The Analyst's Gambit

The Analyst's Gambit: A Second Course in Functional Analysis is a textbook written to serve a graduate course in Functional Analysis. It provides a sequel to the author's previous volume, *A First Course in Functional Analysis*, but it is not necessary to have read one in order to make use of the other. As a graduate text, the reader is assumed to have taken undergraduate courses in set theory, calculus, metric spaces and topology, complex analysis, measure theory (or, alternatively, have enough mathematical maturity to carry on without having seen every particular fact that is used).

A particular strength of the book is that it includes numerous applications. Besides being engaging and interesting in their own right, these applications also illustrate how functional analysis is used in other parts of mathematics. The applications to problems from varied fields (PDEs, Fourier series, group theory, neural networks, topology, etc.) constitute an enticing external motivation for studying functional analysis. There are also applications of the material to functional analytic problems (Lomonosov's invariant subspace theorem, the spectral theorem, Stone's theorem), showcasing the power of the results as well as the elegance and unity of the theory.

Features

- Can be used as the primary textbook for a graduate course in functional analysis
- Rich variety of exercises
- Emphasis on substantial and modern applications

Orr Moshe Shalit is a professor of mathematics at the Technion Israel Institute of Technology, where he teaches and conducts research in operator theory, operator algebras, functional analysis and function theory. His first book, *A First Course in Functional Analysis*, was published by Chapman & Hall / CRC in 2017.

The Analyst's Gambit
A Second Course in Functional Analysis

Orr Moshe Shalit

CRC Press
Taylor & Francis Group
Boca Raton London New York

CRC Press is an imprint of the
Taylor & Francis Group, an **informa** business
A CHAPMAN & HALL BOOK

First edition published 2026
by CRC Press
2385 NW Executive Center Drive, Suite 320, Boca Raton FL 33431

and by CRC Press
4 Park Square, Milton Park, Abingdon, Oxon, OX14 4RN

CRC Press is an imprint of Taylor & Francis Group, LLC

ISBN: 978-1-032-28657-0 (hbk)
ISBN: 978-1-032-28659-4 (pbk)
ISBN: 978-1-003-29786-4 (ebk)

DOI: 10.1201/9781003297864

Typeset in CMR10 font
by KnowledgeWorks Global Ltd.

Publisher's note: This book has been prepared from camera-ready copy provided by the authors.

To Nohar

Contents

Preface

This is a textbook to accompany the graduate course *Functional Analysis* offered by the Department of Mathematics at the Technion. It is a sequel to my book *A First Course in Functional Analysis* [ ], which is the accompanying textbook for the introductory undergraduate course in functional analysis (mostly Hilbert spaces and Fourier analysis) at the Technion.

The student/reader is assumed to have taken undergraduate courses in set theory, calculus, metric spaces and topology, complex analysis, and measure theory, or, alternatively, have enough mathematical maturity to carry on without having seen every particular fact that is used. Note that the introductory undergraduate course in functional analysis is **not** an indispensable prerequisite for taking the "second" course or reading this book. However, readers who have taken the first course, or a similar course on Hilbert spaces and Fourier analysis, will probably cope better with the pace and get more out of some of the examples.

The book includes significant applications of functional analysis to other parts of mathematics. I see these applications as an essential part of the course. To be honest, I feel that the selection of applications, their careful presentation, and measured timing are among the highlights of this book. Besides being beautiful and fun, they illustrate how the theory is used in other parts of mathematics. The applications to problems from varied fields (PDEs, Fourier series, group theory, neural networks, topology, etc.) constitute a gratifying external motivation for studying functional analysis. There are also applications of the material to functional analytic problems (Lomonosov's invariant subspace theorem, the spectral theorem, Stone's theorem), showcasing the power of the results as well as the elegance and unity of the theory. A common theme is that, by employing the abstract framework of functional analysis to solve problems, the mathematician executes a maneuver by which a difficult task is traded for a more tractable one. For example: the task of proving existence might be exchanged for the task of bounding a quantity, and so forth.

How to use this book for a course

The book is designed to align closely with the course. There are thirteen chapters, roughly one chapter per week (three fifty-minute lectures). In practice, one or two sections in every chapter will be left for self-study. (Alternatively, one might proceed a bit more leisurely, at the expense of the final

two chapters). The high pace is compensated for by the logical coherence of
the material and the fact that the key results are used in significant ways in
later chapters.

Instructors of courses on functional analysis face a dilemma: how much
measure theory to assume? One approach is to require that the students take
a course on measure theory before taking functional analysis, allowing the
instructor to use the material freely. Another approach is to devote a large
portion of the course to measure theory, treating measure theory as a part of
functional analysis.[1] I took a middle road: define and use whatever is needed,
but prove only results that illustrate key notions and are used often, such as
the completeness of L^p spaces or the identification of their dual spaces.

Exercises are located throughout the text and at the end of each chapter.
Solving exercises is a crucial component of the course. I recommend reading all
of them because there are some important facts that appear only as exercises.
Some of the most rewarding applications of the material were left for the
students to work out on their own, so take care not to miss them!

Sources

I took the course Functional Analysis as a graduate student in 2003–2004. The
instructor was Baruch Solel, who later became my PhD thesis advisor. I kept
my notes and homework assignments from that course for years until I started
teaching functional analysis, and relied on them as a basis for the courses
that I taught. Solel's course on functional analysis, together with his course
on operator algebras that I took later as a PhD student, have substantially
shaped my knowledge of functional analysis and, ultimately, this book. Other
sources are the books [14], [28] and [34] which were recommended literature
in the courses that I took as a student, and [2], [16], [31], [40], [42], as well as
many others which I encountered later in life, as a teacher and researcher.

Acknowledgments

This book grew out of my lecture notes, which developed over years of teaching
functional analysis at Ben-Gurion University and the Technion. I am indebted
to many colleagues and students who provided valuable feedback on previous
versions. A big thank-you goes to Ram Band, Nick Crawford, Erel Firer-
Frischer, Malte Gerhold, Mikhail Mironov, Zuly Salinas, Guy Salomon, Jeet
Sampat, Itai Shafrir, Ofri Shahaf, Tama Shalit, Eli Shamovich, Michael Skeide
and Ami Viselter for their help in reviewing and improving these notes.

I would not have been able to complete this project without the support
of my beloved wife and children through a series of truly difficult years. What
I owe them, words cannot express.

Rosh Pina

---

[1]When teaching the undergraduate course, I use a different approach (that would not
be appropriate for a graduate course), which is to avoid measure theory entirely; see [11].

Chapter 1

Basic notions and first examples of Banach spaces

1.1 Introduction

Functional analysis is a branch of mathematics that was born at the beginning of the 20th century when mathematicians sought a unified conceptual framework for studying problems in analysis. The observation that many naturally occurring collections of functions (such as the collection of continuous functions, continuously differentiable functions, or integrable functions, etc.) are vector spaces, which, in addition, carry a natural metric, has led to an elegant theory with profound consequences. Over the years the field was fashioned to be a rich blend of linear algebra, abstract algebra, topology, geometry, and (of course!) analysis.

The subject has grown vast. Very general and abstract approaches are available. In this book, we focus on what many would agree to be the core of the subject. We shall learn methods and notions that are used throughout applications of abstract analysis, and we will give concrete examples of applications to learn how these tools work in practice.

In this chapter, we need to review some material that should be familiar to most readers. There is significant overlap with [44, Chapter 7].

Conventions. All vector spaces will be either over the real numbers $\mathbb{R}$ or over the complex numbers $\mathbb{C}$. In many cases, there will be no need to be specific, and then we shall refer to elements of the underlying field as *scalars*. In cases where it is important to specify over which field we are working a space over the real numbers will be referred to as a real space, etc.

DOI: 10.1201/9781003297864-1

1.2   Normed spaces

1.2.1   Basic concepts

Definition 1.2.1. Let X be a vector space. A ***semi-norm*** on X is a function
$\|\cdot\| : X \to [0, \infty)$ that satisfies

1. $\|x + y\| \le \|x\| + \|y\|$ for all $x, y \in X$,

2. $\|ax\| = |a|\|x\|$ for all $x \in X$ and every scalar a.

A semi-norm is called a ***norm*** if it has the additional property:

$$\|x\| = 0 \implies x = 0.$$

A vector space X with a (semi-)norm $\|\cdot\|$ defined on it is called a ***(semi-)
normed space***. Sometimes one refers to the pair $(X, \|\cdot\|)$ for definiteness.

Every normed space is a metric space[1], where the metric d is given by

$$d(x, y) = \|x - y\|.$$

We then say that the metric d is induced by the norm. Hence every normed
space becomes a topological space with a basis consisting of the ***open balls***
$B_r(x) = \{y \in X : \|x - y\| < r\}$, where (x, r) runs over all $x \in X$ and all $r > 0$.

We write $\overline{B}_r(x)$ for the ***closed ball*** $\{y \in X : \|x - y\| \le r\}$ centered at x
with radius r. This is equal to the closure of $B_r(x)$. The ***closed unit ball*** of
a normed space X, denoted by X_1, is the set

$$X_1 = \{x \in X : \|x\| \le 1\} = \overline{B}_1(0).$$

A normed space is said to have a certain property if the underlying metric
space has that property. For example, a normed space is said to be ***separable***
if it is separable as a metric space, and it is said to be ***complete*** if it is
complete as a metric space.

Definition 1.2.2. A ***Banach space*** is a normed space which is complete.

There is a familiar completion procedure that justifies (to a large extent)
putting an emphasis on Banach spaces.

Proposition 1.2.3. *Every normed space X can be linearly and isometrically
embedded in a unique Banach space which contains X as a dense subspace.*

Proof. The completion $\tilde{X}$ is the unique metric space completion of the metric
space X, which the reader is assumed to know from a previous introductory
course in metric spaces. Since addition, scalar multiplication, and the norm are
uniformly continuous on any ball, they extend uniquely to the completion.   □

---

[1]For a crash introduction to metric spaces, see the appendix of [ ]. For a thorough
treatment of metric and topological spaces, see [ ].

1.2.2 Examples of Banach spaces

Example 1.2.4. If X is a topological space, then the space $C_b(X)$ of complex valued bounded continuous functions equipped with the sup norm $\|f\|_\infty = \sup_{x \in X} |f(x)|$ is a Banach space. In particular, if X is a compact space, then the space $C(X)$ of all continuous functions equipped with the norm $\|\cdot\|_\infty$ is a Banach space.

Example 1.2.5. Let X be a locally compact Hausdorff space. A function $f \in C(X)$ is said to have **compact support** if the **support** supp(f) of f, defined to be

$$\mathrm{supp}(f) = \overline{\{x : f(x) \neq 0\}},$$

is a compact subset of X. A function f is said to **vanish at infinity** if for all $\epsilon > 0$, there is a compact K such that $|f(x)| < \epsilon$ for all $x \notin K$. Let $C_c(X)$ denote the space of all compactly supported continuous functions, and let $C_0(X)$ denote the space of all continuous functions vanishing at infinity. If X is not compact then $C_c(X)$ is an incomplete normed space, and the completion of $C_c(X)$ is the Banach space $C_0(X)$.

Example 1.2.6. Let $p \in [1, \infty)$. For any finite or infinite sequence $x = (x_k)_k$ of complex numbers, we define

$$\|x\|_p = \left(\sum_k |x_k|^p \right)^{1/p}.$$

We also define

$$\|x\|_\infty = \sup_k |x_k|.$$

An important family of finite dimensional Banach spaces is given by the spaces commonly denoted by ℓ_n^p (for $p \in [1, \infty]$ and $n \in \mathbb{N}$), which are the vector space $\mathbb{C}^n$ endowed with the norm $\|\cdot\|_p$ (there are also obvious versions of ℓ^p spaces over the reals).

The infinite dimensional ℓ^p spaces ($p \in [1, \infty]$) are given by

$$\ell^p = \{x = (x_k)_{k=0}^\infty \in \mathbb{C}^\mathbb{N} : \|x\|_p < \infty\}.$$

For $p = 2$ we get the familiar Hilbert space ℓ^2. For $p = 1$ or $p = \infty$, it is also clear that $\|\cdot\|_p$ is a norm, but if $p \notin \{1, 2, \infty\}$ this is not apparent. However, once one shows that $\|\cdot\|_p$ is a norm, it is straightforward to show that ℓ^p is complete (completeness of the spaces ℓ^p also follows from Proposition 1.5.20). Thus, the spaces ℓ^p and ℓ_n^p are Banach spaces.

In order to keep the presentation self-contained, we present the details showing that $\|\cdot\|_p$ is a norm (the reader who is familiar with the details can skip to Example 1.2.9). The nontrivial part is to show that $\|\cdot\|_p$ satisfies the triangle inequality. We will need a few preparations first. Two extended real numbers $p, q \in [1, \infty]$ are said to be **conjugate exponents** if

$$\frac{1}{p} + \frac{1}{q} = 1.$$

If $p = 1$, then we understand this to mean that $q = \infty$, and vice versa.

Lemma 1.2.7 (Hölder's inequality for sequences). *Let $p, q \in [1, \infty]$ be conjugate exponents. For any two (finite or infinite) sequences $x_1, x_2, \ldots$ and $y_1, y_2, \ldots$*

$$\sum_k |x_k y_k| \leq \|x\|_p \|y\|_q.$$

Proof. The heart of the matter is to prove the inequality for finite sequences (we leave the case of infinite sequences to the reader). Therefore, we need to prove that for every $x = (x_k)_{k=1}^n$ and $y = (y_k)_{k=1}^n$ in $\mathbb{C}^n$,

$$\sum_{k=1}^n |x_k y_k| \leq \left(\sum_{k=1}^n |x_k|^p \right)^{1/p} \left(\sum_{k=1}^n |y_k|^q \right)^{1/q}. \tag{1.1}$$

The case $p = 1$ (or $p = \infty$) is immediate. The right-hand side of (1.1) is continuous in p when x and y are held fixed, so it is enough to verify the inequality for a dense set of values of p in $(1, \infty)$. Define

$$S = \left\{ \frac{1}{p} \in (0, 1) \,\middle|\, p \text{ satisfies Equation (1.1) for all } x, y \in \mathbb{C}^n \right\}.$$

Now our task reduces to that of showing that S is dense in $(0, 1)$. By the Cauchy-Schwarz inequality, we know that $\frac{1}{2} \in S$. Also, the roles of p and q are interchangeable, so $\frac{1}{p} \in S$ if and only if $1 - \frac{1}{p} \in S$.

Set $a = \frac{q}{2p+q}$, so that $2ap = (1 - a)q$. Now, if $\frac{1}{p} \in S$, we apply (1.1) to the sequences $(|x_k||y_k|^a)_k$ and $(|y_k|^{1-a})_k$, and then we use the Cauchy-Schwarz inequality, to obtain

$$\sum_{k=1}^n |x_k y_k| = \sum_{k=1}^n |x_k||y_k|^a |y_k|^{1-a}$$

$$\leq \left(\sum_{k=1}^n |x_k|^p |y_k|^{ap} \right)^{1/p} \left(\sum_{k=1}^n |y_k|^{(1-a)q} \right)^{1/q}$$

$$\leq \left(\left(\sum_{k=1}^n |x_k|^{2p} \right)^{1/2} \left(\sum_{k=1}^n |y_k|^{2ap} \right)^{1/2} \right)^{1/p} \left(\sum_{k=1}^n |y_k|^{(1-a)q} \right)^{1/q}$$

$$= \left(\sum_{k=1}^n |x_k|^{p'} \right)^{\frac{1}{p'}} \left(\sum_{k=1}^n |y_k|^{q'} \right)^{\frac{1}{q'}},$$

where $\frac{1}{p'} = \frac{1}{2p}$ and $\frac{1}{q'} = \frac{1}{2p} + \frac{1}{q}$. Therefore, if $s = \frac{1}{p} \in S$, then $\frac{s}{2} = \frac{1}{2p} \in S$; and if $s = \frac{1}{q} \in S$, then $\frac{s+1}{2} = \frac{1}{2}\frac{1}{q} + \frac{1}{2} = \frac{1}{q} + \frac{1}{2}\frac{1}{p}$ is also in S.

Since $\frac{1}{2}$ is known to be in S, it follows by induction that for every $n \in \mathbb{N}$ and $m \in \{1, 2, \ldots, 2^n - 1\}$, the fraction $\frac{m}{2^n}$ is in S. Hence S is dense in $(0, 1)$, and the proof is complete. □

Lemma 1.2.8 (Minkowski's inequality). *For every $p \in [1, \infty]$ and every two (finite or infinite) sequences $x_1, x_2, \ldots$ and $y_1, y_2, \ldots$*

$$\|x + y\|_p \leq \|x\|_p + \|y\|_p.$$

In particular, if $x, y \in \ell^p$, then $x + y \in \ell^p$.

Proof. Again, the cases $p = 1, \infty$ are easy, and the result for infinite sequences is a consequence of the result for finite sequences. Therefore, let $p \in (1, \infty)$, and let $x, y \in \mathbb{C}^n$. With $q = \frac{p}{p-1}$, we use Hölder's inequality and the identity $q(p-1) = p$ to find

$$\|x + y\|_p^p = \sum_k |x_k + y_k|^p = \sum_k |x_k + y_k||x_k + y_k|^{p-1}$$

$$\leq \sum_k |x_k||x_k + y_k|^{p-1} + \sum_k |y_k||x_k + y_k|^{p-1}$$

$$\leq \|x\|_p \left(\sum_k |x_k + y_k|^p \right)^{1/q} + \|y\|_p \left(\sum_k |x_k + y_k|^p \right)^{1/q}$$

$$= (\|x\|_p + \|y\|_p) \|x + y\|_p^{p-1}.$$

Minkowski's inequality follows. $\qquad\qquad\square$

Example 1.2.9. One can define $\ell^p(S)$ for any set S. For $p = \infty$, we define $\ell^\infty(S)$ to be the space of all bounded functions on S with supremum norm. For $p \in [1, \infty)$, we define

$$\ell^p(S) = \left\{ a : S \to \mathbb{C} : \|a\|_p := \sup_{F \subseteq S, |F| < \infty} \left(\sum_{s \in F} |a(s)|^p \right)^{1/p} < \infty \right\}.$$

Then $\ell^p(S)$ is a Banach space for all $p \in [1, \infty]$.

Example 1.2.10 (Spaces of convergent sequences). Define

$$c = \{ x = (x_k)_{k=0}^\infty \in \ell^\infty : \lim_{k \to \infty} x_k \text{ exists} \},$$

and

$$c_0 = \{ x = (x_k)_{k=0}^\infty \in \ell^\infty : \lim_{k \to \infty} x_k = 0 \}.$$

These spaces are closed subspaces of ℓ^∞, so when equipped with the norm $\| \cdot \|_\infty$, they are Banach spaces in their own right.

Example 1.2.11. No course in functional analysis is complete without mention of the so-called "L^p spaces". Since the traditional definition of these spaces requires some material from measure theory, we postpone their introduction to Section 1.5. However, in certain circumstances these spaces can be defined in a functional analytic manner that is equivalent to the traditional definition, as follows.

Fixing $p \in [1, \infty)$, define a norm on the space $C_c(\mathbb{R})$ of continuous compactly supported functions on the real line by $\|f\|_p = (\int_{\mathbb{R}} |f(x)|^p dx)^{1/p}$. One can show that $\|\cdot\|_p$ is a norm on $C_c(\mathbb{R})$. Here too the nontrivial thing to prove is that the triangle inequality holds, that is, that there is an integral analogue of Lemma 1.2.8 — for the moment the reader can take this on faith (see Section 1.5). Then one can define the space $L^p = L^p(\mathbb{R})$ to be the completion of $C_c(\mathbb{R})$ with respect to the norm $\|\cdot\|_p$.

Similarly, for every interval $[a, b] \subset \mathbb{R}$, one can define the space $L^p([a, b])$ to be the completion of $C([a, b])$ with respect to the norm $\|f\|_p = (\int_a^b |f(x)|^p dx)^{1/p}$. We shall use the notation $L^p[a, b]$, for brevity.

1.3 Bounded operators

1.3.1 Basic definitions

Definition 1.3.1. Let X and Y be normed spaces. A linear operator $T \colon X \to Y$ is said to be **bounded** if the **operator norm** $\|T\|$ of T, defined as

$$\|T\| = \sup_{x \in X_1} \|Tx\|,$$

satisfies $\|T\| < \infty$. The space of all bounded operators from X to Y is denoted by $B(X, Y)$. If $Y = X$ we abbreviate $B(X) = B(X, X)$.

It follows from the definition that $\|Tx\| \leq \|T\| \|x\|$ for all $x \in X$. One can check that the "operator norm" is indeed a norm, so the space $B(X, Y)$ becomes a normed space.

Let X, Y, Z be normed spaces. If $T : X \to Y$ and $S : Y \to Z$ are linear operators, then their composition $S \circ T$ is the map $S \circ T : X \to Z$ given by $S \circ T(x) = S(T(x))$. The composition $S \circ T$ is usually denoted by ST. It is immediate that

$$\|ST\| \leq \|S\| \|T\|.$$

In particular, for every normed space X, the normed space $B(X)$ is an *algebra*. If $T \in B(X)$, then $T^n = T \circ T \circ \cdots \circ T$ (the composition of T with itself, n times) is also in $B(X)$, and moreover $\|T^n\| \leq \|T\|^n$. If $p(z) = \sum_{k=0}^n a_k z^k$ is a polynomial, then we can naturally define $p(T)$ to be the operator

$$p(T) = \sum_{k=0}^n a_k T^k,$$

and it is clear that $p(T) \in B(X)$, because it is the sum of bounded operators. The "zeroth power" T^0 is understood as the identity operator: $T^0 = I$.

We have the familiar notions of **kernel** and **image** of operators between normed spaces, defined as

$$\ker T = \{x \in X : Tx = 0\}, \quad \mathrm{Im}(T) = \{Tx : x \in X\}.$$

The proofs of the following propositions are basic.

Proposition 1.3.2. *For a linear transformation* $T : X \to Y$ *mapping between two normed spaces, the following are equivalent:*

1. *T is bounded.*

2. *T is continuous.*

3. *T is continuous at some* $x_0 \in X$.

Proposition 1.3.3. *If* Y *is a Banach space then* $B(X, Y)$ *is a Banach space.*

Exercise 1.3.4. Make sure that you have proven the above two propositions at least once in your life.

Definition 1.3.5. Let X be a normed space. A linear map from X into the scalar field is called a **linear functional**. We let X^* denote the space of all bounded linear functionals on X. The space X^* is called the **dual space** (or the **conjugate space**) of X.

In some parts of the literature, the dual space is denoted by X', but we will not use this notation. Another common notation that is used sometimes, is to write the action of X^* on X as an inner product (also called **pairing**)

$$\langle x, x^* \rangle := x^*(x) \quad \text{for all } x \in X, x^* \in X^*.$$

This notation is very useful when one has to keep track of which element belongs to which space . Since we will later discuss C*-algebras we shall avoid writing elements of X^* as little xs with stars, but the pairing (the "inner product" notation) is useful and we shall use it.

Remark 1.3.6. The dual space X^*, being a space of bounded operators on X, is a normed space with the norm

$$\|f\| = \sup_{x \in X_1} |f(x)| = \sup_{x \in X_1} |\langle x, f \rangle| \quad , \quad f \in X^*.$$

By Proposition 1.3.3, X^* is complete. So X^* is a Banach space. Is every Banach space the dual space of some other normed space? We'll return to this question later.

1.3.2 Examples of dual spaces

Example 1.3.7. By the Riesz representation theorem [11, Theorem 5.2.3], every Hilbert space can be identified (isometrically and anti-linearly) with its dual in the following way

$$H^* = \{\langle \cdot, g \rangle : g \in H\}.$$

Example 1.3.8. The dual of ℓ^1 can be isometrically identified with ℓ^∞; this is usually summarized by writing $(\ell^1)^* = \ell^\infty$. To be precise: every $b \in \ell^\infty$ gives rise to a linear functional $\Phi_b \in (\ell^1)^*$ by way of $\Phi_b(a) = \sum_n a_n b_n$. We will now show that $b \mapsto \Phi_b$ is an isometric (i.e., norm preserving) linear isomorphism. Clearly $|\Phi_b(a)| \leq \|b\|_\infty \|a\|_1$ for every $a \in \ell^1$, so Φ_b is well defined, bounded, and $\|\Phi_b\| \leq \|b\|_\infty$. To complete the identification of $(\ell^1)^*$ with ℓ^∞, we will show that the map $b \mapsto \Phi_b$ is also isometric and surjective.

For every n, let e_n denote the element of ℓ^1 with 1 in the nth slot and 0s elsewhere. It is easy to see that for every $a = (a_n)_n \in \ell^1$, we have the norm convergent series

$$a = \lim_{N \to \infty} \sum_{n=0}^{N} a_n e_n = \sum_{n \in \mathbb{N}} a_n e_n.$$

Given $\Psi \in (\ell^1)^*$, we define a sequence $b = (b_n)_n$ by $b_n = \Psi(e_n)$. Then $\|b\|_\infty \leq \|\Psi\|$. By linearity and continuity,

$$\Psi(a) = \lim_{N \to \infty} \sum_{n=0}^{N} a_n \Psi(e_n) = \lim_{N \to \infty} \sum_{n=0}^{N} a_n b_n = \Phi_b(a),$$

whence, $\Psi = \Phi_b$. Thus $b \mapsto \Phi_b$ is surjective. Finally, $\|b\|_\infty \leq \|\Psi\| = \|\Phi_b\| \leq \|b\|_\infty$, so $b \mapsto \Phi_b$ is isometric.

Example 1.3.9. Similarly to the above example, one can show that every $b \in \ell^1$ gives rise to a bounded functional Γ_b on ℓ^∞ given by

$$\Gamma_b(a) = \sum a_n b_n.$$

The map $b \mapsto \Gamma_b$ is an isometric linear map of ℓ^1 into $(\ell^\infty)^*$. We currently do not have enough tools to determine whether this map is surjective.

Example 1.3.10. If $1 < p, q < \infty$ are conjugate exponents, then $(\ell^p)^* = \ell^q$.

Exercise 1.3.11. Show that if $1 < p, q < \infty$ are conjugate exponents, then $(\ell^p)^* = \ell^q$, where the isometric isomorphism is similar to the one given in Example 1.3.8.

Remark 1.3.12. Note the subtle difference between the case $p = 2$ of the above exercise and Example 1.3.7.

1.3.3 Isomorphisms of Banach spaces

In Example 1.3.8 we showed that $(\ell^1)^* = \ell^\infty$. However, these two spaces are not *the same*. What we really established is that there is a bijective linear map between ℓ^∞ and $(\ell^1)^*$ that preserves the norm, and this allows us to conveniently identify the spaces. This kind of identification is very useful and common, and we take a moment here to formalize it. In the following discussion, let X and Y be normed spaces.

Definition 1.3.13. An operator $T \in B(X, Y)$ is called an *isomorphism* if it is bijective and has a bounded inverse[2]. The spaces X and Y are said to be *isomorphic* if there is an isomorphism between them.

Definition 1.3.14. An operator $T \in B(X, Y)$ is said to be an *isometry* if $\|Tx\| = \|x\|$ for all $x \in X$. The spaces X and Y are said to be *isometrically isomorphic* (or simply *isometric*) if there is an isometric isomorphism between them.

Two Hilbert spaces are isomorphic as Banach spaces if and only if there is a unitary map from one onto the other and this happens if and only if they have orthonormal bases of the same cardinality (see [14, Section 3.4]). In contrast, Banach spaces are much more diverse. Mapping this complex landscape is a central theme in the advanced study of Banach spaces. While we are not planning to delve into the classification theory of Banach spaces, we cannot completely ignore the natural and compelling questions that arise.

Exercise 1.3.15. Prove that ℓ^p is not isomorphic to ℓ^∞ for all $1 \le p < \infty$. Harder: what can you say if we replace ℓ^∞ by ℓ^q, for some $q \in [1, \infty)$?

The notion of isomorphism is significantly weaker than that of isometric isomorphism. For example, a Hilbert space can never be isometrically isomorphic to a Banach space that is not Hilbert (why?), but it certainly can be isomorphic to one.

Exercise 1.3.16. Let H and K be Hilbert spaces.

1. Prove that if $T : H \to K$ is an isometric isomorphism, then it is a unitary map.

2. Prove that if H is isomorphic to K as Banach spaces, then they are isometrically isomorphic, hence they are also isomorphic as Hilbert spaces in the sense that there exists a unitary map of H onto K.

3. Give an example of a Banach space isomorphism between Hilbert spaces that is not a Hilbert space isomorphism.

---

[2] We shall see later that the inverse of a bounded bijective linear map between Banach spaces is also bounded, automatically.

Definition 1.3.17. Let X be a normed space with a norm $\|\cdot\|$, and let $\|\cdot\|'$ be another norm defined on X. The norms $\|\cdot\|$ and $\|\cdot\|'$ are said to be **equivalent** if there exist two positive constants c, C such that

$$c\|x\| \leq \|x\|' \leq C\|x\| \quad \text{for all } x \in X.$$

Clearly, the norms $\|\cdot\|$ and $\|\cdot\|'$ are equivalent if and only if the identity map is an isomorphism from $(X, \|\cdot\|)$ onto $(X, \|\cdot\|')$.

Exercise 1.3.18. Let $(X, \|\cdot\|_X)$ and $(Y, \|\cdot\|_Y)$ be normed spaces, and let $T : X \to Y$ be a bijective linear map. Define $\|x\|' = \|Tx\|_Y$. Prove that $\|\cdot\|'$ is a norm on X. Prove that T is an isomorphism from $(X, \|\cdot\|_X)$ onto $(Y, \|\cdot\|_Y)$ if and only if $\|\cdot\|$ and $\|\cdot\|'$ are equivalent.

Exercise 1.3.19. Let $\ell_0 \subseteq c_0$ be the space of all finitely supported sequences. Prove that the $\|\cdot\|_p$ norms on ℓ_0 are all mutually nonequivalent.

We close this section with a few words on finite dimensional spaces. For brevity, we treat complex spaces — obvious analogues hold for real spaces. Recall that every finite dimensional vector space over $\mathbb{C}$ is linearly isomorphic to $\mathbb{C}^n$ for some n. By the construction in Exercise 1.3.18, every finite dimensional normed space is isometric to $(\mathbb{C}^n, \|\cdot\|')$ where $\|\cdot\|'$ is some norm on $\mathbb{C}^n$. The classification of n-dimensional spaces thus reduces to the classification of norms on $\mathbb{C}^n$.

Theorem 1.3.20. *All norms on $\mathbb{C}^n$ are equivalent. In other words, every two normed spaces of the same dimension over $\mathbb{C}$ are isomorphic.*

Proof. We leave the proof as an exercise for readers who do not recall having seen this fact earlier in their studies. $\qquad\square$

1.4   Quotient spaces

1.4.1   The algebraic setting

We begin by recalling the construction of the quotient of a vector space by a linear subspace. Let X be a vector space, and let $M \subseteq X$ be a linear subspace. We define an equivalence relation on X:

$$x \sim y \iff x - y \in M.$$

The equivalence class of an element $x \in X$ is simply $x + M = \{x + m : m \in M\}$. The **quotient space** X/M is defined to be the set of all equivalence classes, with the structure of a linear space given by

$$(x + M) + (y + M) = (x + y) + M,$$

and $a(x + M) = ax + M$. We write $\dot{x}$ for the equivalence class of x, that is, $\dot{x} = x + M$. The map $\pi : X \to X/M$, given by $\pi(x) = \dot{x}$, is called the **quotient map**. The quotient map is a surjective linear map, and $\ker \pi = M$. Thus,

$$\dim(X) = \dim(M) + \dim(X/M).$$

The dimension of X/M is called the **codimension** of M (in X).

Lemma 1.4.1. *Let M be a linear subspace of a vector space X. The following are equivalent.*

1. *M has codimension 1.*

2. *M is a maximal subspace (i.e., there exists no proper subspace of X that properly contains M).*

3. *There exists a nonzero linear functional F on X such that $M = \ker F$.*

Proof. If $\dim(X/M) = 1$, there is an isomorphism ϕ from X/M onto the scalar field. Then $F = \phi \circ \pi$ is nonzero functional with $\ker F = M$. Conversely, given a functional with $\ker F = M$, let $v = F^{-1}(1)$. Then $v \notin \ker F = M$, so $\dot{v} \neq 0$. Moreover, for every $x \in X$, $x - F(x)v \in \ker F$, so $\dot{x} = F(x)\dot{v}$. It follows that $\{\dot{v}\}$ is a basis for X/M.

We have established (1) $\iff$ (3). The equivalence (1) $\iff$ (2) is no harder and is left as an exercise for the reader. $\qquad\square$

1.4.2 The semi-normed setting

If $\| \cdot \|$ is a single semi-norm on a space X, then $\ker \| \cdot \| := \{x : \|x\| = 0\}$ is a subspace, and one can, and usually one does, "quotient out" by this subspace to obtain a normed space. This allows us to ignore semi-normed spaces and deal instead only with normed spaces.

Proposition 1.4.2. *Let X be a vector space, and let $\| \cdot \|$ be a semi-norm on X. The set $M := \ker \| \cdot \|$ is a subspace of X. If on X/M we define*

$$\|\dot{x}\|_{X/M} = \|x\|,$$

then $\| \cdot \|_{X/M}$ is a norm on X/M.

Proof. The proof is left to the readers. $\qquad\square$

In Section 1.5.3 we will apply this proposition to define the L^p spaces.

Remark 1.4.3. The procedure of quotienting away the kernel of a semi-norm and passing to a normed space is useful only when the space is equipped with a *single* semi-norm. However, it is important for us to keep in mind that a very common situation in analysis is when we have a vector space X together with a whole family of semi-norms $\{\| \cdot \|_\alpha\}_{\alpha \in \mathbb{A}}$, and that this family is used to introduce a topology on X that is often *fundamentally not normable*. We will return to this issue in Chapter 6.

1.4.3 The normed setting

We now proceed to study the quotients of a normed space $(X, \|\cdot\|)$ by a closed subspace M. We define

$$\|\dot{x}\|_{X/M} = d(x, M) = \inf\{\|x - m\| : m \in M\}. \tag{1.2}$$

Let π denote the quotient map $X \to X/M$. We usually simply write $\|\dot{x}\|$ for $\|\dot{x}\|_{X/M}$, when no confusion can occur.

Theorem 1.4.4. *Let X be a normed space, $M \subseteq X$ a closed subspace, and define the quotient space X/M as above. With the norm defined as in (1.2), the following hold:*

1. $(X/M, \|\cdot\|_{X/M})$ *is a normed space.*

2. π *is a contraction:* $\|\pi(x)\| \leq \|x\|$ *for all x.*

3. *For every $y \in X/M$ such that $\|y\| < 1$, there is an $x \in X$ with $\|x\| < 1$ such that $\pi(x) = y$.*

4. *If U is open in X then $\pi(U)$ is open in X/M.*

5. *If X is complete, then so is X/M.*

6. *If F is a closed subspace containing M, then $\pi(F)$ is closed.*

Proof. For every $x \in X$ and every scalar a, the identity $\|a\dot{x}\| = |a|\|\dot{x}\|$ is trivial. Next, $\|\dot{x}\| = 0 \iff \dot{x} = 0$, (i.e., $\|\dot{x}\| = 0 \iff x \in M$) follows from the fact that M is closed. For every $x, y \in X$, let $m, n \in M$ be such that

$$\|x - m\| < \|\dot{x}\| + \epsilon \quad \text{and} \quad \|y - n\| < \|\dot{y}\| + \epsilon.$$

Then, $m + n \in M$, so

$$\|\dot{x} + \dot{y}\| \leq \|x + y - m - n\| \leq \|x - m\| + \|x - n\| \leq \|\dot{x}\| + \|\dot{y}\| + 2\epsilon.$$

It follows that $\|\dot{x} + \dot{y}\| \leq \|\dot{x}\| + \|\dot{y}\|$. Thus (1.2) defines a norm.

The following assertions are not hard to prove in the order in which they appear, and we skip to the one before last.

Suppose that $\{y_n\}$ is a Cauchy sequence in X/M. To show that the sequence converges, it suffices to show that a subsequence converges. Passing to a subsequence we may assume that $\|y_n - y_{n+1}\| < 2^{-n}$. Let $x_1 \in X$ be a representative of y_1. By assertion (3) in the statement of the theorem, we can find x_2 such that $\pi(x_2) = y_2$ and $\|x_1 - x_2\| < \frac{1}{2}$. Continuing inductively, we find a sequence $\{x_n\}$ such that $\pi(x_n) = y_n$ and $\|x_n - x_{n+1}\| < 2^{-n}$ for all n. If X is complete, the Cauchy sequence $\{x_n\}$ converges to some $x \in X$. By (2) we have that $y_n = \pi(x_n) \to \pi(x)$. We conclude that X/M is complete. $\square$

Exercise 1.4.5. Complete the proof of the Theorem 1.4.4.

Exercise 1.4.6. A map is called **open** (**closed**) if it maps open (closed) sets to open (closed) sets. By Theorem 1.4.4(4), the quotient map π is open. Is the quotient map closed?

1.5 Measure theory and L^p spaces

In this section we will briefly review the rudiments of measure theory in order to carefully define L^p spaces, which form a prominent class of examples of Banach spaces in theory and applications. Measure theory plays several other important roles in functional analysis, some of which we shall see later in this book. For a full treatment of the subject, and in particular for proofs of statements made below, the reader is referred to the literature. Excellent references for this material are [43] and [19], which take an abstract approach. Alternatively one may consult [27], which starts with a concrete and hands-on construction of the Lebesgue measure and integral on $\mathbb{R}^n$, and after proving all the big theorems in this concrete setting the results are quickly generalized to abstract measure spaces.

1.5.1 Measure spaces

Definition 1.5.1. Let X be a set. A **σ-algebra** in X is a collection Σ of subsets of X with the following properties:

1. $X \in \Sigma$.

2. Σ is closed under taking complements: if $A \in \Sigma$ then $X \setminus A \in \Sigma$.

3. Σ is closed under countably infinite unions: if $A_n \in \Sigma$ for all $n \in \mathbb{N}$ then $\cup_{n \in \mathbb{N}} A_n \in \Sigma$.

A **measurable space** is a pair (X, Σ) consisting of a set and a σ-algebra in it. Elements of Σ are referred to as **measurable sets**.

It follows easily that $\emptyset \in \Sigma$ and that Σ is also closed under finite unions as well as countable (finite or infinite) intersections. Two easy examples of σ-algebras are (i) $\Sigma = \{\emptyset, X\}$ consisting of X and the empty set; and (ii) $\Sigma = 2^X$, that is, the σ-algebra in which *all* subsets of X are measurable. These two examples are referred to as trivial. One has to think a bit to come up with nontrivial examples.

Example 1.5.2. If X is a set, then the collection of all subsets of X that are either countable or have a countable complement in X is a σ-algebra.

When X is uncountable, the above example provides a nontrivial σ-algebra. Still, this is not a very substantial example. The next example is our first interesting example of a σ-algebra. It is not hard to show for every collection Λ in 2^X there exists a smallest σ-algebra containing Λ; we call this **the σ-algebra generated by** Λ. The σ-algebra of Example 1.5.2 is the σ-algebra generated by finite subsets of X.

Definition 1.5.3. If X is a topological space and τ is the topology (i.e. the collection of all open sets) then the σ-algebra generated by τ is called **the Borel σ-algebra.**

It is not straightforward, but it can be shown that the cardinality of the Borel σ-algebra on $\mathbb{R}^k$ is $2^{\aleph_0}$, therefore it is a nontrivial σ-algebra.

Definition 1.5.4. If (X_1, Σ_1) and (X_2, Σ_2) are measurable spaces, a function $f \colon X_1 \to X_2$ is said to be **measurable** if $f^{-1}(B) \in \Sigma_1$ for all $B \in \Sigma_2$. A scalar valued function on a measurable space (X, Σ) is said to be **measurable** if it is measurable as a function from the measurable space (X, Σ) into the scalar field with the Borel σ-algebra.

For an example of a measurable function, consider the characteristic function χ_A of a measurable set A, defined by $\chi_A(x) = 1$ if $x \in A$ and $\chi_A(x) = 0$ otherwise. More generally, a **simple function** is a function of the form $f = \sum_{k=1}^{n} a_k \chi_{A_k}$ where the A_ks are measurable sets and the a_ks are scalars; simple functions are measurable.

Exercise 1.5.5. Prove that if f is a measurable complex valued function then $|f|$ is measurable. Show that the converse is false. Prove also that f is measurable if and only if its real and imaginary parts are; also, a real function f is measurable if and only if $f^+ = \max\{f, 0\}$ and $f^- = (-f)^+$ are measurable.

It is convenient to consider also **extended real valued** functions, that is, functions that might possibly attain the value ∞ or $-\infty$ at some point. We think of $[0, \infty]$ as the one-point compactification of $[0, \infty)$, and then a function $f \colon X \to [0, \infty]$ is said to be measurable if $f^{-1}(B)$ is measurable for every Borel set in $[0, \infty]$.

Definition 1.5.6. Let (X, Σ) be a measurable space. A **measure** on (X, Σ) is a function $\mu \colon \Sigma \to [0, \infty]$ that has the following properties

1. $\mu(A) \geq 0$ for all $A \in \Sigma$.

2. $\mu(\emptyset) = 0$.

3. If $\{A_n\}$ is a collection of disjoint measurable sets then $\mu(\cup_n A_n) = \sum_n \mu(A_n)$.

A triple (X, Σ, μ) or a pair (X, μ) (when Σ is understood) is referred to as a **measure space.**

Sometimes, one uses the terminology **positive measure**, to highlight that the above defined function μ takes values in $[0, \infty]$. If $\mu(X) = 1$, then μ is referred to as a **probability measure**. A **Borel measure** is a measure that is defined on the Borel σ-algebra (or on a larger σ-algebra).

Example 1.5.7. Let X be any set, let $\Sigma = 2^X$, and define $\mu(A) = |A|$ for finite subsets of X and $\mu(A) = \infty$ for infinite subsets. Then μ is a measure, called the **counting measure.**

The above is a rather trivial example of a measure. Exhibiting that there exists a rich and useful supply of measures is a formidable task, that is addressed in a course on measure theory. The most important measure is the Lebesgue measure, which extends the notion of k-dimensional volume. Recall that the **volume** of a rectangle $R = [a_1, b_1] \times \cdots \times [a_k, b_k]$ is defined to be $\text{Vol}(R) = \Pi_{i=1}^{k} |b_i - a_i|$.

Example 1.5.8. For every $k \in \mathbb{N}$, there is a σ-algebra of sets $\mathcal{M}_k$ containing the Borel sets in $\mathbb{R}^k$ and a Borel measure m_k on $\mathcal{M}_k$, with the following properties:

1. $m_k(R) = \text{Vol}(R)$ for every rectangle $R \subset \mathbb{R}^k$.

2. $m_k(A) = \inf\{m_k(U) : A \subseteq U, U \text{ is open }\}$ and also $m_k(A) = \sup\{m_k(K) : K \subseteq A, K \text{ is compact }\}$ for every $A \in \mathcal{M}_k$.

3. If $A \in \mathcal{M}_k$, $B \subset A$ and $m_k(A) = 0$ then $B \in \mathcal{M}_k$.

The sets in $\mathcal{M}_k$ are said to be **Lebesgue measurable**, and m_k is called the **Lebesgue measure**. The construction of the Lebesgue measure takes up an important part in a first course in measure theory; see [27].

1.5.2   Integration on measure spaces

Definition 1.5.9. Let (X, μ) be a measure space and $f \colon X \to [0, \infty]$ a measurable function. The **integral** of f (with respect to μ) is defined to be

$$\int f d\mu = \sup \sum_k a_k \mu(A_k) \tag{1.3}$$

where the supremum is taken over all simple function $s = \sum_{k=1}^{n} a_k \chi_{A_k}$ such that $a_k \geq 0$ for all k and $s \leq f$ (meaning that $s(x) \leq f(x)$ for all $x \in X$).

In particular, the above definition implies that for any simple nonnegative function $s = \sum_k a_k \chi_{A_k}$, the integral of s is defined to be $\int s d\mu = \sum_k a_k \mu(A_k)$. Here, if $\mu(A_k) = \infty$ then $a_k \mu(A_k)$ is understood as ∞ if $a_k > 0$ and 0 if $a_k = 0$. We do not worry about the case in which $a_k = \infty$ because this does not fit our definition of a simple function (the a_k are supposed to be scalars).

Definition 1.5.10. Let (X, μ) be a measure space. A measurable function $f \colon X \to \mathbb{C}$ is said to be **integrable** if $\int |f| d\mu < \infty$. If f is integrable then we define the **integral** of f to be

$$\int f d\mu = \int u^+ d\mu - \int u^- d\mu + i \int v^+ d\mu - i \int v^- d\mu,$$

where u and v are the real and imaginary parts of f.

Finally, if A is a measurable subset of X then the integral of f on A is defined to be

$$\int_A f d\mu = \int f \chi_A d\mu.$$

This simple definition of integral is very versatile and powerful. When X is a nice subset of $\mathbb{R}^k$ and μ is the restriction of the Lebesgue measure to X, then the integral is referred to as **the Lebesgue integral**. It turns out that every function that is integrable in the sense of Riemann on X is also Lebesgue integrable (and in particular, Lebesgue measurable), and that the value of the Lebesgue integral $\int f d\mu$ is equal to the Riemann integral $\int_X f(x) dx$ of f on X. It is a good idea to use any intuition acquired from experience with the Riemann integral to the above general definition of integral. Many of the nice properties of the Riemann integral persist, for example:

Proposition 1.5.11. *Let (X, μ) be a measure space. The integral has the following properties:*

1. *Linearity: $\int (f + g) d\mu = \int f d\mu + \int g d\mu$.*

2. *Additivity: If A and B are disjoint then $\int_{A \cup B} f d\mu = \int_A f d\mu + \int_B f d\mu$.*

3. *Monotonicity: if $f \leq g$ then $\int f d\mu \leq \int g d\mu$.*

4. *Triangle inequality: $|\int f d\mu| \leq \int |f| d\mu$.*

The proof of the above proposition is elementary, though things have to be interpreted appropriately; e.g. linearity actually means that if f and g are integrable then so is $f + g$ and that the equality holds. As for monotonicity, it is enough to assume that $f \leq g$ **almost everywhere** (usually abbreviated as **a.e.**), meaning there is a set $A \in \Sigma$ such that $\mu(X \setminus A) = 0$ and $f(x) \leq g(x)$ for all $x \in A$. The notion of "almost everywhere" is very useful in measure theory. For example, we say that a sequence of functions $\{f_n\}$ of complex or extended real valued functions converges almost everywhere (a.e.) to f if there is a set $A \in \Sigma$ such that $\mu(X \setminus A) = 0$ and $\lim_n f(x) = f(x)$ for all $x \in A$. The main advantages of the notion of integral we gave above are the nice closure and continuity properties it possesses, in particular with respect to almost everywhere convergence.

Exercise 1.5.12. Show that if $\{f_n\}$ is a sequence of measurable functions, then $\sup_n f_n$ is measurable. Conclude that $\lim_n f_n$, if it exists almost everywhere, defines a measurable function.

Now we present two very significant and useful convergence theorems for the integral.

Theorem 1.5.13 (Monotone convergence theorem). *Let $0 \leq f_1 \leq f_2 \leq \ldots$ be a monotone sequence of nonnegative measurable functions on a measure space (X, μ). Then*

$$\lim_n \int f_n d\mu = \int (\lim_n f_n) d\mu.$$

Theorem 1.5.14 (Dominated convergence theorem). *Let* (X, μ) *be a measure space, and let* $\{f_n\}$ *be a sequence of measurable functions that converge almost everywhere to a function* f. *Assume that there exists an integrable* g *such that* $|f_n| \leq g$ *almost everywhere. Then* f *is integrable and*

$$\lim_n \int f_n d\mu = \int f d\mu.$$

1.5.3 Definition of L^p spaces

We finally have enough measure theoretic machinery to define L^p spaces and show that they are Banach spaces.

Example 1.5.15 (L^p-spaces, $p \in [1, \infty)$). Let (X, μ) be a measure space. For $p \in [1, \infty)$, one may define the Lebesgue space $L^p(X, \mu)$ as the space of all measurable functions on X for which $\int_X |f|^p d\mu < \infty$, equipped with the so-called p-norm $\| \cdot \|_p$:

$$\|f\|_p = \left(\int_X |f|^p d\mu \right)^{1/p}.$$

Actually, the p-norm is a semi-norm, because functions that are zero almost everywhere have zero "norm". The persnickety way of defining L^p is as follows.

Let $\mathcal{L}^p = \mathcal{L}^p(X, \mu)$ be the space of measurable functions $f : X \to \mathbb{C}$, which satisfy

$$\int_X |f|^p d\mu < \infty.$$

Then $\|f\|_p = \left(\int_X |f|^p d\mu \right)^{1/p}$ is a semi-norm on $\mathcal{L}^p$ (details regarding the triangle inequality are relegated to Subsection 1.5.4). The space $L^p = L^p(X, \mu)$ is defined to be the quotient space $\mathcal{L}^p / \ker \| \cdot \|_p$. By Proposition 1.4.2, L^p is a normed space. In fact, L^p is a Banach space (details are provided in Subsection 1.5.5). If $\dot{f}$ is an element of L^p with representative $f \in \mathcal{L}^p$, then we define $\|\dot{f}\|_p = \|f\|_p$, and one never runs into any problems by identifying functions, their equivalence class, and all other functions in that equivalence class.

Every measure space gives us a different instance of an L^p space. For example, if $X = \{1, 2, \dots, \}$ and μ is the counting measure, then we get the spaces ℓ^p discussed in Example 1.2.6. Of course, in this case, no quotient is required. On the other hand, if $X = \mathbb{R}$ or $X = [a, b]$ and μ is the Lebesgue measure, then one can show that we obtain the same spaces $L^p(\mathbb{R})$ and $L^p[a, b]$ which we defined in Example 1.2.11.

Remark 1.5.16. We now have two definitions of $L^p(\mathbb{R})$. By Example 1.2.11 $L^p(\mathbb{R})$ is a completion of $C_c(\mathbb{R})$ with respect to a certain norm. By Example 1.5.15 specialized to Lebesgue measure on the line, $L^p(\mathbb{R})$ is the space of p-integrable measurable functions. Each definition has its merits.

Peter Lax wrote in the foreword to [31] that "it is an accident of history that measure theory was invented before functional analysis", and in Appendix A of [31] he shows how starting from the Banach space L^1 — "the object of our desire" — one can obtain the Lebesgue measure on $\mathbb{R}$. Also, in [44] L^p spaces are introduced as completion spaces, and it is shown that for many purposes and applications, the definition of L^p as a completion serves just as well as the traditional one, with the added bonus that it requires no measure theoretic prerequisites. Moreover, the facts that $L^p(\mathbb{R})$ is complete and that $C_c(\mathbb{R})$ is dense in $L^p(\mathbb{R})$ require proof in the traditional approach, while if defined as in Example 1.2.11 we get them for free.

On the other hand, the traditional path has its advantages. First, obviously, the definition of the integral given in this section is incredibly abstract and can be used in spaces with no underlying geometry or even topology, spaces in which no analogue of the Riemann integral exists. Second, there are areas in mathematics in which "the object of our desire" is actually the measure and not the space, for example in probability. But the biggest advantage of the traditional path for us is that in order to robustly apply functional analysis to analysis, we can't be satisfied working with abstract spaces and we want to know what our spaces are made of. In the traditional approach, it is easier to think of L^p spaces as spaces of functions.

Example 1.5.17 (The space $L^\infty(X, \mu)$). Again, let (X, μ) be a measure space. The space $L^\infty = L^\infty(X, \mu)$ is defined to be the space of all essentially bounded functions on X, with the essential supremum norm. In more detail, let $\mathcal{L}^\infty$ be the vector space of all ***essentially bounded*** measurable functions defined by

$$\mathcal{L}^\infty = \mathcal{L}^\infty(X, \mu) = \{f : X \xrightarrow{measurable} \mathbb{C} : \|f\|_{e.s.} < \infty\}$$

where $\|f\|_{e.s.}$ is the ***essential supremum*** of f:

$$\|f\|_{e.s.} = \inf\{M : \mu(\{x : |f(x)| > M\}) = 0\}$$
$$= \sup\{|z| : \mu(\{x : |f(x) - z| < \epsilon\}) > 0\} \text{ for all } \epsilon > 0\}.$$

If X has no nonempty measurable sets of measure zero, then the essential supremum is the supremum norm $\|f\|_\infty = \sup_{x \in X} |f(x)|$. If there are nonempty sets of zero measure, then $\|\cdot\|_{e.s.}$ is a semi-norm, and we define

$$L^\infty = \mathcal{L}^\infty / \ker \|\cdot\|_{e.s.}.$$

The norm on L^∞ is denoted $\|f\|_\infty$ and this usually doesn't lead to confusion.

Remark 1.5.18. It is worth noting that

$$\ker \|\cdot\|_p = \{f : X \xrightarrow{measurable} \mathbb{C} : f = 0 \text{ , a.e.}\},$$

and in particular, it is the same space of functions for all $p \in [1, \infty]$.

1.5.4   The triangle inequality for the p-norm

In this section, we explain why the p-norm satisfies the triangle inequality. Now, if we replace sums with integrals in the proof of Minkowski's inequality for sequences (Lemma 1.2.8), then we obtain Minkowski's inequality for integrals, hence the triangle inequality holds for the p-norm. The only missing technical piece is the integral analogue of Hölder's inequality for sequences (Lemma 1.2.7), which we now prove.

Lemma 1.5.19 (Hölder's inequality for integrals). *For every two measurable functions f and g on a measure space (X, μ) we have*

$$\int |fg| d\mu \leq \left(\int |f|^p d\mu \right)^{1/p} \left(\int |g|^q d\mu \right)^{1/q}.$$

Proof. It suffices to prove the inequality for nonnegative functions. By the definition of the integral and the monotone convergence theorem, it is enough to prove the inequality for simple functions, i.e. for functions that are a finite linear combination of characteristic functions $f = \sum a_n \chi_{A_n}$ and $g = \sum b_n \chi_{A_n}$ (note that we assumed, without loss of generality, that f and g are defined by the same sequence of disjoint sets). Then, using Hölder's inequality for sequences (Lemma 1.2.7) and writing $\mu(A_n) = \mu(A_n)^{1/p} \mu(A_n)^{1/q}$, we find

$$\int fg d\mu = \sum a_n b_n \mu(A_n) \leq \left(\sum a_n^p \mu(A_n) \right)^{1/p} \left(\sum a_n^q \mu(A_n) \right)^{1/q}$$

$$= \left(\int f^p d\mu \right)^{1/p} \left(\int g^q d\mu \right)^{1/q}$$

as required. $\qquad\qquad\qquad\qquad\qquad\qquad\qquad\qquad\qquad\qquad\qquad\qquad\quad$ $\square$

1.5.5   Completeness of L^p spaces

To conclude our presentation of L^p spaces, we prove that they are complete.

Proposition 1.5.20. *For every measure space (X, μ) and every $p \in [1, \infty]$ the normed space $L^p = L^p(X, \mu)$ is complete.*

Proof. We shall prove the case $p < \infty$ and leave the easier case $p = \infty$ as an exercise. Let $\{f_n\}$ be a Cauchy sequence in L^p. It suffices to show that a subsequence converges in L^p. By passing to a subsequence, we may assume that $\|f_n - f_{n+1}\|_p < 2^{-n}$ for all n. Now consider

$$g_n = \sum_{k=1}^{n} |f_{k+1} - f_k| \in L^p,$$

and let $g : X \to [0, \infty]$ be the extended real valued function $g = \lim_{n \to \infty} g_n$. Then $\|g_n\|_p \leq \sum_{k=1}^{n} \|f_{k+1} - f_k\|_p \leq 1$, and therefore $\int |g_n|^p d\mu \leq 1$. By the monotone convergence theorem, $\int |g|^p d\mu \leq 1 < \infty$, so $g \in L^p$. In particular, it follows that g must have finite values a.e., so the series $\sum_{k=1}^{\infty}(f_{k+1} - f_k)$ converges absolutely a.e., and we use the limit to define

$$f = \lim_{n \to \infty} f_1 + \sum_{k=1}^{n}(f_{k+1} - f_k) = \lim_{n \to \infty} f_n$$

almost everywhere. Crashing through with absolute values we have $|f| \leq g + |f_1| \in L^p$, so $f \in L^p$. We claim that $f_n \to f$ in L^p. Indeed,

$$|f - f_n| = |f - f_1 - \sum_{k=1}^{n}(f_{k+1} - f_k)|$$
$$\leq |f| + |f_1| + |g_n| \leq |f| + |f_1| + |g| \in L^p.$$

Invoking the dominated convergence theorem we conclude that $\int |f - f_n|^p d\mu \to 0$, which means that $\|f - f_n\|_p \to 0$, as required. $\qquad\square$

Exercise 1.5.21. Proposition 1.5.20 shows that the L^p spaces are complete. But does it, really? The L^p spaces were defined to be quotient spaces (see Example 1.5.15), whereas here we worked with functions; that is, we showed that the *semi-normed space* $(\mathcal{L}^p, \|\cdot\|_p)$ is complete (in a certain sense). Explain in precise terms why Proposition 1.5.20 implies completeness of L^p.

Exercise 1.5.22. Prove that L^∞ is complete.

1.6   Additional exercises

Exercise 1.6.1. Let X be a normed space. We say that a series $\sum_{n=1}^{\infty} x_n$ *converges* if the limit $\lim_{N \to \infty} \sum_{n=1}^{N} x_n$ exists. In other words, $\sum_{n=1}^{\infty} x_n$ converges if there exists $x \in X$ such that $\lim_{N \to \infty} \|x - \sum_{n=1}^{N} x_n\| = 0$.

1. Prove that if $\sum x_n$ exists, then $\|\sum x_n\| \leq \sum \|x_n\|$.

2. Prove that X is complete if and only if $\sum \|x_n\| < \infty$ implies that $\sum x_n$ converges. In other words: a normed space is complete if and only if every "absolutely convergent" series is convergent.

Exercise 1.6.2. Prove that c_0 and c are closed subspaces of ℓ^∞. Prove that for every $a = (a_n)_0^\infty \in \ell^\infty$, the norm of the coset $a + c_0$ in the quotient space ℓ^∞/c_0 is equal to

$$\|a + c_0\| = \limsup_{n \to \infty} |a_n|.$$

Can you say something about the norms on ℓ^∞/c and c/c_0?

Exercise 1.6.3. Let H be a Hilbert space and let M be a closed subspace of H. Prove that the quotient space H/M is isometrically isomorphic to $M^{\perp}$. Provide two proofs (a) give a description of an isometric isomorphism (b) use orthonormal bases.

Exercise 1.6.4. Let X be a locally compact Hausdorff space. Prove that $C_0(X)$ is complete and that it is the completion of $C_c(X)$. (Ponder: what is the role of Hausdorffness? What is the role of local-compactness?).

Exercise 1.6.5. Discuss the continuity of the p-norm as a function of p (in L^p spaces or ℓ^p spaces, according to your taste and the amount of time you have).

Exercise 1.6.6. Prove that a Hamel basis[3] in an infinite dimensional Banach space must be uncountable. (**Hint:** if $\{v_1, v_2, \dots\}$ is a Hamel basis for X, then X is the countable union $\cup_n \mathrm{span}\{v_1, \dots, v_n\}$; now use Baire's theorem.)

Exercise 1.6.7. Prove that a finite dimensional subspace of a normed space is closed and that every finite dimensional normed space is a Banach space.

Exercise 1.6.8. Let $C^k([a, b])$ denote the space of functions real valued that are k times differentiable, such that $f, f', \dots f^{(k)} \in C([a, b])$. For every $f \in C^k([a, b])$, define $\|f\|_{C^k} = \|f\|_{\infty} + \dots + \|f^{(k)}\|_{\infty}$, that is,

$$\|f\|_{C^k} = \sup_{t \in [a,b]} |f(t)| + \sup_{t \in [a,b]} |f'(t)| + \dots + \sup_{t \in [a,b]} |f^{(k)}(t)|.$$

Prove that $(C^k([a, b]), \|\cdot\|_{C^k})$ is a Banach space.

Exercise 1.6.9. For any $r, p \in [1, \infty]$ and every $m, n \in \mathbb{N}$, the space $M_{m,n}(\mathbb{C})$ of $m \times n$ matrices can be identified with the space of bounded operators $B(\ell_n^p, \ell_m^r)$. Find a formula, in terms of the coefficients of a matrix $A = (a_{ij})_{i,j=1}^{m,n}$, for the operator norms

$$\|A\|_{\ell_n^1 \to \ell_m^p} \quad \text{and} \quad \|A\|_{\ell_n^p \to \ell_m^{\infty}} \quad \text{for } p \in [1, \infty].$$

Exercise 1.6.10. Prove that an infinite matrix $A = (a_{ij})_{i,j=0}^{\infty}$ gives rise to a bounded operator on ℓ^1 by matrix multiplication, if and only if

$$\sup \left\{ \sum_{i=0}^{\infty} |a_{ij}| : j \in \mathbb{N} \right\} < \infty.$$

Prove that the expression above is then equal to the operator norm $\|A\|$.

Exercise 1.6.11 (Best approximation in normed spaces). This exercise should be solved with the analogous Hilbert space result in mind.

---

[3] A Hamel basis $\mathcal{B}$ for a vector space V is just a basis in the sense of linear algebra, i.e. every $v \in V$ can be written in a unique way as a linear combination of elements in $\mathcal{B}$.

1. Let X be a finite dimensional normed space, let $S \subseteq X$ be a closed convex subset, and let $x \in X$. Prove that there exists a point $s \in S$, such that
$$\|x - s\| = d(x, S) = \inf\{\|x - y\| : y \in S\}.$$

2. Give an example of a Banach space X, a closed convex subset $S \subseteq X$, and a point $x \in X$, such that there is more than one $s \in S$ for which
$$\|x - s\| = d(x, S) = \inf\{\|x - y\| : y \in S\}.$$

   Does such an example exist if S is assumed to be a linear subspace?

3. Construct an example of a Banach space X, a closed convex subset $S \subseteq X$, and an element $x \in X$, such that there is no $s \in S$ for which
$$\|x - s\| = d(x, S) = \inf\{\|x - y\| : y \in S\}.$$

   (**Hint:** Let X be $C_{\mathbb{R}}([0,1])$ with the supremum norm, and let $S = \{f \in X : \int_0^1 f(t)dt = \frac{1}{2}$ and $f(0) = 1\}$.)

Exercise 1.6.12 (Existence and uniqueness of best approximation in uniformly convex Banach spaces). A normed space X is said to be ***uniformly convex*** if for all $\epsilon > 0$, there exists $\delta > 0$, such that for all $x, y \in X_1$
$$\|x - y\| > \epsilon \text{ implies that } \left\|\frac{x + y}{2}\right\| < 1 - \delta.$$

1. Prove that every inner product space is uniformly convex.

2. Give an example of a Banach space, in which the norm is not induced by an inner product, that is uniformly convex.

3. Prove that in a uniformly convex Banach space X, for every closed and convex set $S \subseteq X$ and every $x \in X$, there exists a unique $s \in S$ such that
$$\|x - s\| = d(x, S) = \inf\{\|x - y\| : y \in S\}.$$

Exercise 1.6.13. On the space $C^1([a, b])$ we define three norms:

1. $\|f\| = \|f\|_\infty + \|f'\|_\infty$ (recall Exercise 1.6.8),

2. $\|f\|' = |f(0)| + \|f'\|_\infty$,

3. $\|f\|^* = \|f\|_\infty + |f'(0)|$.

Determine which two of these norms are equivalent.

Chapter 2

The Hahn–Banach theorems and duality

2.1 The Hahn–Banach theorems

The Hahn–Banach theorems are a cornerstone of functional analysis. There are a number of theorems that go by the name "the Hahn–Banach theorem", all of them are concerned with the existence of well behaved linear functionals on vector spaces, and in particular with the existence of continuous linear functionals on normed spaces. This chapter contains essentially the same material as [, Chapter 13].

2.1.1 The Hahn–Banach extension theorems

Definition 2.1.1. Let X be a real vector space. A **sublinear functional** is a function $p : X \to \mathbb{R}$ such that

1. $p(x + y) \leq p(x) + p(y)$ for all $x, y \in X$.

2. $p(cx) = cp(x)$ for all $x \in X$ and $c \geq 0$.

For example, a semi-norm is a sublinear functional. We will see other interesting examples.

Theorem 2.1.2 (Hahn–Banach extension theorem, sublinear functional version). *Let X be a real vector space, and let p be a sublinear functional on X. Suppose that $Y \subseteq X$ is a subspace, and that f is a linear functional on Y such that $f(y) \leq p(y)$ for all $y \in Y$. Then, there exists a linear functional F on X such that $F\big|_Y = f$ and $F(x) \leq p(x)$ for all $x \in X$.*

Proof. Let Y be a subspace, $f : Y \to \mathbb{R}$ a linear functional dominated by p as in the theorem, and let $x \notin Y$. Define

$$W = \{y + cx : y \in Y, c \in \mathbb{R}\}.$$

Our first goal is to show that one can always extend f to a functional F on W such that $F(w) \leq p(w)$ for all $w \in W$. We will later use this fact to show that f can be extended all the way up to X.

DOI: 10.1201/9781003297864-2

Since x is independent of Y, we are free to define $F(x) = t$ for any real number t, and that determines uniquely a linear extension F of f, given by $F(y+cx) = f(y)+ct$. The issue here is to choose t so that F is dominated by p on W. Now, having chosen the value $F(x) = t$, the requirement $F(w) \le p(w)$ for all $w \in W$ is equivalent to

$$f(y) + ct = F(y + cx) \le p(y + cx) \tag{2.1}$$

for all $y \in Y, c \in \mathbb{R}$. If $c = 0$, then (2.1) becomes $f(y) \le p(y)$, which is satisfied by assumption. By dividing (2.1) by $|c|$, and replacing y with a multiple of y, we see that the validity of (2.1) for all $c \ne 0$ and all $y \in Y$, is equivalent to

$$t \le p(x - y) + f(y) \text{ and } f(y) - p(y - x) \le t \tag{2.2}$$

being true for all $y \in Y$. Therefore, if there exists any $t \in \mathbb{R}$ such that

$$\sup\{f(z) - p(z - x) : z \in Y\} \le t \le \inf\{p(x - y) + f(y) : y \in Y\}, \tag{2.3}$$

then we can define $F(x) = t$, and that determines a linear extension F defined on W, which is dominated by p. To see that the supremum appearing in (2.3) is smaller than the infimum, we let $y, z \in Y$, and estimate

$$f(z) - f(y) = f(z - y) \le p(z - y) \le p(z - x) + p(x - y),$$

or $f(z) - p(z - x) \le f(y) + p(x - y)$. We conclude that a t satisfying (2.3) exists, thus we can always extend a linear functional as in the statement of the theorem to a subspace spanned by Y and by another vector.

To complete the proof of the theorem, let $\mathcal{P}$ be the collection of all pairs (Z, g) such that

1. Z is a linear subspace of X containing Y.

2. g is a linear functional on Z that extends f.

3. $g(z) \le p(z)$ for all $z \in Z$.

The pair (Y, f) is in $\mathcal{P}$, so $\mathcal{P}$ is not empty. Let $\mathcal{P}$ be partially ordered by the rule $(Z, g) \le (Z', g')$ if and only if $Z' \supseteq Z$ and $g'|_Z = g$. It is easy to see that every chain in $\mathcal{P}$ has an upper bound, thus by Zorn's lemma $\mathcal{P}$ has a maximal element $(\hat{Z}, \hat{g})$. Now, $\hat{Z}$ must be X, otherwise, there would exist $x \notin \hat{Z}$, and then, by the first part of the proof, we would be able to extend $\hat{g}$ to the space $\{z + cx : z \in \hat{Z}, c \in \mathbb{R}\}$. But that would contradict maximality, thus we conclude that $\hat{Z} = X$, and $F = \hat{g}$ is the required extension of f. □

Remark 2.1.3. It is sometimes interesting to keep track of just how much our proofs are nonconstructive, or whether they depend on the Axiom of Choice in an essential way. If X is finite dimensional, or if X is a separable normed space and p is continuous, then the Axiom of Choice can be avoided in the proof.

Recall that for any normed space X, the dual space X^* is defined to be the space of all bounded linear functionals from X into the scalar field, with norm given by

$$\|F\| = \sup_{x \in X_1} |F(x)|, \quad F \in X^*,$$

where $X_1 = \{x \in X : \|x\| \leq 1\}$.

Theorem 2.1.4 (Hahn–Banach extension theorem, bounded functional version). *Let Y be a subspace of a normed space X, and let $f \in Y^*$. Then, there exists $F \in X^*$ such that $F|_Y = f$ and $\|F\| = \|f\|$.*

Proof. We prove the theorem first for the case where X is a vector space over the reals. Define $p(x) = \|f\|\|x\|$ for all $x \in X$. This is easily seen to be a sublinear functional on X, which dominates f on Y. By Theorem 2.1.2, there exists F extending f such that $|F(x)| \leq p(x) = \|f\|\|x\|$ for all x, thus $\|F\| \leq \|f\|$. Since F extends f, it follows that $\|F\| = \|f\|$.

Suppose now that X is a normed space over the complex numbers. Then X and Y are also normed spaces over the reals, and Y is a subspace of X also when these are considered as real linear spaces. Define $g = \operatorname{Re} f$. Then g is a bounded real functional on Y, and $\|g\| \leq \|f\|$. By the previous paragraph, g extends to a bounded real functional G on X such that $\|G\| = \|g\|$.

We now define $F(x) = G(x) - iG(ix)$. Some computations show that F is linear and extends f. To get $\|F\| = \|f\|$, it suffices to show that $\|F\| \leq \|G\|$.

Fix $x \in X$, and write $F(x) = re^{it}$. Then $|F(x)| = r = e^{-it}F(x) = F(e^{-it}x) = G(e^{-it}x) - iG(ie^{-it}x)$. But this is a real number, so its imaginary part vanishes and we get $|F(x)| = G(e^{-it}x) \leq \|G\|\|x\|$, which proves $\|F\| \leq \|G\|$. □

Exercise 2.1.5. Show that in the above proof, the functional F is a linear functional that extends f.

Exercise 2.1.6. Let X be a vector space over the complex numbers. X can also be considered as a real vector space. Show that for every complex linear functional $F : X \to \mathbb{C}$, the real part of F (the functional $x \mapsto \operatorname{Re} F(x)$) is a real linear functional on the real space X, and that

$$F = \operatorname{Re} F(x) - i \operatorname{Re} F(ix), \quad x \in X.$$

Deduce that $F \mapsto \operatorname{Re} F$ is a bijection between complex linear functionals and real linear functionals on X.

2.1.2 The Hahn–Banach separation theorems in a normed space

Theorem 2.1.7 (Hahn–Banach separation theorem, subspace/point). *Let M be a linear subspace of a normed space X, and let $x \in X$. Put*

$$d = d(x, M) = \inf\{\|x - m\| : m \in M\}.$$

Then, there exists $F \in X_1^* = \{G \in X^* : \|G\| \leq 1\}$, such that $F(x) = d$ and $F\big|_M = 0$.

Proof. On span$\{x, M\}$ we define a linear functional by $f(cx + m) = cd$. By definition, $f(x) = d$ and $f\big|_M = 0$. For every $c \neq 0$ and every $m \in M$,

$$\|cx + m\| = |c|\|x - (-c^{-1}m)\| \geq |c|d,$$

thus

$$|f(cx + m)| = |cd| \leq \|cx + m\|.$$

This shows that $\|f\| \leq 1$. Applying the Hahn–Banach extension theorem to f provides $F \in X_1^*$ that satisfies all the requirements. $\qquad\square$

Corollary 2.1.8. *Let X be a normed space and $x \in X$. Then, there exists $F \in X_1^*$ for which $F(x) = \|x\|$.*

Proof. This follows Theorem 2.1.7 applied to $M = \{0\}$. $\qquad\square$

Corollary 2.1.9. *Let M be a subspace in a normed space. A point $x \in X$ is in $\overline{M}$ if and only if $F(x) = 0$ for all $F \in X^*$ that vanishes on M.*

Proof. One implication follows from the continuity of F. For the other implication, we use Theorem 2.1.7 to see that if $F(x) = 0$ for all F that vanishes on M, then $d(x, M) = 0$, in other words $x \in \overline{M}$. $\qquad\square$

Corollary 2.1.10. *A subspace M is dense in X if and only if for all $F \in X^*$, $F\big|_M = 0$ implies $F = 0$.*

Theorem 2.1.7 says that there is a functional $F \in X_1^*$ giving witness to the fact that the distance of x from M is d. This is sometimes restated by saying that F "separates" M from x. We now turn our attention to using functionals for separating convex sets. We recall a couple of definitions.

Definition 2.1.11. Let X be a vector space. A subset $C \in X$ is said to be *convex* if for all $x, y \in C$,

$$(1 - t)x + ty \in C, \quad \text{for all } t \in [0, 1].$$

Definition 2.1.12. A *hyperplane* in X is a subset of the form

$$F^{-1}(c) = \{x \in X : F(x) = c\},$$

where F is a nonzero linear functional on X and c is a scalar.

Note that a subset H of a vector space X is a hyperplane if and only if there exists a nonzero linear functional F and some $x_0 \in X$ such that $H = x_0 + \ker F$.

The Hahn–Banach separation theorems are most conveniently stated for real vector spaces. Since every complex space is also a real space, the separation theorems below can be applied to complex spaces — the statements and definitions should be modified by replacing every linear functional F with its real part $\operatorname{Re} F$, and understanding that by a separating hyperplane we mean a separating **real hyperplane** $(\operatorname{Re} F)^{-1}(c)$.

Definition 2.1.13. If A, B are subsets of a real vector space, we say that the hyperplane $F^{-1}(c)$ **separates** A from B if for all $x \in A, y \in B$

$$F(x) \le c \le F(y).$$

We say that this hyperplane **strictly separates** A from B if there is some $\epsilon > 0$ such that for all $x \in A, y \in B$

$$F(x) \le c - \epsilon < c + \epsilon \le F(y).$$

Draw two convex bodies A and B in the plane, so that their boundaries are disjoint. No matter how you drew A and B, so long as the bodies are convex and do not touch, you will be able to draw a straight line that cuts the plane into two halves, one containing A and the other containing B. A straight line in the plane is given by an equation of the form $ax + by = c$, so this picture illustrates the fact that whenever we are given two disjoint convex bodies A and B in the plane, we can find a hyperplane that separates A from B. The Hahn–Banach separation theorems that we will prove below give precise formulations of this separation phenomenon in the setting of any real normed space X.

In infinite dimensional spaces, a hyperplane is not necessarily a nice set. To make our geometric vocabulary complete, we note that if F is bounded, then $F^{-1}(c)$ is closed. The converse is also true.

Proposition 2.1.14. *A hyperplane $F^{-1}(c)$ in a normed space is closed if and only if it is not dense, and this happens if and only if F is bounded.*

Proof. It suffices to treat the case $c = 0$, because $F^{-1}(c) = \ker F + x_0$ for some x_0 with $F(x_0) = c$. Now, if F is bounded then it is continuous, thus $\ker F = F^{-1}(0)$ is closed. If $\ker F$ is closed then it cannot be dense because $F \ne 0$ (part of the definition of hyperplane is that F is not trivial). If F is not bounded, then there is a sequence $x_n \to 0$ such that $F(x_n) = 1$. Now for $x \notin \ker F$, we have $\ker F \ni x - F(x)x_n \to x$, so $\ker F$ is dense. $\square$

To obtain the separation theorems, we will make use of the following device.

Definition 2.1.15 (The Minkowski functional). Let X be a normed space and let $C \subseteq X$ be a convex and open set containing 0. Define $p : X \to [0, \infty)$ by

$$p(x) = \inf\{t > 0 : t^{-1}x \in C\}.$$

p is called the **Minkowski functional** of C.

The reader is invited to work out the Minkowski functional of the open unit ball.

Lemma 2.1.16. *Let X be a normed space, let $C \subseteq X$ be a convex and open set containing 0, and let p be the Minkowski functional of C defined as above. Then, p is a sublinear functional, $C = \{x \in X : p(x) < 1\}$, and there exists some $M > 0$ such that $p(x) \le M\|x\|$ for all $x \in X$.*

Proof. There is some r such that the closed ball $\overline{B}_r(0)$ is contained in C, thus for every $x \ne 0$, $\frac{r}{\|x\|}x \in C$, hence $p(x) \le \frac{1}{r}\|x\|$. That takes care of the last assertion.

We now show that $x \in C$ if and only if $p(x) < 1$. Assume that $x \in C$. The set C is open, so $(1+r)x \in C$ for sufficiently small $r > 0$, whence $p(x) \le (1+r)^{-1} < 1$. Conversely, assume that $p(x) < 1$. Thus there is some $0 < t < 1$ for which $t^{-1}x \in C$. But since x is on the ray connecting 0 and $t^{-1}x$, and since C is convex, $x \in C$. We conclude that $C = \{x \in C : p(x) < 1\}$.

Clearly, $p(0) = 0$. For $c > 0$, $p(cx)$ can be written as

$$\inf\{ct > 0 : (ct)^{-1}cx \in C\} = c\inf\{t > 0 : t^{-1}x \in C\} = cp(x).$$

We proceed to prove that p is subadditive. Let $x, y \in X$ and $r > 0$. From the definition of p together with convexity, $(p(x) + r)^{-1}x$ and $(p(y) + r)^{-1}y$ are in C. Every convex combination

$$t(p(x) + r)^{-1}x + (1 - t)(p(y) + r)^{-1}y \ , \ t \in [0, 1] \tag{2.4}$$

is also in C. We now choose the value of t cleverly, so that the coefficients of x and y will be the same, and in that way, we will get something times $x + y$. The solution of the equation

$$t(p(x) + r)^{-1} = (1 - t)(p(y) + r)^{-1}$$

is $t = \frac{p(x)+r}{p(x)+p(y)+2r}$, and plugging that value of t in (2.4) gives $(p(x) + p(y) + 2r)^{-1}(x + y) \in C$. Thus, $p(x + y) \le p(x) + p(y) + 2r$. Letting r tend to 0 we obtain the result. $\qquad\square$

Theorem 2.1.17 (Hahn–Banach separation theorem, convex/open). *Let X be a real normed space and let $A, B \subseteq X$ be two nonempty disjoint convex sets, and suppose that B is open. Then there exists a closed hyperplane which separates A and B.*

Proof. Let us first treat the case where $0 \in B$ and A consists of one point, say $A = \{a\}$. Let p be the Minkowski functional of B. Define a linear functional $f : \operatorname{span}\{a\} \to \mathbb{R}$ by

$$f : \lambda a \mapsto \lambda.$$

By Lemma 2.1.16, $p(a) \ge 1$. Therefore if $\lambda \ge 0$, then $f(\lambda a) = \lambda \le \lambda p(a) = p(\lambda a)$. If $\lambda < 0$, then $f(\lambda a) < 0 \le p(\lambda a)$. We see that $f(\lambda a) \le p(\lambda a)$ for all λ.

By Theorem 2.1.2, f can be extended to a functional F on X such that $F(x) \leq p(x)$ for all $x \in X$. By the last part of Lemma 2.1.16, F is bounded. This F satisfies $F(a) = 1$ and $F(x) \leq p(x) < 1$ for all $x \in B$, so the closed hyperplane $F^{-1}(1)$ separates A and B, and in fact

$$F(b) < 1 = F(a) \quad \text{for all } b \in B.$$

If $A = \{a\}$ and B is open but does not contain 0, then we choose some $b_0 \in B$ and apply the result from the previous paragraph to the sets $\{a - b_0\}$ and $B - b_0 = \{b - b_0 : b \in B\}$.

Now let us treat the general case. The set $A - B = \{a - b : a \in A, b \in B\}$ is convex, open, and is disjoint from the set $\{0\}$. By the previous paragraph, there is a closed hyperplane that separates $A - B$ and $\{0\}$ as follows

$$F(a - b) < F(0) = 0 \quad \text{for all } a \in A, b \in B,$$

so $F(a) < F(b)$ for all $a \in A$ and $b \in B$. If c satisfies $\sup F(A) \leq c \leq \inf F(B)$, then $F^{-1}(c)$ separates A and B. $\qquad\Box$

Exercise 2.1.18. Prove that if A is convex and if B is convex and open, then the set $A - B$ is convex and open.

Theorem 2.1.19 (Hahn–Banach separation theorem, compact/closed). *Let X be a real normed space and let $A, B \subseteq X$ be two nonempty disjoint convex sets, such that A is closed and B is compact. Then there exists a closed hyperplane which strictly separates A and B.*

Proof. As above, we consider $A - B$, call this set C. As above, C is convex and does not contain 0. Since B is compact, it follows that C is closed. Let $B_r(0)$ be a small ball disjoint from C. By Theorem 2.1.17, there is a functional $F \in X^*$ and $c \in \mathbb{R}$ such that

$$F(a) - F(b) = F(a - b) \leq c \leq \inf F(B_r(0)) = -\|F\| r$$

for all $a \in A$ and $b \in B$. It follows that A and B can be strictly separated by some hyperplane defined by F. $\qquad\Box$

Exercise 2.1.20. Prove that if A is closed and B is compact, then the set $A - B$ is closed. What happens if B is not assumed to be compact?

2.2 Application: the Banach limit

Let ℓ^∞ be the Banach space consisting of all bounded sequences of complex numbers, equipped with the supremum norm

$$\ell^\infty = \left\{ a = (a_n)_{n=0}^\infty : \|a\|_\infty = \sup_{n \geq 0} |a_n| < \infty \right\}.$$

The space ℓ^∞ contains a closed subspace

$$c = \left\{ a \in \ell^\infty : \lim_{n \to \infty} a_n \text{ exists} \right\}.$$

The functional $f : a \mapsto \lim_n a_n$ is a bounded linear functional on c. By the Hahn–Banach theorem, this functional can be extended to a bounded linear functional F defined on all of ℓ^∞. We will show that F can be chosen to have some of the nice properties that f has. For example, if we let S denote the backward shift operator

$$S\,(a_0, a_1, a_2, \ldots) = (a_1, a_2, \ldots),$$

then f is translation invariant, in the sense that $f(Sa) = f(a)$ for all $a \in c$.

Theorem 2.2.1 (Banach limit). *There exists a bounded linear functional L on ℓ^∞, such that for all $a = (a_n)_{n=0}^\infty$,*

1. *$L(Sa) = L(a)$,*

2. *If $m \le a_n \le M$ for all n, then $m \le L(a) \le M$,*

3. *If $\lim_n a_n$ exists, then $L(a) = \lim_n a_n$.*

Remark 2.2.2. A functional on ℓ^∞ satisfying the above conditions is called a **Banach limit**.

Proof. Let M be the range of the operator $S - I$, that is

$$M = \{ Sa - a : a \in \ell^\infty \}.$$

M is a subspace, and every $b \in M$ has the form

$$b = (b_n)_{n=0}^\infty = (a_1 - a_0, a_2 - a_1, \ldots).$$

Such an element b satisfies $\sum_{k=0}^n b_k = a_{n+1} - a_0$. In particular

$$\sup_n \left| \sum_{k=0}^n b_k \right| < \infty. \tag{2.5}$$

Thus, if we let $\mathbf{1} = (1, 1, 1, \ldots)$, then $\mathbf{1} \notin M$. We claim that

$$d(\mathbf{1}, M) = \inf\{\|\mathbf{1} - b\| : b \in M\} = 1.$$

Indeed, if there was an element $b \in M$ such that $\|\mathbf{1} - b\| < 1$, this would contradict (2.5). By Theorem 2.1.7, there exists $L \in (\ell^\infty)^*$ such that $\|L\| = 1$, $L(\mathbf{1}) = 1$ and $L\big|_M = 0$. Since $L(b) = 0$ for all $b \in M$, we see that $L(Sa) = L(a)$ for all $a \in \ell^\infty$.

Next, suppose that $a_n \ge 0$ for all $n \in \mathbb{N}$. We claim that $L(a) \ge 0$. First, let us show that $L(a) \in \mathbb{R}$. Suppose to the contrary that $L(a) = u + iv$ with

$u, v \in \mathbb{R}$ and $v \neq 0$; without loss of generality assume that $v > 0$. Then for $t > 0$ we have that

$$|L(1 - ita)| = |1 - it(u + iv)| \geq 1 + tv.$$

On the other hand, since $a_n \geq 0$,

$$\|1 - ita\|_\infty = \sup_n \sqrt{1 + t^2 a_n^2} = 1 + \frac{t^2 \|a\|_\infty^2}{2} + \text{ higher order terms}$$

and we find that $\|1 - ita\| < |L(1 - ita)|$ for small $t > 0$, contradicting $\|L\| = 1$. Thus $L(a)$ must be real. Now, consider the sequence

$$x = \|a\| \mathbf{1} - a = (\|a\| - a_0, \|a\| - a_1, \|a\| - a_2, \dots).$$

By definition, $x_n \geq 0$ for all n, so as above $L(x)$ must be real. Since $\|x\| \leq \|a\|$ and $\|L\| = 1$, we get

$$\|a\| - L(a) = L(x) \leq \|L\| \|x\| \leq \|a\|,$$

and it follows that $L(a) \geq 0$. From here, it is easy to prove that L satisfies the second property stated in the theorem.

Finally, to prove that $L(a) = \lim_n a_n$ when the limit exists it suffices to prove this for bounded sequences a such that $a_n \in \mathbb{R}$ for all n. We will show that for any such sequence,

$$\liminf_n a_n \leq L(a) \leq \limsup_n a_n.$$

We will show the second inequality; the first one is shown in a similar manner. For this, consider the sequence $y = (a_k, a_{k+1}, a_{k+2}, \dots)$. Then $y_n \leq \sup_{m \geq k} a_m$ for all n, and therefore

$$L(a) = L(y) \leq \sup_{m \geq k} a_m.$$

Letting $k \to \infty$ we obtain $L(a) \leq \limsup_n a_n$. The other inequality is proved similarly. □

We can now settle a question that was raised in the first chapter. Recall that in Example 1.3.9, we noted that there exists an isometric map

$$\ell^1 \ni b \mapsto \Gamma_b \in (\ell^\infty)^*,$$

where Γ_b is given by

$$\Gamma_b(a) = \sum a_n b_n.$$

Motivated by the case where the roles of ℓ^1 and ℓ^∞ are reversed and by the case of ℓ^p spaces where $1 < p < \infty$, one is naturally led to ask whether the map $b \mapsto \Gamma_b$ is surjective.

Exercise 2.2.3. Show that the map $b \mapsto \Gamma_b$ discussed above is not surjective.

Note that the above exercise only settles the question of whether a particular map is an isomorphism between ℓ^1 and $(\ell^\infty)^*$, that is, the exercise shows that the map given by $b \mapsto \Gamma_b$ is not an isomorphism. But it leaves open the question of whether ℓ^1 is isometrically isomorphic (or maybe just isomorphic) to the dual of ℓ^∞. We will answer this question in the next section.

2.3   The dual space and the double dual

In the first chapter, we defined the dual space X^* of a Banach space X, but we did not prove anything significant regarding dual spaces. We could not determine whether or not $(\ell^\infty)^* \cong \ell^1$. In fact, we could not even say that a general normed space has *any* nonzero bounded functional defined on it.

Now with the Hahn–Banach theorems at our hands, we know that every normed space has lots of bounded functionals on it — enough to separate points. One can learn many things about a Banach space X from its dual X^*. For example:

Exercise 2.3.1. Prove that if X^* is separable, then X is separable, too.

Exercise 2.3.2. Prove that $(\ell^\infty)^*$ is not isomorphic to ℓ^1.

Definition 2.3.3. Let X be a normed space. Then X^* is a Banach space in its own right and therefore has a dual space

$$X^{**} = (X^*)^*.$$

The space X^{**} is called the **double dual** (or **bidual**) of X.

Every $x \in X$ gives rise to a function $\hat{x} \in X^{**} = (X^*)^*$ by way of

$$\hat{x}(f) = f(x), \quad f \in X^*.$$

Proposition 2.3.4. *The map $x \mapsto \hat{x}$ is an isometry from X into X^{**}.*

Proof. Linearity is trivial. To see that the map is norm preserving,

$$\|\hat{x}\| = \sup_{f \in X_1^*} |f(x)| = \|x\|,$$

where the first equality is true by definition of the norm of a functional, and the second equality follows from Corollary 2.1.8.   □

Definition 2.3.5. A Banach space X is said to be **reflexive** if the map $x \mapsto \hat{x}$ is an isometry of X onto X^{**}.

In Exercise 2.4.14 you will explore some examples and non-examples of reflexive spaces.

Exercise 2.3.6 (Completions revisited). In Proposition 1.2.3 we stated the fact every normed space X can be embedded isometrically as a dense subspace of a Banach space. The isometric embedding of X in its double dual given in Proposition 2.3.4 suggests a different, more conceptual proof of this fact. Let $\widehat{X}$ denote the image of X in X^{**} under the map $x \mapsto \hat{x}$. Prove that $\overline{\widehat{X}}$ is the unique completion of X. Prove that if X is an inner product space, then the norm of $\overline{\widehat{X}}$ is induced by an inner product, thus it is a Hilbert space.

2.4 Additional exercises

Exercise 2.4.1. Let $H = \mathbb{C}^n$ be the standard n-dimensional Hilbert space. We norm the algebra $M_n = M_n(\mathbb{C})$ of $n \times n$ matrices in the following two ways. First, we give M_n the operator norm obtained by identifying it with the space $B(H)$ of bounded operators:

$$\|A\|_{op} = \sup_{h \in H_1} \|Ah\|.$$

Second, we can norm M_n with the following **trace norm**:

$$\|A\|_{tr} = \mathrm{trace}((A^*A)^{1/2}).$$

Prove that the dual space of $(M_n, \|\cdot\|_{op})$ is $(M_n, \|\cdot\|_{tr})$, with pairing:

$$\langle A, B \rangle = \mathrm{trace}(AB).$$

What about the dual of $(M_n, \|\cdot\|_{tr})$?

Exercise 2.4.2. Find the dual space of c_0. Find the dual space of c. You should find that c_0^* is isometrically isomorphic to c^*. Ponder: does this imply that c_0 is isometrically isomorphic to c?

Exercise 2.4.3. Consider the space $X = \ell_3^1$ (i.e., $X = (\mathbb{R}^3, \|\cdot\|_1)$).

1. Let $F \in X^*$ be given by $F(x, y, z) = ax + by + cz$. Prove that $\|F\| = \max\{|a|, |b|, |c|\}$.

2. Let $Y = \{(x, y, z) : z = 0 = x - 3y\}$ be a linear subspace, and let $f : Y \to \mathbb{R}$ be given by $f(x, y, z) = x$, for all $(x, y, z) \in Y$. Find the general form of $F \in X^*$ such that $F\big|_Y = f$ and $\|F\| = \|f\|$.

Exercise 2.4.4. Give a proof of the Hahn–Banach extension theorem (which one? you decide) that works for separable normed spaces, and does not make use of Zorn's lemma.

Exercise 2.4.5. Let X be a real vector space with two norms $\|\cdot\|$ and $\|\cdot\|'$. Let f be a linear functional on X, and suppose that for every $x \in X$, either $f(x) \leq \|x\|$, or $f(x) \leq \|x\|'$. Prove that there is some $t \in [0, 1]$ such that

$$f(x) \leq t\|x\| + (1 - t)\|x\|' \quad \text{for all } x \in X.$$

(**Hint:** consider the convex sets $\{(a, b) : a < 0, b < 0\}$ and $\mathrm{conv}\{(\|x\| - f(x), \|x\|' - f(x)) : x \in X\}$ in $\mathbb{R}^2$, where $\mathrm{conv}(A)$ denotes the smallest convex set containing A.)

Exercise 2.4.6. Prove that *any* two convex sets in a finite dimensional normed space can be separated by a hyperplane.

Exercise 2.4.7. In $H = L^2[-\pi, \pi]$ (with the usual inner product) consider the subspaces

$$M = \left\{ a_0 + \sum_{n=1}^{\infty} \left(a_n \sin(nx) + \frac{a_n}{n} \cos(nx) \right) : (a_n) \in \ell^2 \right\},$$

and

$$N = \left\{ \sum_{n=1}^{\infty} b_n \sin(nx) : (b_n) \in \ell^2 \right\}.$$

Let $f(x) = \sum_{n=1}^{\infty} \frac{1}{n} \cos(nx)$.

1. Prove that $\overline{M + N} = L^2$, but that $f \notin M + N$.

2. Prove that $f + N$ and M are convex, disjoint and closed sets, but that they cannot be separated by a closed hyperplane.

3. Can they be separated by *some* hyperplane?

Exercise 2.4.8 (Amenability of $\mathbb{Z}$). Let G be a group. A function $\mu : 2^G \mapsto [0, 1]$ such that

- $\mu(A_1 \cup A_2) = \mu(A_1) + \mu(A_2)$ for every disjoint $A_1, A_2 \in 2^G$,

- $\mu(G) = 1$.

is called a ***finitely additive probability measure*** (as opposed to a probability measure, which is countably additive). A group G is said to be ***amenable*** if there exists a finitely additive probability measure on G which is also ***shift invariant***, in the sense that $\mu(gA) = \mu(A)$ for all $g \in G$ and $A \subseteq G$. (Here we use the notation $gA = \{ga : a \in A\}$).

1. Prove that every finite group is amenable[1].

2. Prove that for every n, the free group $\mathbb{F}_n$ generated by n generators is not amenable[2].

3. Prove that $\mathbb{Z}$ is amenable (**Hint:** get inspiration from the Banach limit construction.)

Exercise 2.4.9 (Non-uniqueness of Hahn–Banach extensions). Let $C_\mathbb{R}([0,1])$ be the space of all continuous real valued functions on $[0,1]$, and let $\mathcal{B}_\mathbb{R}([0,1])$ be the space of all bounded Borel measurable functions equipped with the supremum norm. Let $f : C_\mathbb{R}([0,1]) \to \mathbb{R}$ be the linear functional $f(\varphi) = \int_0^1 \varphi(t)dt$.

1. Prove that for all $a \in [0,1]$, there is a linear functional $F : \mathcal{B}_\mathbb{R}([0,1]) \to \mathbb{R}$, that extends f and such that $\|F\| = 1$ and $F(\chi_{[0,1]\cap\mathbb{Q}}) = a$.

2. Let $c,d \in (0,1)$. For F as above, find all possible values of $F(\chi_{[c,d]})$ (**Hint:** if $\varphi \le \psi$, what can you say about $F(\varphi)$ and $F(\psi)$?)

Exercise 2.4.10. True or false: there exists an element $f \in \ell^\infty \setminus c$ of norm 1, such that for all $c \in [0,1]$, there exists a Banach limit L (i.e., a norm one, positive, shift invariant extension of $a \mapsto \lim_n a_n$) such that $L(f) = c$.

Exercise 2.4.11 (Application of Banach limits). Let T be an operator on a real Hilbert space $(H, \langle \cdot, \cdot \rangle)$. Suppose that there exist $0 < c < C < \infty$ such that
$$c\|h\| \le \|T^n h\| \le C\|h\| \quad , \quad \text{for all } n \in \mathbb{N}.$$
Prove that T is similar to an isometry; that is, prove that there exists an inner product space G and a bounded invertible linear map $S : H \to G$ with bounded inverse such that STS^{-1} is an isometry.

Exercise 2.4.12. Let X be a reflexive Banach space, and M a closed subspace of X. True or false: M is also a reflexive Banach space?

Exercise 2.4.13. True or false: a Banach space X is reflexive if and only if X^* is reflexive?

Exercise 2.4.14. Determine which of the following spaces is reflexive: a finite dimensional normed space, c_0, c, $C([0,1])$, ℓ^1, ℓ^p $(1 < p < \infty)$, ℓ^∞, $L^\infty[0,1]$.

Exercise 2.4.15. Let X be a real normed space. A function $\phi: X \to \mathbb{R}$ is said to be *lower semicontinuous* if
$$\phi^{-1}((\infty, t]) = \{x \in X : \phi(x) \le t\}$$

---

[1] This is not an exercise in functional analysis.
[2] Neither is this.

is closed for all $t \in \mathbb{R}$. It is said to be **convex** if

$$\phi(tx + (1 - t)y) \le t\phi(x) + (1 - t)\phi(y)$$

for all $x, y \in X$ and all $t \in [0, 1]$. A **continuous affine functional** is a function $g: X \to \mathbb{R}$ of the form $g(x) = f(x) + b$ where $f \in X^*$ and $b \in \mathbb{R}$.

1. Prove that a function $\phi: X \to \mathbb{R}$ is lower semicontinuous if and only if $f(x) \le \liminf_{n\to\infty} f(x_n)$ whenever $x_n \to x$.

2. Prove that function ϕ is lower semicontinuous and convex if and only if it is equal to the supremum of a family of continuous affine functionals, that is, there exists a family $\mathcal{F}$ of continuous affine functionals such that $\phi = \sup_{g \in \mathcal{F}} g$ in the sense that

$$\phi(x) = \sup\{g(x) : g \in \mathcal{F}\}$$

for all $x \in X$.

Chapter 3

The dual spaces of L^p and $C_0(X)$

Our goal in this chapter is to illustrate the utility of the theory accumulated thus far by exhibiting a couple of applications of the Hahn–Banach theorem to problems in analysis. Before presenting the applications we shall need to identify the dual spaces of L^p and of $C_0(X)$, and for this we will require some additional material from measure theory.

3.1 Some measure theoretic preliminaries

In this section, we review notions and results that we need in order to find the dual of L^p. Good references for this material are [19] or [13].

Definition 3.1.1. A measure space (X, Σ, μ) is said to be ***finite*** if $\mu(X) < \infty$; it is ***σ-finite*** if there is a sequence of measurable sets $E_n \in \Sigma$ such that $X = \cup_{n=1}^{\infty} E_n$ and such that $\mu(E_n) < \infty$ for all n.

Usually, the measure *space* is not referred to and one just says that the measure μ is finite or σ-finite. Most measure spaces that arise in reasonable applications to analysis are σ-finite. Every σ-finite space has the useful property that it is ***semi-finite***; which means that for every $E \in \Sigma$ with $\mu(E) > 0$, there exists $F \in \Sigma$, such that $F \subseteq E$ and $0 < \mu(F) < \infty$. To facilitate the discussion, while losing practically nothing, we shall assume below that the measures we work with are σ-finite. We shall also need the notion of a complex measure.

Definition 3.1.2. Let (X, Σ) be a measurable space. A ***complex measure*** on (X, Σ) is a function $\nu : \Sigma \to \mathbb{C}$ such that whenever $\{A_n\}$ is a collection of disjoint measurable sets then $\nu(\cup_n A_n) = \sum_n \nu(A_n)$.

Example 3.1.3. Let (X, Σ, μ) be a measure space and let $h \in L^1(X, \mu)$. Then the set function

$$\nu(A) = \int_A h \, d\mu$$

is a complex measure.

DOI: 10.1201/9781003297864-3

Definition 3.1.4. Let ν be a complex measure on a measure space (X, Σ). The **variation** of ν is the set function $|\nu|$ given by[1]

$$|\nu|(A) = \sup \left\{ \sum_n |\nu(A_n)| : \{A_n\} \text{ is a measurable partition of } A \right\}$$

for all $A \in \Sigma$.

One can show that the variation of a complex measure is a finite measure. A complex measure ν is said to be σ-finite if the (positive) measure $|\nu|$ is σ-finite.

Definition 3.1.5. Suppose that μ and ν are two measures defined on the same measurable space. We say that ν is **absolutely continuous** with respect to the measure μ if $\mu(A) = 0$ implies $\nu(A) = 0$ for every measurable set A. One then writes $\nu \ll \mu$.

The following important and nontrivial theorem will be used below.

Theorem 3.1.6 (Radon–Nikodym theorem). *Let (X, Σ, μ) be a σ-finite measure space. If ν is a complex measure that is absolutely continuous with respect to μ, then there exists a unique $h \in L^1(X, \mu)$, such that*

$$\nu(A) = \int_A h \, d\mu, \quad \text{for all } A \in \Sigma.$$

The function h is called the **Radon–Nikodym derivative** of ν with respect to μ. The conclusion of the theorem is often written as $d\nu = h \, d\mu$ or $h = \frac{d\nu}{d\mu}$.

3.2   The dual of L^p

Let (X, Σ, μ) be a σ-finite measure space. Let $p \in [1, \infty)$ and let $q = \frac{p}{p-1}$ be its conjugate exponent. We will abbreviate $L^p = L^p(X, \mu)$. Our goal in this section is to prove that $(L^p)^* = L^q$, where the pairing between the spaces is given by

$$\langle f, g \rangle = \int f g \, d\mu \quad \text{for all } f \in L^p, g \in L^q.$$

We shall prove this under the assumption that μ is σ-finite. One can show that if $p \in (1, \infty)$ then $(L^p)^* = L^q$ is true for any measure, not necessarily a σ-finite one. In particular, it follows that for such p the space L^p is always

---

[1] By a **measurable partition** of A, we mean a family $\{A_n\}$ of disjoint measurable sets such that A is equal to the disjoint union $\cup_n A_n$.

reflexive. On the other hand, if the measure is not semi-finite then it might happen that $(L^1)^* \neq L^\infty$. See [19, Chapter 6] for a complete treatment. We have already seen that $(L^\infty)^* \neq L^1$ when μ is the counting measure on $\mathbb{N}$ (see Exercise 2.3.2).

3.2.1 The inclusion $L^q \subseteq (L^p)^*$

We will begin by showing that L^q is contained in the dual of L^p. Given any function g, let $\mathrm{sign}\, g$ be the function that is equal to 0 when $g = 0$, and otherwise it is defined so that $g = (\mathrm{sign}\, g)|g|$. Clearly, if g is measurable then so is $\mathrm{sign}\, g$.

Let $g \in L^q$. By Hölder's inequality (Lemma 1.5.19), we have that for every $f \in L^p$,

$$\left| \int fg d\mu \right| \leq \|f\|_p \|g\|_q.$$

Therefore, we can define $\Phi_g \in (L^p)^*$ by $\Phi_g : f \mapsto \int fg$, which is clearly linear and has $\|\Phi_g\| \leq \|g\|_q$.

Proposition 3.2.1. $\|\Phi_g\| = \|g\|_q$.

Proof. Suppose that $g \neq 0$, for otherwise there is nothing to prove.

Case 1: $q = \infty$. It suffices to find, for every $\epsilon > 0$, a function $f \in (L^p)_1$ such that $\Phi_g(f) \geq \|g\|_q - \epsilon$. Let $E = \{x \in X : |g(x)| > \|g\|_\infty - \epsilon\}$. By the definition of essential supremum, $\mu(E) > 0$. By semi-finiteness, there exists $F \in \Sigma$ such that $F \subseteq E$ and $0 < \mu(F) < \infty$. We may therefore define

$$f = \mu(F)^{-1} \chi_F \overline{\mathrm{sign}\, g}.$$

Then $f \in (L^1)_1$, and

$$\Phi_g(f) = \frac{1}{\mu(F)} \int_F |g| d\mu \geq \|g\|_\infty - \epsilon.$$

Case 2: $q < \infty$. This case is simpler: we simply define

$$f = \|g\|_q^{1-q} \overline{\mathrm{sign}\, g} |g|^{q-1}.$$

Then

$$\int |f|^p = \|g\|_q^{(1-q)p} \int |g|^{(q-1)p} d\mu = \|g\|_q^{-q} \|g\|_q^q = 1,$$

so $f \in (L^p)_1$. We check that

$$\Phi_g(f) = \|g\|_q^{1-q} \int \overline{\mathrm{sign}\, g} |g|^{q-1} g d\mu = \|g\|_q^{1-q} \int |g|^q d\mu = \|g\|_q.$$

$\qquad\qquad\qquad\qquad\qquad\qquad\qquad\qquad\qquad\qquad\qquad\qquad\qquad\qquad\qquad\quad \square$

Remark 3.2.2. The above proposition also holds if $p = \infty$ and $q = 1$ (with easy proof).

Exercise 3.2.3. Give an example of a measure space in which the map $g \in L^\infty \mapsto (L^1)^*$ is not injective (of course, you have to change the standing assumptions).

3.2.2 The equality $L^q = (L^p)^*$

In the previous section, we showed that the linear map $g \mapsto \Phi_g$ is an isometry of L_q into $(L^p)^*$. In this section we will show that (in the case $p < \infty$) it is surjective, establishing the following theorem.

Theorem 3.2.4. *Let (X, μ) be a σ-finite measure space and let $p \in [1, \infty)$. Then the map $g \mapsto \Phi_g$ is an isometric surjection of L^q onto $(L^p)^*$.*

Let us write $\mathcal{S}$ for the space of all simple functions $\sum a_n \chi_{A_n}$ where every A_n is measurable of finite measure. A basic fact we shall require is that $\mathcal{S}$ is dense in L^p, for all $p \in [1, \infty)$.

If g is a function such that $gs \in L^1$ for all $s \in \mathcal{S}$, then we may define

$$M_q(g) = \sup\left\{\left|\int gs d\mu\right| : s \in \mathcal{S} \text{ and } \|s\|_p \le 1\right\}$$

and we may also write down the following prescription, which a priori may not be well defined:

$$N_q(g) = \sup\left\{\left|\int gf d\mu\right| : f \in L^p \text{ and } \|f\|_p \le 1\right\}.$$

Proposition 3.2.5. *Let g be a function such that $gs \in L^1$ for all $s \in \mathcal{S}$, and write $M_q = M_q(g)$ and $N_q = N_q(g)$. If $M_q < \infty$, then $g \in L^q$ and $\|g\|_q = M_q = N_q$.*

Proof. Suppose that $M_q < \infty$. To see that gf is integrable for all $f \in L^p$, approximate f with a sequence of simple functions s_n such that $|s_n| \le |f|$ and $s_n \to f$ pointwise (a routine task in measure theory) and apply the dominated convergence theorem. This also shows that $N_q \le M_q$, but now that N_q is seen to be well defined it is clear that $M_q \le N_q$.

By Hölder's inequality, $M_q = N_q \le \|g\|_q$ (with natural interpretation when $g \notin L^q$). The proof will be complete once we show the reverse inequality.

Case 1: $q = \infty$. To show that $\|g\|_\infty \le N_\infty = M_\infty$, we will show that the assumption $M_\infty < \|g\|_\infty$ leads to a contradiction. Indeed, if this holds, then there is some $\epsilon > 0$ such that $\mu\{x \in X : |g(x)| > M_\infty + \epsilon\} > 0$. But then, by semi-finiteness, there is some $F \in \Sigma$ such that $0 < \mu(F) < \infty$ and $|g| > M_\infty + \epsilon$ on F. Now we may define

$$f = \mu(F)^{-1}\chi_F \overline{\operatorname{sign} g},$$

and we find that $f \in L^1$, $\|f\|_1 = 1$ and

$$N_\infty \geq \left| \int f g d\mu \right| = \frac{1}{\mu(F)} \int_F |g| d\mu \geq M_\infty + \epsilon,$$

a contradiction.

Case 2: $q < \infty$. Assume that $g \neq 0$ (for otherwise there is nothing to prove), and let $\{s_n\}$ be a sequence in S such that $|s_n| \leq g$ for all n and $s_n \to g$ pointwise. Define

$$f_n = \|s_n\|_q^{1-q} |s_n|^{q-1} \overline{\text{sign } g}.$$

Then $|f_n|^p = \|s_n\|_q^{-q} |s_n|^q$ and so $f_n \in (L^p)_1$. By the dominated convergence theorem $\|g\|_q = \lim \|s_n\|_q$, but

$$\lim_n \|s_n\|_q = \lim_n \int |f_n s_n| d\mu$$

$$\leq \sup_n \int |f_n g| d\mu$$

$$= \sup_n \int f_n g d\mu \leq N_q$$

and we conclude that $\|g\|_q \leq N_q$, as required. $\square$

Proof of Theorem 3.2.4. It remains to prove the surjectivity of the map $g \mapsto \Phi_g$. Given $\Phi \in (L^p)^*$, we need to find $g \in L^q$ such that $\Phi = \Phi_g$.

Case 1: $\mu(X) < \infty$. In this case, every characteristic function is in L^p. We define a complex measure $\nu : \Sigma \to \mathbb{C}$ by

$$\nu(A) = \Phi(\chi_A).$$

To see that ν is a measure, let A_n be a disjoint sequence of measurable sets. Then $\chi_{\cup A_n} = \sum \chi_{A_n}$ and the series converges in L^p; indeed,

$$\left\| \chi_{\cup A_n} - \sum_{n=1}^N \chi_{A_n} \right\|_p = \|\chi_{\cup_{n>N} A_n}\|_p = \mu(\cup_{n>N} A_n)^{1/p} \xrightarrow{N \to \infty} 0,$$

where the last limit follows from $\mu(\cup_{n>N} A_n) = \sum_{n>N} \mu(A_n)$, the latter being the tail of a convergent sum. Because Φ is linear and bounded, we find that

$$\nu(\cup A_n) = \Phi\left(\sum \chi_{A_n}\right) = \sum \Phi(\chi_{A_n}) = \sum \nu(A_n).$$

This shows that ν is a (complex) measure.

Next, note that ν is absolutely continuous with respect to μ. Indeed, if $\mu(A) = 0$, then $\chi_A = 0$ as an element of L^p, so $\nu(A) = \Phi(\chi_A) = 0$.

By the Radon–Nikodym theorem, there exists $g \in L^1$ such that $\nu(A) = \int_A g d\mu$ for all $A \in \Sigma$. It follows that

$$\Phi(s) = \int s g d\mu, \quad \text{for all } s \in \mathcal{S}.$$

We see that $M_q(g) \leq \|\Phi\|$. By Proposition 3.2.5, $g \in L^q$, and also $\|g\|_q = \|\Phi\|$. So $\Phi(s) = \Phi_g(s)$ for all $s \in \mathcal{S}$, and thanks to the fact that $\mathcal{S}$ is dense in L^p, we conclude that $\Phi = \Phi_g$, as required.

Case 2: $\mu(X) = \infty$, but μ is σ-finite. Let $E_n \subseteq E_{n+1}$ be a sequence of subsets of finite measure such that $X = \cup E_n$. Let $\Phi \in (L^p)^*$. We consider $L^p(E_n, \mu)$, as a closed subspace of $L^p = L^p(X, \mu)$ consisting of functions in L^p that are supported in E_n. Then $\Phi_n := \Phi\big|_{L^p(E_n)}$ is bounded, and by the first case considered there exists $g_n \in L^q(E_n)$, such that $\Phi_n = \Phi_{g_n}$ and $\|g_n\|_q = \|\Phi_n\| \leq \|\Phi\| < \infty$. Proposition 3.2.1 implies that the equivalence class of g_n is determined by Φ_n, and it follows that if $m < n$, then $g_m = g_n$ almost everywhere in E_m. We may therefore define $g = \lim g_n \chi_{E_n}$ almost everywhere. An application of the monotone convergence theorem 1.5.13 shows that $g \in L^q$. Finally, applying the dominated convergence theorem 1.5.14 twice, we see that if $f \in L^p$, then $f \chi_{E_n} \to f$ in L^p, and

$$\Phi(f) = \lim_n \Phi(f \chi_{E_n}) = \lim_n \int f g \chi_{E_n} d\mu = \int f g d\mu,$$

and we conclude that $\Phi = \Phi_g$.

Remark 3.2.6. Let us consider a point that is usually glossed over. Strictly speaking, we defined a map $L\colon g \mapsto \Phi_g$ on the space of functions $\mathcal{L}^q(X, \mu)$ and not on the quotient space $L^q(X, \mu) = \mathcal{L}^p(X, \mu)/\ker \|\cdot\|_p$ (see Example 1.5.15). This map is evidently linear, but it is not truly the map we are after. By Proposition 3.2.1, $\|\Phi_g\| = \|g\|_q$. This has two consequences. First, from $\|\Phi_g\| \leq \|g\|_q$ we obtain $\ker \|\cdot\|_q \subseteq \ker L$, which means that Φ_g depends only on the equivalence class $\dot{g}$ of g, and so we can promote L to a well defined linear map $\dot{g} \mapsto \Phi_g$ from L^q into $(L^p)^*$ that satisfies $\|\Phi_g\| \leq \|\dot{g}\|_q$. Second, from $\|\Phi_g\| \geq \|g\|_q$ it follows that the functional Φ_g determines the values of the function g almost everywhere (see Remark 1.5.18). Therefore we really do have an isometric isomorphism $L^q \cong (L^p)^*$.

3.3   Application: weak solutions of PDEs

3.3.1   The notion of a weak solution

Partial differential equations are one of the most important and useful branches of mathematics. It is a huge subject. When working in PDEs one requires an arsenal of different tools, and functional analysis is just one of the many tools that PDE specialists use. Since our only goal here is to give an example of how the Hahn–Banach theorem can be used in the theory of PDEs[2], and so as to be very concrete, we'll discuss only one PDE, namely

$$\text{div}(u) = F. \tag{3.1}$$

Here the function $u = (u_1, u_2) \colon \mathbb{R}^2 \to \mathbb{R}^2$ is a vector valued function on the plane, $F \colon \mathbb{R}^2 \to \mathbb{R}$ is a scalar valued function on the plane, and div is the divergence operator

$$\text{div}(u) = \frac{\partial u_1}{\partial x} + \frac{\partial u_2}{\partial y}.$$

In its simplest form, the problem is: given a specified function F, does there exist a solution u that satisfies $\text{div}(u) = F$?

Classically, a **solution** to equation (3.1) means a differentiable function u (meaning that both u_1 and u_2 are differentiable functions) such that

$$\text{div}(u)(x,y) = \frac{\partial u_1}{\partial x}(x,y) + \frac{\partial u_2}{\partial y}(x,y) = F(x,y)$$

holds for every $(x, y) \in \mathbb{R}^2$. The question of whether a classical solution exists or not is a respectable mathematical question, but as noted above, PDEs arise in applications, and applications may require to allow for solutions not differentiable or even continuous.[3] So one is led to consider weak solutions, that is, functions u which are not differentiable, but which solve the PDE (3.1) in some sense.

In what sense? Assume that $F \in C = C(\mathbb{R}^2)$ and that $u \in C^1 = C^1(\mathbb{R}^2)$ is a solution to (3.1). It then follows that for every smooth function $w \in C_c^\infty(\mathbb{R}^2)$ (i.e., w is an infinitely differentiable compactly supported function) the following holds:

$$\int \left(\frac{\partial u_1}{\partial x}(x,y) + \frac{\partial u_2}{\partial y}(x,y) \right) w(x,y) dx dy = \int F(x,y) w(x,y) dx dy. \tag{3.2}$$

In fact, if $F \in C$, then $u \in C^1$ is a classical solution to (3.1) if and only if the above equality of integrals holds for every $w \in C_c^\infty$.

Now, if we integrate (3.2) by parts, we find that u is a classical solution to (3.1) if and only if

$$-\int (u_1 w_x + u_2 w_y) = \int Fw,$$

---

[2]The author first encountered the following application of the Hahn–Banach theorem to PDEs in a seminar by Haim Brezis at the Technion, where it was also mentioned in connection to the paper [ ].

[3]There is another reason to consider weak solutions besides the need that arises in applications: sometimes the existence of a classical solution is established in two steps. First step: a weak solution is shown to exist. Second step: the weak solution is shown to enjoy some regularity properties and is proved to be a solution in the classical case. See, e.g., [,].

or

$$\int u \cdot \mathrm{grad}(w) = - \int Fw, \quad \text{for all } w \in C_c^\infty \tag{3.3}$$

where $\mathrm{grad}(w) = (w_x, w_y)$ is the gradient of w. So (3.1) is equivalent to (3.3) for $u \in C^1$ and $F \in C$. But (3.3) makes sense also if u and F are merely *locally integrable*, in the sense that their restriction to every bounded open set is integrable. Thus, for a locally integrable F, we say that a locally integrable function u is a **weak solution** to (3.1) if it satisfies (3.3). Experience has shown that this is a reasonable notion of solution to the original PDE.

Now we are free to study (3.3) where F belongs to a certain class of functions and ask whether a solution u in a given class of functions exists. The main result of this section is that for every $F \in L^2$, there exists a $u \in L^\infty$ such that u is a weak solution to $\mathrm{div}(u) = F$.

Remark 3.3.1. Before moving forward, it is worth stressing that finding *some* solution for $\mathrm{div}(u) = F$ is not a difficult problem — the difficulty is in finding a *bounded* solution. Indeed, it is natural to seek a solution of the form $u = (u_1, 0)$, and one easily sees that such a u will satisfy (3.1) if one sets

$$u_1(x, y) = \int_0^x F(t, y)dt.$$

Another solution for (3.1) is given by $u = \mathrm{grad}\,\phi$, where ϕ is the weak solution of the *Poisson equation* $\Delta\phi = F$. By the theory of distributions, a weak solution ϕ can be obtained convolving F with the so-called *fundamental solution* of the Poisson equation (see [17]). This gives rise to a closed form expression for a weak solution in terms of F; however, this solution will again not necessarily be bounded.

3.3.2 The existence of L^∞ solutions to $\mathrm{div}(u) = F$

Let us fix some notation. For simplicity, let all our functions be real valued. We let $L^1 \oplus L^1$ denote the space of all pairs (f, g), where $f, g \in L^1(\mathbb{R}^2)$. We equip this space with the norm

$$\|(f, g)\| = \|f\|_1 + \|g\|_1.$$

Likewise, $L^\infty \oplus L^\infty$ is the space of pairs of bounded functions with the norm

$$\|(f, g)\| = \max\{\|f\|_\infty, \|g\|_\infty\}.$$

Exercise 3.3.2. $L^1 \oplus L^1$ is a Banach space, and $(L^1 \oplus L^1)^* = L^\infty \oplus L^\infty$ (see also Exercise 3.6.3).

Theorem 3.3.3. *For every $F \in L^2$, there exists a weak solution $u = (u_1, u_2) \in L^\infty \oplus L^\infty$ to the equation $\mathrm{div}(u) = F$.*

Proof. Let $M \subset L^1 \oplus L^1$ be the space

$$M = \{(f,g) \in L^1 \oplus L^1 : \exists w \in C_c^\infty.(f,g) = \text{grad}(w)\}.$$

Since M is the range of a linear map, it is a linear subspace of $L^1 \oplus L^1$.
We will require the following lemma.

Lemma 3.3.4. *If $(f,g) \in M$, then there is a unique $w \in C_c^\infty$ for which $(f,g) = \text{grad}(w)$. The map $(f,g) \mapsto w$ is linear and bounded as a map from $M \subset L^1 \oplus L^1$ into L^2.*

Assume the lemma for now, and let us proceed with the proof of the theorem. On M we define the linear functional

$$\phi : M \to \mathbb{R}$$

by

$$\phi(f,g) = -\int Fw,$$

where $w \in C_c^\infty$ is such that $(f,g) = \text{grad}(w)$. Now since $F \in L^2$ by assumption, the map $w \mapsto -\int Fw$ is a bounded functional on L^2. Using this fact together with Lemma 3.3.4 we conclude that ϕ, which is nothing but the composition of the map $M \ni \text{grad}(w) \mapsto w \in L^2$ with the map $L^2 \ni w \mapsto -\int Fw$, is a well defined, linear and bounded functional on $M \subset L^1 \oplus L^1$. By the Hahn–Banach extension theorem, ϕ extends to a bounded functional Φ on $L^1 \oplus L^1$. By Exercise 3.3.2, there exists a $u = (u_1, u_2) \in L^\infty \oplus L^\infty$ such that $\Phi(f,g) = \int (u_1 f + u_2 g)$ for all $f, g \in L^1 \oplus L^1$. Restricting only to elements of the form $(f,g) = \text{grad}(w) \in M$, we find that

$$\int u \cdot \text{grad}(w) = \phi(\text{grad}(w)) = -\int Fw$$

for all $w \in C_c^\infty$. In other words, $u \in L^\infty \oplus L^\infty$ is a weak solution to the equation $\text{div}(u) = F$, as required.   □

This may seem a little magical, but don't forget that we have yet to prove Lemma 3.3.4.

Proof of Lemma 3.3.4. Since the gradient operator $\text{grad}: C^\infty \to C^\infty \oplus C^\infty$ annihilates only constant functions, its restriction to C_c^∞ has a trivial kernel. Therefore, the linear transformation $\text{grad}: C_c^\infty \to M$ has a linear inverse $\text{grad}^{-1}: M \to C_c^\infty$, which sends every $(f,g) \in M$ to the unique $w \in C_c^\infty$ such that $\text{grad}(w) = (f,g)$. Since $(f,g) = \text{grad}(w) \implies f_y = g_x$, the linear transformation grad^{-1} is given by an explicit formula

$$w(x,y) = \text{grad}^{-1}(f,g)(x,y) = \int_{-\infty}^{x} f(t,y)dt.$$

The only nontrivial issue is boundedness with respect to the appropriate norms. For $w \in C_c^\infty$,

$$w(x, y) = \int_{-\infty}^{x} \frac{dw}{dx}(t, y) dt.$$

We obtain the estimate

$$|w(x, y)| \leq \int_{-\infty}^{\infty} \left| \frac{dw}{dx}(t, y) \right| dt.$$

Similarly,

$$|w(x, y)| \leq \int_{-\infty}^{\infty} \left| \frac{dw}{dy}(x, s) \right| ds.$$

Multiplying the two estimates above, we obtain

$$|w(x, y)|^2 \leq \int_{-\infty}^{\infty} \left| \frac{dw}{dx}(t, y) \right| dt \times \int_{-\infty}^{\infty} \left| \frac{dw}{dy}(x, s) \right| ds.$$

Integrating with respect to x and y, we conclude that

$$\|w\|_2^2 \leq \left\| \frac{dw}{dx} \right\|_1 \left\| \frac{dw}{dy} \right\|_1 \leq \frac{1}{2} \left(\left\| \frac{dw}{dx} \right\|_1 + \left\| \frac{dw}{dy} \right\|_1 \right)^2 = \frac{1}{2} \| \operatorname{grad}(w) \|_{L^1 \oplus L^1}^2$$

as required.

Remark 3.3.5. An application of the Hahn–Banach theorem to a PDE as above is an instance of a maneuver that may be called *the analyst's gambit*: by invoking duality, we have traded an existence problem for the problem of bounding an explicit integral expression. As mentioned above, the standard methods for solving linear PDEs do not give rise to a bounded u solving (3.1), so it seems like we have to pull a solution out of thin air. By using duality, we shifted the burden from the task of finding a solution, to the task of proving that the operator grad^{-1} — for which we have an explicit formula — is bounded. Bounding a quantity may still be a hard analytical problem, but now there is an arsenal of tools for approaching it directly. Note that the above proof of Theorem 3.3.3 is nonconstructive and provides no clue as to what the solution is or how to find it. However, existence results for PDEs can be instrumental in guiding research efforts toward deriving closed-form analytical solutions or numerical schemes.

Remark 3.3.6. Lemma 3.3.4 is a typical example of an estimate that one has to prove in order to apply functional analysis to PDEs, and falls under the wide umbrella of the *Sobolev–Nirenberg inequalities*. The reader who is interested in further applications of functional analysis to PDEs is referred to the book [7].

3.4 The dual of $C_0(X)$

We now describe a very important and useful class of dual spaces — the duals of Banach spaces of continuous functions. We review the definitions and results that we shall need below, without proofs and without attempting to cover the greatest generality; for full details, see [19] or [43].

3.4.1 The Banach space $M(X)$

Let X be a locally compact Hausdorff topological space[4]. A (positive) Borel measure μ on X is said to be **regular** if

$$\mu(A) = \inf\{\mu(U) : \ U \text{ is open and contains } A\}$$

and

$$\mu(A) = \sup\{\mu(K) : \ K \text{ is compact and contained in } A\}$$

for every measurable A. A complex Borel measure ν is said to be **regular** if its variation $|\nu|$ (recall Definition 3.1.4) is regular. Regular Borel measures play an important role in analysis because they tie between topology and measure theory.

Let $M(X)$ denote the space of regular complex Borel measures on X. For every two measures $\nu_1, \nu_2 \in M(X)$ and every scalar a we define a new measure Borel $a\nu_1 + \nu_2$ by

$$(a\nu_1 + \nu_2)(A) = a\nu_1(A) + \nu_2(A)$$

for all Borel sets A. These operations make $M(X)$ into a vector space.

If we define the norm of a complex measure ν to be its **total variation**, given by

$$\|\nu\| = |\nu|(X)$$

then $M(X)$ becomes a normed space and in fact it is a Banach space.

3.4.2 The action of $M(X)$ on $C_0(X)$

The space $M(X)$ acts on $C_0(X)$ by integrating continuous functions against measures:

$$\langle f, \nu \rangle = \int f d\nu.$$

One can make sense of the right hand side by first defining what it means for simple functions (there can only be one meaning) and then obtaining a

---

[4]A **locally compact** space is a topological space in which every point has a compact neighborhood. A space is **Hausdorff** if for every two points $x \neq y$ there exist two disjoint neighborhoods $U_x \ni x$ and $U_y \ni y$ such that $U_x \cap U_y = \emptyset$. Most function spaces of interest in analysis live on locally compact Hausdorff spaces.

definition for continuous functions by some approximation procedure. Here is an equivalent way to define the integral of a continuous function against a complex measure. If ν is a regular complex Borel measure, then it is evidently absolutely continuous with respect to its variation $|\nu|$. By the Radon–Nikodym theorem (Theorem 3.1.6), there exists therefore a function $h \in L^1(X, |\nu|)$ such that $\nu(A) = \int h\, d|\nu|$ for every measurable set A. In fact, one can show that $|h| = 1$ a.e. (we shall need this below). Therefore it makes sense to define

$$\int f\, d\nu = \int f h\, d|\nu|$$

for every $f \in C_0(X)$, and this definition agrees with the one that we would have gotten had we first defined the integral for simple functions and then extended it to other functions by approximation.

Every regular complex Borel measure ν therefore gives rise to a functional $\Psi_\nu : C_0(X) \to \mathbb{C}$ given by

$$\Psi_\nu(f) = \int f\, d\nu, \quad f \in C_0(X).$$

This functional is evidently linear, and moreover

$$|\Psi_\nu(f)| = \left| \int f\, d\nu \right| \leq \int |f| |h| d|\nu| \leq \|f\|_\infty |\nu|(X) = \|f\|_\infty \|\nu\|.$$

Thus $\Psi_\nu \in C_0(X)^*$ and $\|\Psi_\nu\| \leq \|\nu\|$.

3.4.3   The Riesz representation theorem

Up to now, we described a map $\nu \mapsto \Psi_\nu$ from $M(X)$ into $C_0(X)^*$. It is easy to see that this map is linear. The nontrivial and useful fact is that this map is an isometric isomorphism.

Theorem 3.4.1 (Riesz representation theorem). *Let X be a locally compact Hausdorff space. Then the map $\nu \mapsto \Psi_\nu$ described above is an isometric isomorphism of $M(X)$ onto $C_0(X)^*$. In other words, for every bounded linear functional $\Psi \in C_0(X)^*$ there exists a unique regular complex Borel measure ν on X such that*

$$\Psi(f) = \int f\, d\nu, \quad f \in C_0(X).$$

Briefly, the Riesz representation theorem says that $C_0(X)^* = M(X)$.

Remark 3.4.2. There is a closely related result, also referred to as "the Riesz representation theorem", which states that if $\Psi \colon C_0(X) \to \mathbb{C}$ is a *positive* linear functional, in the sense that $f \geq 0$ implies $\Psi(f) \geq 0$, then there exists a *positive* regular Borel measure μ on X such that Ψ is given by integration against μ. Theorem 3.4.1 can be deduced from the Riesz representation theorem for positive functionals by proving first that every bounded linear functional on $C_0(X)$ can be written as a linear combination of positive functionals.

The Riesz representation theorem is a central result in measure theory and is typically proved (in one form or another) in a course or in a textbook on that subject, thus we shall not discuss the proof here. The theorem is presented here to serve as an example of a dual space so that we can illustrate in the sequel how duality is used in applications of functional analysis.

3.5 Application: universality of neural networks

In this section, we illustrate how the Hahn–Banach theorem can be used to establish approximation-theoretic results, in a context of widespread contemporary interest: neural networks. We will prove the so-called "universality theorem" for neural networks, which states roughly that any continuous function can be uniformly approximated by neural networks. But first, we must begin by reviewing what neural networks are.

3.5.1 Neural networks

Let $\sigma \colon \mathbb{R} \to \mathbb{R}$ be some fixed function. If $x \in \mathbb{R}^k$, then we write $\sigma(x)$ for the entry-wise application of σ to $x = (x_1, \dots, x_k)$

$$\sigma(x) = (\sigma(x_1), \dots, \sigma(x_n)).$$

An **affine map** between vector spaces V and W is a map $F \colon V \to W$ of the form $F(v) = Av + w$ where A is a linear map $V \to W$ and w is a vector in W. A **neural network** (with activation function σ) is a function $N \colon \mathbb{R}^k \to \mathbb{R}^m$ that is obtained as the composition of a finite number of affine maps and entry-wise applications of σ. In other words, we can use the following recursive definition: a neural network is either an affine map, or it has the form

$$N(x) = F(\sigma(M(x))$$

where M is a neural network[5].

Neural networks are just another way of representing functions, or, better: another model of computation. Neural networks are used in practice to represent any kind of function: the input can be a digital recording of a voice (represented as a vector) and the output a transcript of the spoken text; or

---

[5]The class of functions that we call here "neural networks" is sometimes referred to as "feedforward neural networks" or "multilayer feedforward perceptron" (MLP) to distinguish it from other, more refined, or sophisticated models.

the input may be a blurry and noisy image and the output will be a clean and de-blurred version of the image; etc. etc. A friendly introduction to neural networks can be found in [37].

This model of computation is inspired by some models of how the brain works (hence the name), though one should not take this analogy too literally. Computations in this model are extremely efficient because all operations are either matrix multiplications or entry-wise applications of a single fixed function, and modern processors can perform such operations extremely fast on very large inputs. We avoid any attempt to go into an overview of the widespread and profound achievements of neural networks — whatever we write will be outdated by the time the ink is dry.

3.5.2   Approximation by single layer neural networks

The remarkable success of neural networks raises an important theoretical question: which functions can neural networks represent? We will show that the simplest kind of neural networks — single layer neural networks — already form a space of functions that is dense in the space of continuous functions. Several researchers proved results along these lines during the 1980s and 1990s (see [39]), the proof we present here is due to Cybenko [13].

A *single layer neural network* (with activation function σ) is a neural network of the form

$$N(x) = F(\sigma(G(x)))$$

where F and G are affine functions. To show that for any compact set $K \subset \mathbb{R}^k$ the space of single layer neural networks is dense in the space of continuous functions from K to $\mathbb{R}^m$, it suffices to treat the case where $m = 1$ and $K = R$ is some fixed rectangle

$$R = \Pi_{i=1}^{k}[a_i, b_i];$$

it will be convenient to assume that

$$0 < a_i < b_i < 1 \quad \text{for all} \quad i = 1, \ldots, k, \tag{3.4}$$

and we do so. Thus, we fix R, and our task is to show that the span of the functions $x \mapsto \sigma(y \cdot x + b)$, where $y \in \mathbb{R}^k$ and $b \in \mathbb{R}$, is dense in the space $C_{\mathbb{R}}(R)$ of continuous real valued functions on R. We define a subspace $\mathcal{M}(\sigma) \subseteq C_{\mathbb{R}}(R)$ by

$$\mathcal{M}(\sigma) = \text{span}\{\sigma(y \cdot x + b) : y \in \mathbb{R}^k, b \in \mathbb{R}\}.$$

The question of whether or not $\mathcal{M}(\sigma)$ is dense in $C_{\mathbb{R}}(R)$ obviously depends on the function σ. If σ is a linear function, then $\mathcal{M}(\sigma)$ will contain only affine functionals; in fact, adding more layers will not improve the situation. Moreover, if σ is a polynomial then $\mathcal{M}(\sigma)$ will consist only of polynomials of degree no larger than the degree of σ. Remarkably, being a polynomial is the only obstruction that stands between $\mathcal{M}(\sigma)$ and density (see Section 7.5). We shall prove a somewhat weaker result, which still covers almost all cases of practical interest.

Definition 3.5.1. A function $\sigma\colon \mathbb{R} \to \mathbb{R}$ is said to be a **sigmoidal** if it is continuous and satisfies

$$\lim_{t \to -\infty} \sigma(t) = 0, \quad \lim_{t \to \infty} \sigma(t) = 1.$$

For several decades activation functions in neural networks were usually taken to be sigmoidal functions, such as the **logistic function**

$$\sigma(t) = \frac{e^t}{1 + e^t}.$$

Today there are other activation functions of interest, see Exercise 3.5.3.

Theorem 3.5.2. *If σ is sigmoidal then $\mathcal{M}(\sigma)$ is dense in $C_{\mathbb{R}}(R)$.*

Proof. By the Hahn–Banach theorem (Corollary 2.1.10), to show that $\mathcal{M}(\sigma)$ is dense, it suffices to show that every bounded linear functional on $C_{\mathbb{R}}(R)$ that vanishes on $\mathcal{M}(\sigma)$ is the zero functional. By the Riesz representation theorem, we have to show that if ν is a regular complex Borel measure such that

$$\int \sigma(y \cdot x + b)d\nu = 0 \tag{3.5}$$

for all $b \in \mathbb{R}$ and $y \in \mathbb{R}^k$, then $\nu = 0$. This is another version of *the analyst's gambit* mentioned in Remark 3.3.5. Here, instead of solving an existence problem (the problem of finding an element in $\mathcal{M}(\sigma)$ that approximates an element in $C_{\mathbb{R}}(R)$), we now have to solve a uniqueness problem (the problem of showing that ν has to be zero). As always in serious applications, there is still analytical work to be done.

To show that ν is zero, it suffices to show that $\int f d\nu = 0$ for all continuous functions f on R. By the density of trigonometric polynomials in the space $C_{per}([0,1]^k)$ of continuous periodic functions in the unit cube (see [44, Corollary 4.1.4]), and thanks to assumption (3.4), it suffices to show that $\int e^{2\pi i n \cdot x} d\nu = 0$ for all $n \in \mathbb{Z}^k$, or equivalently, that

$$\int \sin(2\pi n \cdot x)d\nu = \int \cos(2\pi n \cdot x)d\nu = 0.$$

Note that the integrands are functions that have the specific form $x \mapsto g(n \cdot x)$ where g is a bounded continuous function on the real line. Now, every continuous function on the line can be approximated uniformly on every bounded subinterval by a finite sum of indicator functions of half-open intervals, therefore it suffices to show that $\int h(n \cdot x)d\nu = 0$ for every h that is the indicator function of a half-open interval $a < t \leq b$. In fact, it suffices to consider $h = \chi_{(b,\infty)}$ for some $b \in \mathbb{R}$. For such h we have

$$\int h(n \cdot x)d\nu = \nu\{x \in R : n \cdot x > b\}. \tag{3.6}$$

On the other hand, by the dominated convergence theorem we have

$$0 = \lim_{t \to \infty} \int \sigma(t(n \cdot x - b) + c) d\nu$$
$$= \nu\{x \in R : n \cdot x > b\} + \sigma(c)\nu\{x \in R : n \cdot x = b\},$$

where the first equality follows from the assumption (3.5) that ν annihilates $\mathcal{M}(\sigma)$. Letting $c \to -\infty$ we find that $\nu\{x \in R : n \cdot x > b\} = 0$, and plugging this back in (3.6) we get $\int h(n \cdot x) d\nu = 0$, which is what we were after, and the proof is complete.   □

Exercise 3.5.3. Let $\sigma(x) = x^+ := \max\{0, x\}$. This is the so-called Rectified Linear Unit function (ReLU for short) which is a popular activation function at the time of writing. Show that $\mathcal{M}(\sigma)$ is dense in $C_{\mathbb{R}}(R)$.

3.6   Additional exercises

Exercise 3.6.1. For $p \in (1, \infty)$, prove that $(L^p)^* = L^q$ in general (an alternative exercise for the measure-lazy: prove $\ell^p(S)^* = \ell^q(S)$ for any set S).

Exercise 3.6.2. Find the dual space of ℓ^∞.

Exercise 3.6.3. Let X and Y be Banach spaces. Define a direct sum vector space

$$X \oplus Y = \{(x, y) : x \in X, y \in Y\},$$

with the obvious addition and multiplication by scalar operations. We introduce norms on $X \oplus Y$; for $p \in [1, \infty)$ we define

$$\|(x, y)\|_{\oplus,p} = (\|x\|_X^p + \|y\|_Y^p)^{1/p}.$$

We also define

$$\|(x, y)\|_{\oplus,\infty} = \max\{\|x\|_X, \|y\|_Y\}.$$

Prove that for every p, the space $(X \oplus Y, \|\cdot\|_{\oplus,p})$ is a Banach space. Prove that for every pair of conjugate exponents $p, q \in [1, \infty]$,

$$(X \oplus Y, \|\cdot\|_{\oplus,p})^* = (X^* \oplus Y^*, \|\cdot\|_{\oplus,q}).$$

Exercise 3.6.4. Cybenko's proof of the universality theorem actually has a mistake in it. Read the proof of Theorem 1 in the paper [13]; find the mistake.

Exercise 3.6.5 (Runge approximation theorem[6]). Let $K \subset \mathbb{C}$ be a compact subset of the complex plane that is simply connected (the reader should take

---

[6]I took this Exercise from the textbook [31] by Lax, who attributes this elegant proof to Hormander.

"simply connected" to mean that $\mathbb{C} \setminus K$ is connected). Let $A(K)$ denote the closure in $C(K)$ of all restrictions to K of functions that are analytic in a neighborhood of K. Runge's approximation theorem states that the analytic polynomials are dense in $A(K)$.

1. It might be natural to quickly conclude that Runge's theorem follows immediately from Stone–Weierstrass or from the existence of power series; explain why this is not so.

2. Explain why Runge's theorem would follow from the following statement: *if $\phi \in C(K)^*$ is such that $\phi(p) = 0$ for every polynomial p, then $\phi(f) = 0$ for every $f \in A(K)$.*

3. Use Cauchy's integral formula to show that it suffices to consider only functions of the form $f = f_w(z) = \frac{1}{w-z}$ for $w \notin K$.

4. Let $\phi \in C(K)^*$. Prove that $g(w) = \phi(f_w)$ is analytic in the complement of K.

5. Now assume that $\phi(p) = 0$ for every polynomial p. Prove Runge's theorem by showing that $g(w) = 0$ for large $|w|$ and then deducing that $g(w) = 0$ for all $w \notin K$.

Exercise 3.6.6 (Muntz approximation theorem). Let $q_1 < q_2 < q_3 < \cdots$ be a sequence of positive numbers converging to ∞. Consider the space

$$M = \operatorname{span}\{1, x^{q_1}, x^{q_2}, \ldots\} \subseteq C([0,1]).$$

The Muntz approximation theorem is the following beautiful result:

$$M \text{ is dense in } C([0,1)) \iff \sum_{k}^{\infty} \frac{1}{q_k} = \infty.$$

In this exercise, we will prove that the divergence of the sum is a sufficient condition for the density of M[7].

1. Explain why it suffices to prove the following statement: *for every bounded linear functional ϕ in the unit ball of $C([0,1])^*$ that vanishes on M, if there exists some $n > 0$ such that $\phi(x^n) \neq 0$, then $\sum_{k}^{\infty} q_k^{-1} < \infty$.*

2. Now fix a functional ϕ in the unit ball of $C([0,1])^*$ and define a function $F(z) = \phi(x^z)$ on the right half plane $\mathbb{H} = \{z \in \mathbb{C} : \operatorname{Re} z > 0\}$. Prove that F is analytic in $\mathbb{H}$.

3. Assume that $\phi\big|_M = 0$ and that $F(n) \neq 0$ for some $n > 0$. We prove the required statement as follows.

---

[7]This application of the Hahn-Banach theorem is also from [31]. For a classical-analytical proof of Muntz's theorem (including the necessity of the condition) see [11].

4. For every N we define a function

$$B_N(z) = \prod_{k=1}^{N} \frac{z - q_k}{z + q_k}, \quad z \in \mathbb{H}.$$

Prove that $B_N(z) = 0$ if and only if $z \in \{q_1, \ldots, q_N\}$, and that $|B_N(z)| \to 1$ as $\operatorname{Re} z \to 0$ and as $|z| \to \infty$.

5. Define $g_N = \frac{F}{B_N}$ on $\mathbb{H}$. Prove that g_N is analytic and that $|g_N| \leq 1$.

6. Conclude that for the integer $n > 0$ for which we assumed that $F(n) \neq 0$, it holds that

$$\prod_{k=1}^{N} \left| \frac{p_k + n}{p_k - n} \right| \leq \frac{1}{|F(n)|}.$$

7. Conclude now that $\lim_N \prod_{k=1}^{N} \left| \frac{q_k + n}{q_k - n} \right| < \infty$ and so $\sum_k^{\infty} q_k^{-1} < \infty$.

8. Recap: what did we prove?

Chapter 4

The open mapping, uniform boundedness and closed graph theorems

4.1 The fundamental consequences of completeness

In this chapter, we shall learn three important consequences of the completeness of Banach spaces, which have deep and surprising implications and have come to be known as "the three big theorems of Banach space theory". We begin by recalling some notions from the theory of metric spaces.

Definition 4.1.1. A set F in a metric space is said to be *nowhere dense* if the interior of $\overline{F}$ is empty (here $\overline{F}$ is the closure of F).

Definition 4.1.2. A set in a metric space is said to be *of the first category* (or *meager*), if it is the union of countably many nowhere dense sets. A set is *of the second category* if it is not of the first category.

Definition 4.1.3. A set in a metric space is said to be *generic* (or *residual*) if it is the complement of a set of the first category.

Let us recall *Baire's category theorem* and its immediate consequence, which the readers should have seen in a course in metric spaces (see the appendix of [44] for proofs).

Theorem 4.1.4 (Baire category theorem). *If X is a complete metric space, then X is of the second category.*

Corollary 4.1.5. *Let X be a complete metric space, and let F_i be a sequence of closed subsets of X. If $X = \cup_i F_i$, then for some j, F_j has nonempty interior.*

A Banach space X is a complete metric space, so the Baire category theorem applies; this has profound consequences for the operator theory on X.

4.1.1 The uniform boundedness principle

Theorem 4.1.6 (The uniform boundedness principle). *Let X be a Banach space, let Y be a normed space, and let $\mathcal{F} \subseteq B(X,Y)$ be a family of operators.*

DOI: 10.1201/9781003297864-4

If for all $x \in X$,

$$\sup_{T \in \mathcal{F}} \|Tx\| < \infty,$$

then $\sup_{T \in \mathcal{F}} \|T\| < \infty$.

The uniform boundedness principle is often referred to as the *Banach–Steinhaus theorem*. Since X is of the second category, Theorem 4.1.6 follows immediately from the following stronger version of the uniform boundedness principle. It is worth noting that Theorem 4.1.6 can also be proved without recourse to Baire's category theorem; see Exercise 4.5.1.

Theorem 4.1.7. *Let X be a Banach space, Y a normed space, and $\mathcal{F} \subseteq B(X, Y)$ a family of operators. Let $A \subseteq X$ be the set*

$$A = \{x \in X : \sup_{T \in \mathcal{F}} \|Tx\| < \infty\}.$$

If A is of the second category, then $\sup_{T \in \mathcal{F}} \|T\| < \infty$.

Proof. For all $T \in \mathcal{F}$ and $n \in \mathbb{N}$, let $A_{n,T} = \{x \in X : \|Tx\| \leq n\}$. This set is closed, hence the set

$$A_n = \{x \in X : \sup_{T \in \mathcal{F}} \|Tx\| \leq n\} = \cap_{T \in \mathcal{F}} A_{n,T}$$

is closed, too. Note that $A = \cup_{n=1}^{\infty} A_n$. Since A is of the second category, there is some integer N such that A_N has a nonempty interior.

Suppose that the open ball of radius r centered at y is contained in A_N. In other words, for all $z \in X$, $\|z - y\| < r$ implies that $\|Tz\| \leq N$ for all T. Rearranging this information, we find that for all $x \in B_r(0)$,

$$\|Tx\| \leq \|T(x + y)\| + \|T(y)\| \leq 2N.$$

We conclude that for all $T \in \mathcal{F}$, $\|T\| \leq 2Nr^{-1}$. That completes the proof. $\square$

4.1.2 The open mapping theorem

We now come to the second "big theorem". Recall that a map $T : X \to Y$ is said to be **open** if $T(U)$ is open in Y, whenever U is open in X.

Theorem 4.1.8 (The open mapping theorem). *Let X and Y be Banach spaces, and let $T \in B(X, Y)$. If T is surjective, then it is an open mapping.*

Remark 4.1.9. The open mapping theorem is sometimes referred to as the *Banach–Schauder theorem*.

Proof. It suffices to show that if $x \in X$, and $B_r(x)$ is the open ball of radius r around x, then $T(B_r(x))$ contains an open ball centered at Tx (indeed, this will show that every point in the image of an open set has a neighborhood in the image of that open set). But by linearity, this reduces to showing that $T(B_1(0))$ contains an open ball centered at 0.

Since T is surjective, we have that $Y = \cup_{n=1}^{\infty} T(B_n(0))$. By the Baire category theorem one of the sets $\overline{T(B_n(0))}$ has nonempty interior. But

$$\overline{T(B_n(0))} = 2n\overline{T(B_{1/2}(0))}.$$

Thus, the set $\overline{T(B_{1/2}(0)}$ contains an open ball around some point, say

$$B_r(y) \subseteq \overline{T(B_{1/2}(0))}. \tag{4.1}$$

We want to show, as an intermediate step, that there is an open ball centered at 0 that is contained in $\overline{T(B_1(0))}$. In fact, we will show that $B_{r/2}(0) \subseteq \overline{T(B_1(0))}$.

Let y be as in (4.1). There exists $x \in B_{1/2}(0)$ such that $\|Tx - y\| < r/2$. Now for all $z \in B_{r/2}(0)$, $\|Tx + z - y\| < r$, so

$$Tx + z \in B_r(y) \subseteq \overline{T(B_{1/2}(0))}.$$

It follows that

$$z = Tx + z - Tx \in \overline{T(B_{1/2}(0))} + T(B_{1/2}(0))$$
$$\subseteq \overline{T(B_{1/2}(0))} + \overline{T(B_{1/2}(0))}$$
$$= \overline{T(B_1(0))},$$

thus,

$$B_{r/2}(0) \subseteq \overline{T(B_1(0))}.$$

Now, to prove the theorem, we have to show that $T(B_1(0))$ — not its closure — contains an open ball around the origin. Since T is open if and only if $\frac{2}{r}T$ is open, we may as well scale T and assume that we proved that

$$B_1(0) \subseteq \overline{T(B_1(0))}. \tag{4.2}$$

We will show that (4.2) implies that $B_{1/2}(0) \subseteq T(B_1(0))$. That will complete the proof.

First, note that from (4.2) it follows that

$$B_{2^{-k}}(0) \subseteq \overline{T(B_{2^{-k}}(0))} \tag{4.3}$$

for all $k = 1, 2, \ldots$.

Now let $u \in B_{1/2}(0)$. We will construct a sequence $x_1, x_2, \ldots$ such that $\sum x_k$ converges to an element in $B_1(0)$ and such that $T(\sum x_k) = u$. Choose $x_1 \in B_{1/2}(0)$ such that $\|u - Tx_1\| < 1/4$. This is possible thanks to (4.3). Using (4.3) again, choose $x_2 \in B_{1/4}(0)$ such that $\|(u - Tx_1) - Tx_2\| < 1/8$. Continuing this way, we construct a sequence $x_1, x_2, \ldots$ such that $\|x_k\| < 2^{-k}$ and $\|u - T(\sum_{k=1}^{n} x_k)\| < 2^{-n-1}$. It follows that $x := \sum x_k \in B_1(0)$, and from the continuity of T it follows that $Tx = u$. $\qquad \square$

Remark 4.1.10. Note that, unlike the uniform boundedness principle, this theorem requires both X and Y to be complete. Where did we use the completeness of X? Of Y?

The following is an important application of the open mapping theorem. Recall that if T is a bijective linear map, then T has a (unique) linear inverse. It is a nontrivial fact that the inverse of a bounded invertible map is bounded.

Corollary 4.1.11 (The inverse mapping theorem). *Let X and Y be Banach spaces. Let $T : X \to Y$ be a bijective linear map which is bounded. Then the inverse of T is bounded.*

Proof. By the open mapping theorem, the inverse of T is continuous. $\qquad\square$

Corollary 4.1.12. *Let X be a Banach space with a norm $\| \cdot \|$. Suppose that there is another norm $\| \cdot \|'$ defined on X which induces a complete metric, and suppose that there exists a constant C such that*

$$\|x\|' \leq C\|x\|,$$

for all $x \in X$. Then $\| \cdot \|$ and $\| \cdot \|'$ are equivalent.

Proof. The identity mapping from $(X, \| \cdot \|)$ to $(X, \| \cdot \|')$ is bounded and bijective. By the inverse mapping theorem, its inverse is bounded. $\qquad\square$

4.1.3   The closed graph theorem

For any normed spaces X and Y we define the normed space $X \oplus_1 Y$ to be the vector space
$$X \times Y = \{(x, y) : x \in X, y \in Y\}$$
with the natural vector space operations and the norm
$$\|(x, y)\|_{X \oplus_1 Y} = \|x\|_X + \|y\|_Y;$$
if X and Y are Banach spaces then so is $X \oplus_1 Y$ (see Exercise 3.6.3). The projection maps
$$\pi_X : X \oplus_1 Y \to X, \quad (x, y) \mapsto x$$
and
$$\pi_Y : X \oplus_1 Y \to Y, \quad (x, y) \mapsto y$$
are continuous linear maps. One can show that other natural choices of a norm, such as a norm of the form $\|(x, y)\| = \|(\|x\|_X, \|y\|_Y)\|'$, when $\| \cdot \|'$ is any norm on $\mathbb{R}^2$, will lead to an equivalent norm on the vector space $X \oplus Y$.

Definition 4.1.13. Let $T : X \to Y$. The **graph of** T, denoted $\Gamma(T)$, is the set
$$\Gamma(T) = \{(x, y) \in X \times Y : y = Tx\} \subseteq X \oplus_1 Y.$$
T is said to be **closed** if its graph is closed.

It is clear that $\Gamma(T)$ is a linear subspace of $X \oplus Y$. If T is continuous, then $\Gamma(T)$ is closed. Indeed, suppose that $(x_n, Tx_n) \in \Gamma(T)$ and $(x_n, Tx_n) \to (x, y)$. Then, $x_n \to x$, so $y = \lim_n Tx_n = Tx$ and this means that $(x, y) \in \Gamma(T)$.

Theorem 4.1.14 (The closed graph theorem). *Let X and Y be Banach spaces and let $T : X \to Y$ be a linear map. If T is closed, then T is continuous.*

Proof. $\Gamma(T)$ is a closed subspace of $X \oplus_1 Y$, hence it is a Banach space. Let $\pi_X : X \oplus_1 Y \to X$ be the natural projection map on X given by $(x, y) \mapsto x$ and likewise let $\pi_Y : X \oplus_1 Y \to Y$ be the projection map on Y. These are continuous linear maps. Hence the restrictions $p = \pi_X|_{\Gamma(T)} : \Gamma(T) \to X$ and $q = \pi_Y|_{\Gamma(T)} : \Gamma(T) \to Y$ are continuous, and p is invertible. By the inverse mapping theorem, p^{-1} is bounded. Hence $T = q \circ p^{-1}$ is also bounded. $\square$

4.2   Complemented subspaces

Let X be a vector space and let M be a linear subspace. By basic linear algebra, there always exists an **algebraic complement**, meaning a subspace N such that

$$M \cap N = \{0\} \quad \text{and} \quad M + N = X.$$

This is the same as saying that every $x \in X$ can be written as $m + n$ for $m \in M, n \in N$ in a unique way. Note: the decomposition $x = m + n$ is unique once the complement N is chosen; however, the choice of N is not unique. Holding the algebraic complement N fixed, we may define **the projection of X onto M parallel to N** to be the map $P : X \to M$ defined by $P(m + n) = m$. It is clear that $I - P$ is then the projection of X onto N parallel to M. The projection P satisfies

$$P^2 = P, \; \operatorname{Im}(P) = \ker(I - P) = M, \text{ and } \ker(P) = N.$$

Conversely, if Q is a linear operator satisfying $Q^2 = Q$, and if M and N are subspaces of X such that $\operatorname{Im}(Q) = M$ and $\ker(Q) = N$, then M and N are algebraic complements of one another, and Q is the projection onto M parallel to N.

The considerations in the above paragraphs were purely algebraic. Now let X be a Banach space, and let M be a closed subspace. We know that M is complemented algebraically, but this construction is usually not useful or appropriate in the setting of Banach spaces.

Definition 4.2.1. A closed subspace M in a normed space X is said to be **complemented** if there exists a closed subspace $N \subseteq X$ such that $X = M + N$ and $M \cap N = \{0\}$.

Thus, M is complemented if and only if it has a closed algebraic complement. In this situation we write $X = M \oplus N$. Sometimes the term **topologically complemented** is used instead, for emphasis, but we shall mostly stick to Definition 4.2.1.

Theorem 4.2.2. *A closed subspace M of a Banach space X is complemented if and only if M is the range of a continuous projection.*

Proof. Let P be a continuous projection. Then $X = \mathrm{Im}(P) + \ker(P)$, and $\mathrm{Im}(P) \cap \ker(P) = \{0\}$ as in the purely algebraic case. Now $\ker(P)$ is closed, because P is continuous, and $\mathrm{Im}(P) = \ker(I - P)$ so it is closed, too.

Conversely, suppose that $X = M \oplus N$. Then we may define algebraically the projection P onto M parallel to N, and it holds that $\mathrm{Im}(P) = M$. We must show that P is continuous. By the closed graph theorem, it suffices to show that the graph $\Gamma(P)$ of P is closed.

Let $(x_k, Px_k) \in \Gamma(P)$ converge to (x, y). Write $x_k = m_k + n_k$, so $Px_k = m_k \to y$, whence $y \in M$. Therefore $y = Py$. What we need to show is that $y = Px$. The sequences x_k, m_k converge to x and y, respectively, so $n_k = x_k - m_k$ converges to $x - y$. Therefore, since N is closed, we find that $x - y \in N = \ker(P)$. We conclude that $Px = Py = y$. $\square$

Let us compare the general situation with what happens in Hilbert spaces. In a Hilbert space H, every closed subspace M has a unique orthogonal complement $M^\perp$, and therefore every subspace is complemented. Closed subspaces in Hilbert spaces have infinitely many topological complements, but the orthogonal complement is the most useful. If N is a complement of M, then the projection onto M parallel to N has norm 1 if and only if $N = M^\perp$.

In a general Banach space, there is no canonical choice of complement. Since there is no canonical choice, it is perhaps not surprising that there are closed subspaces of Banach spaces that have no topological complement (see Exercise 4.5.8 for an example).

Exercise 4.2.3. Let M be a closed subspace of a Banach space X.

1. If $\dim M < \infty$, then M is complemented in X.

2. If $\dim X/M < \infty$, then M is complemented in X.

4.3 Quotients, revisited

Recall from Section 1.4 the notion of the quotient X/M of a Banach space X by a closed subspace M. Recall also the quotient map $\pi : X \mapsto X/M$, defined by $\pi(x) = \dot{x} = x + M$.

Theorem 4.3.1. *Let X and Y be normed spaces. Every $T \in B(X, Y)$ induces a unique bounded operator $\hat{T} : X/\ker T \to Y$ such that*

$$\hat{T}(\dot{x}) = Tx.$$

The operator $\hat{T}$ is bounded and $\|\hat{T}\| = \|T\|$. If T is surjective and if X and Y are complete, then $\hat{T}$ is an isomorphism of $X/\ker T$ onto Y.

Proof. Let $M = \ker T$. Then M is a closed subspace, so the quotient X/M is a normed space. That T gives rise to a well defined operator $\hat{T}$ given by $\hat{T}(\dot{x}) = Tx$ is left as an exercise. To prove that $\|\hat{T}\| = \|T\|$, we first estimate

$$\|\hat{T}\| = \sup_{\|\dot{x}\|<1} \|\hat{T}\dot{x}\| = \sup\{\|T(x+m)\| : x \in X, m \in M, \|x+m\| < 1\} \le \|T\|.$$

On the other hand, $T = \hat{T} \circ \pi$, so $\|T\| \le \|\pi\|\|\hat{T}\| = \|\hat{T}\|$. Thus, $\|\hat{T}\| = \|T\|$, and it remains to show the last part of the theorem. If T is surjective so is $\hat{T}$. But $\hat{T}$ is clearly injective. Under the assumption that X and Y are complete, we may invoke the inverse mapping theorem to see that $\hat{T}$ is an isomorphism. $\square$

Theorem 4.3.2. *Let X be a Banach space and let M be a complemented subspace of X with complement N. Then*

1. *X is isomorphic to $M \oplus_1 N$.*

2. *N is isomorphic to X/M.*

Proof. The map $T : M \oplus_1 N \to X$ given by $T(m, n) = m + n$ is injective, surjective, and bounded (by the triangle inequality), hence T is an isomorphism.

To see that $N \cong X/M$, invoke Theorem 4.3.1 with $N = Y$ and $T : X \to Y$ equal to the projection onto N parallel to M. $\square$

Remark 4.3.3. The above theorem shows that given two Banach spaces M and N, the direct sum $M \oplus N$ has a well defined meaning as a Banach space, up to isomorphism. Indeed, given M and N, we can form the algebraic direct sum $X = M \oplus N = \{(m, n) : m \in M, n \in N\}$ with the obvious operations. Then *whatever* complete norm we choose on this space with respect to which M and N are closed subspaces, the space X will be isomorphic to $M \oplus_1 N$.

Exercise 4.3.4. Let $T \in B(X, Y)$ be a bounded operator between Banach spaces. Prove that if $\operatorname{Im} T$ has finite codimension (that is, if $\dim Y/\operatorname{Im} T < \infty$) then $\operatorname{Im} T$ is closed.

4.4 Application: Divergence of Fourier series

Recall that for every integer n, the *nth **Fourier coefficient*** of a function $f \in L^1[0, 1]$ is defined to be

$$\hat{f}(n) = \int_0^1 f(x)e^{-2\pi i n x}\, dx.$$

The Nth **partial Fourier sum** of f is defined to be the function

$$S_N(f)(x) = \sum_{n=-N}^{N} \hat{f}(n)e^{2\pi inx}.$$

A fundamental question in Fourier analysis is whether or not, and in what sense, do the partial Fourier sums of a function converge to it, so that we recover f as the **Fourier series**

$$f = \sum_{n=-\infty}^{\infty} \hat{f}(n)e^{2\pi inx}.$$

Here are a couple of easy results (see [11, Chapter 4]). For every $f \in L^2$ the partial sums $S_N(f)$ converge to f in the L^2 norm. For stronger convergence results we shall need to restrict our attention to a nicer space of functions. Let $C_{per}([0,1])$ be the Banach space of continuous functions f on $[0,1]$ for which $f(0) = f(1)$, endowed with the supremum norm. If $f \in C_{per}([0,1])$ is also continuously differentiable on $[0,1]$, then the Fourier series of f converges uniformly to f. It is then natural to ask whether the Fourier series of a continuous periodic function always converges pointwise to the function. In this section, we will use the uniform boundedness principle to see that the answer to this question is a resounding no. As is usual, a serious application requires some analytical preparations.

4.4.1   The Dirichlet kernel

It is convenient to have an alternative representation for the partial sums $S_N(f)$. Note that

$$S_N(f)(x) = \sum_{n=-N}^{N} \left(\int_0^1 f(t)e^{-2\pi int}dt \right) e^{2\pi inx} = \int_0^1 f(t)D_N(x-t)dt,$$

where $D_N(z) = \sum_{n=-N}^{N} e^{2\pi inz}$.

Definition 4.4.1. The sequence of functions $\{D_N\}$ is called the **Dirichlet kernel**.

Definition 4.4.2. If $g, h \in C_{per}([0,1])$, then the **convolution** of g with h is the function

$$g * h(x) = \int_0^1 g(t)h(x-t)dt = \int_0^1 g(x-t)h(t)dt.$$

The first equality above is the definition, while the second is a simple observation using a change of variables. We reached the following result.

Proposition 4.4.3. *For every $f \in C_{per}([0,1])$, the partial Fourier sums are given by*

$$S_N(f) = f * D_N.$$

We shall require the following analytical formula for D_N.

Proposition 4.4.4. *For all N,*

$$D_N(x) = \frac{\sin(2\pi(N+1/2)x)}{\sin(\pi x)}.$$

Proof. This follows from multiplying both sides by $\sin(\pi x)$ and using trigonometric identities. □

4.4.2   Continuous functions with divergent Fourier series

Theorem 4.4.5. *There exists a function $f \in C_{per}([0,1])$ whose Fourier series diverges at 0.*

Theorem 4.4.5 can be proved in a constructive manner, by writing down an example. We will obtain the following much stronger, albeit non-constructive, result, which implies Theorem 4.4.5 immediately.

Theorem 4.4.6. *Given $x_0 \in [0,1]$, the set of functions in $C_{per}([0,1])$ whose Fourier series diverges at x_0 is generic. In fact, given any sequence $x_1, x_2, \ldots$ in $[0,1]$, the set of functions in $C_{per}([0,1])$ whose Fourier series diverges at x_i for all i is generic.*

In particular, we see that for any countable dense subset S of the interval, there is a continuous and periodic function whose Fourier series diverges on that S. Things cannot, however, get much worse, because the Fourier series of any L^2 function converges almost everywhere[1].

Proof. A countable intersection of generic sets is generic, hence it suffices to prove the first assertion. Also one may assume that $x_0 = 0$ (yes, it is obvious, but the readers should make sure they really know how to justify this).
For every N, let F_N be the linear functional

$$F_N(f) = S_N(f)(0) = \int_0^1 f(t) D_N(t) dt.$$

(We made use of $D_N(-t) = D_N(t)$.) We require the following lemma.

Lemma 4.4.7. *For all N, F_N is bounded, and $\|F_N\| \to \infty$.*

---

[1]The proof of this result (which is called *Carleson's theorem* [10]) is beyond our means.

Assuming the lemma for now, let us complete the proof of Theorem 4.4.6. Let $A \subseteq C_{per}([0,1])$ be the set of functions whose Fourier series converges at 0. Then for every $f \in A$, $F_N(f) = S_N(f)(0)$ is a convergent sequence in N, so in particular, we have that for all $f \in A$,

$$\sup\{|F_N(f)| : N \in \mathbb{N}\} < \infty.$$

If A was of the second category, then the uniform boundedness principle (Theorem 4.1.7) would have implied that $\sup_N \|F_N\| < \infty$, contradicting the lemma. Hence A is of the first category, thus its complement is generic (by definition of "generic"). □

Proof of Lemma 4.4.7. The straightforward estimate

$$|F_N(f)| \leq \int |f(t)D_N(t)|dt \leq \|f\|_\infty \int |D_N(t)|dt,$$

gives $\|F_N\| \leq \|D_N\|_1$, and, in particular, we see that F_N is bounded. Let $\{f_n\}$ be a sequence of continuous periodic functions of norm less than 1 converging in the L^1 norm to the function $\text{sign}(D_N)$. Then $F_N(f_n) \to \|D_N\|_1$, while $|F_N(f_n)| \leq \|F_N\|$. This gives $\|F_N\| = \|D_N\|_1$.

We shall now show that $\|D_N\|_1$ diverges to infinity when $N \to \infty$. For this we make the estimates:

$$\begin{aligned}
\|D_N\|_1 &\geq \int_0^1 \frac{|\sin(2\pi(N+1/2)x)|}{\pi x}dx \\
&= \int_0^{2\pi(N+1/2)} \frac{|\sin(u)|}{\pi u}du \\
&\geq \frac{1}{\pi}\sum_{k=1}^N \int_{2\pi(k-1)}^{2\pi k} \frac{|\sin u|}{u}du \\
&\geq \frac{1}{2\pi^2}\int_0^{2\pi} |\sin u|du \sum_{k=1}^N \frac{1}{k} \geq C\log N \to \infty.
\end{aligned}$$

Remark 4.4.8. Here is a question for you to think about. Theorem 4.1.7 doesn't really use completeness of X (check it). In what way does the completeness of $C_{per}([0,1])$ play a role?

Theorem 4.4.6, together with the theorem on uniform convergence recalled at the beginning of this section, imply the following corollary, which roughly says that continuous differentiability is atypical for continuous functions.

Corollary 4.4.9. *The set of continuous functions that are not continuously differentiable is generic in $C_{per}([0,1])$.*

Proof. As we recalled at the beginning of the section, every continuously differentiable function $f \in C_{per}([0,1])$ has a uniformly convergent Fourier series.

Hence, the set of periodic continuous functions that *are not* continuously differentiable contains the set E of functions in $C_{per}([0,1])$ whose Fourier series diverges at 0. By Theorem 4.4.6, the set E is generic, hence so is the set of non-continuously differentiable functions in $C_{per}([0,1])$. □

4.5 Additional exercises

Exercise 4.5.1 (Elementary proof of the uniform boundedness principle[2]). Prove Theorem 4.1.6 without using Baire's category theorem, along the following lines. Suppose that $\sup_{T \in \mathcal{F}} \|T\| = \infty$ and let there be given sequences $M_n, r_n > 0$, to be defined later.

1. Find a sequence of maps $T_n \in \mathcal{F}$ such that $\|T_n\| > M_n$ for all n.

2. Prove that for all $x_0 \in X$,

$$\sup\{\|T_n(x_0 + x)\| : x \in B_r(x_0)\} \geq \|T_n\| r. \tag{4.4}$$

3. Use (4.4) to construct inductively a sequence of points $x_n \in B_{r_n}(0)$ such that $\|T_n(\sum_{k=1}^{n} x_k)\| \geq r_n \|T_n\|$.

4. Show that one can find sequences $M_n, r_n > 0$ such that

$$M_n r_n \to \infty \quad \text{and} \quad \sum_{k=n+1}^{\infty} r_k < \frac{r_n}{2},$$

and that in this case the above construction yields $\lim_n \|T_n x\| = \infty$ for $x = \sum_{n=0}^{\infty} x_n$.

Exercise 4.5.2. Let X be a linear subspace of the linear space $\mathbb{C}^A$ of all complex valued functions on a set A. Suppose that X is also equipped with a norm $\|\cdot\|$, which makes $(X, \|\cdot\|)$ into a Banach space. Suppose, in addition, that for every $a \in A$, the evaluation map $ev_a : X \to \mathbb{C}$, given by

$$ev_a(f) = f(a),$$

is a bounded linear functional on X. Such a space is said to be a ***reproducing kernel Banach space***. A function $\varphi : A \to \mathbb{C}$ is called a ***multiplier*** of X if $\varphi f \in X$ for all $f \in X$ (where φf denotes the usual pointwise product between functions). A multiplier φ gives rise to a linear operator $M_\varphi : X \to X$ by

$$M_\varphi(f) = \varphi f.$$

---

[2]This exercise is inspired by the short paper [16], which contains also some interesting historical comments.

1. Prove that for every multiplier φ of X, the linear operator M_φ is bounded.

2. Now, for an example. Let H^2 be the space of all analytic functions $f(z) = \sum_{n=0}^{\infty} a_n z^n$ with square integrable coefficients, i.e., such that $\sum |a_n|^2 < \infty$. Endow H^2 with the inner product $\langle \sum a_n z^n, \sum b_n z^n \rangle = \sum a_n \bar{b}_n$. Prove that H^2 is a reproducing kernel Hilbert space of functions on $A = \mathbb{D}$ (the open unit disc).

3. Describe all the multipliers φ of H^2, and find $\|M_\varphi\|$. Ponder: did item (1) help you in any way?

Exercise 4.5.3. Prove that for every separable Banach space X, there exists a closed subspace $M \subseteq \ell^1$, such that X is isometrically isomorphic to ℓ^1/M. Show that ℓ^1 cannot be replaced by ℓ^p for any other $p \in (1, \infty)$.

Exercise 4.5.4. True or false: a quotient of a reflexive space by a closed subspace is reflexive.

Exercise 4.5.5. Let X be a Banach space, and M a closed subspace of X. True or false: every bounded operator T from M to another Banach space Y can be extended to a bounded operator $\hat{T} : X \to Y$.

Exercise 4.5.6. Prove that if X is isomorphic to Y, then X is reflexive if and only if Y is reflexive.

Exercise 4.5.7. Let M be a closed subspace of $L^2[0,1]$ such that every function $f \in M$ is essentially bounded, that is $M \subseteq L^\infty$. Prove that M is finite dimensional. (**Guidance**[3]: use one of the big theorems to conclude that there exists a constant C such that $\|f\|_\infty \leq C\|f\|_2$ for all $f \in M$. Show then that for every orthonormal set $\{f_1, \ldots, f_n\} \subset M$ and every a in the unit ball of $\mathbb{C}^n$, $\|\sum_{k=1}^{n} a_k f_k\|_\infty \leq C$. Deduce that $\sum_{k=1}^{n} |f_k(x)|^2 \leq C^2$ for almost all $x \in [0,1]$, and conclude that n must be bounded by C^2.)

Exercise 4.5.8. Recall that c_0 is the closed subspace of ℓ^∞ consisting of sequences that converge to 0. In this exercise, we will show that c_0 is a non–complementable subspace of ℓ^∞.

1. Recall that if $X = M \oplus N$, then X/M is isomorphic to N.

2. Recall that if $f \in X^*$, then f induces a functional $\dot{f}$ on X/M by way of $\dot{f}(\dot{n}) = f|_N(n)$, where n is the unique element in the equivalence class $\dot{n}$ from N.

3. From here on, assume for contradiction that $\ell^\infty = c_0 \oplus N$. Denote the projection from ℓ^∞ to ℓ^∞/c_0 by π.

---

[3]The outline is taken from [12, Chapter 5], which contains many nice applications.

4. Prove the following seemingly unrelated lemma: *There exists a family* $\mathcal{F} \subseteq 2^{\mathbb{N}}$ *consisting of infinite subsets of* $\mathbb{N}$, *such that* $|\mathcal{F}| = 2^{\aleph_0}$, *and such that* $A \cap B$ *is finite for every pair of distinct* $A, B \in \mathcal{F}$.

5. For every $A \subseteq \mathbb{N}$, denote the characteristic function of A by 1_A. Prove that if $A_1, \ldots, A_n$ are distinct elements in $\mathcal{F}$, then for all $c_1, \ldots, c_n \in \mathbb{C}$,

$$\left\| \pi \left(\sum_{k=1}^{n} c_k 1_{A_k} \right) \right\| \leq \max_{1 \leq k \leq n} |c_k|.$$

6. Fix $g \in (\ell^\infty/c_0)^*$. Prove that for every sequence $\{A_k\}_{k=1}^\infty \subset \mathcal{F}$ of distinct sets, $g(\pi 1_{A_k}) \xrightarrow{k \to \infty} 0$, and conclude that for every n, the number of elements $A \in \mathcal{F}$ for which $|g(\pi 1_A)| > 1/n$, is finite. Deduce that there are at most a countable number of As in $\mathcal{F}$ such that $g(\pi 1_A) \neq 0$.

7. Let f_i be the functional on ℓ^∞ given by $f_i(a_1, a_2, \ldots) = a_i$. Assuming that $\ell^\infty = c_0 \oplus N$, prove that if for some $z \in \ell^\infty/c_0$ it holds that $\dot{f}_i(z) = 0$ for all i, then $z = 0$.

8. Contradict the previous item by showing that there exists a nonzero $z \in \ell^\infty/c_0$ such that $\dot{f}_i(z) = 0$ for all i.

Exercise 4.5.9. In this exercise, we will denote by $C(\mathbb{T})$ the continuous functions on the unit circle $\mathbb{T}$. For $f \in C(\mathbb{T})$, let us denote by

$$\hat{f}(n) = \frac{1}{2\pi} \int_0^{2\pi} f(e^{i\theta}) e^{-in\theta} \, d\theta$$

the Fourier coefficients. Prove that given a complex valued sequence $\{\gamma_n\}_{n=-\infty}^\infty$, there exists a regular complex Borel measure μ on $\mathbb{T}$ such that

$$\gamma_n = \int_{\mathbb{T}} e^{-in\theta} \, d\mu(e^{i\theta})$$

for all n (i.e., γ_n are the "Fourier coefficients of the measure μ") if and only if for every $f \in C(\mathbb{T})$ there exists a $g \in C(\mathbb{T})$ with $\hat{g}(n) = \gamma_n \hat{f}(n)$ for all n. (**Hint:** one direction is straightforward; for the other direction, use the closed graph theorem to show that $f \mapsto g$ is a bounded operator on $C(\mathbb{T})$, and then apply the Riesz representation theorem to an appropriate bounded linear functional on $C(\mathbb{T})$.)

Exercise 4.5.10. A sequence $\{x_n\}_{n=1}^\infty$ in a Banach space X is said to be a **Schauder basis** (or sometimes, simply a **basis**) if for every $x \in X$, there exists a unique sequence of scalars $\{c_n\}_{n=1}^\infty$ such that x is equal to the sum of the norm convergent series $\sum c_n x_n$.

1. Prove that if X has a Schauder basis, then it is infinite dimensional and separable.

2. If $x = \sum c_n x_n$, let

$$\|x\|_\infty := \sup_{N \in \mathbb{N}} \left\| \sum_{n=1}^{N} c_n x_n \right\|.$$

Prove that $\| \cdot \|_\infty$ is a norm, and that $(X, \| \cdot \|_\infty)$ is a Banach space.

3. Prove that $\| \cdot \|$ and $\| \cdot \|_\infty$ are equivalent norms.

4. Prove that for every k, the linear functional $f_k : \sum c_n x_n \mapsto c_k$ is bounded.

5. Prove that the maps $P_k(\sum c_n x_n) = \sum_{n=1}^{k} c_n x_n$ are uniformly bounded.

Exercise 4.5.11. The trigonometric system is of course a Schauder basis in $L^2[0, 1]$; this turns out to be true also in $L^p[0, 1]$ for $p \in (1, \infty)$ — a fact that is equivalent to the convergence of Fourier series in L^p (prove the equivalence. For convergence of Fourier series in L^p see [30]). Show that the trigonometric system is not a Schauder basis in $L^1[0, 1]$ nor in $C_{per}([0, 1])$. How do these facts reconcile with Fejér's theorem?

Exercises 4.5.12, 4.5.14 and 4.5.15 illustrate obtaining results on Fourier series using the point of view of functional analysis[4]. Recall the notation used in Section 4.4: For every $f \in L^1[0, 1]$ and $n \in \mathbb{Z}$, we let $\hat{f}(n)$ be the nth Fourier coefficient and $S_N(f)(x) = \sum_{n=-N}^{N} \hat{f}(n)e^{2\pi i n x}$ the N-th partial Fourier sum. The **Cesàro means** of the partial sums are defined to be

$$\sigma_N(f)(x) = \frac{1}{N} \sum_{n=0}^{N-1} S_n(f)(x) = f * K_N(x)$$

where $K_N(t) = \sum_{n=-N}^{N} \left(1 - \frac{|n|}{N+1}\right) e^{2\pi i n t}$ is the **Fejér kernel.** We also define an (evidently linear and bounded) operator $\mathcal{F} : L^1[0, 1] \to \ell^\infty$ by

$$\mathcal{F}(f) = (\hat{f}(n))_{n \in \mathbb{Z}}.$$

We write L^p for $L^p[0, 1]$, and c_0, ℓ^p for $c_0(\mathbb{Z})$, $\ell^p(\mathbb{Z})$, etc.

Exercise 4.5.12. Recall that for all $f \in L^2$, we have that $\mathcal{F}(f) \in \ell^2$, and in particular $\mathcal{F}(f) \in c_0$. Now let's consider $f \in L^1$.

1. Prove that $\mathcal{F}$ maps L^1 into c_0 (this is the *Riemann-Lebesgue lemma,* which states that $\lim_{|n| \to \infty} \hat{f}(n) = 0$, but please don't prove it by cleverly manipulating integrals).

2. Does $\mathcal{F}$ map L^1 *onto* c_0?

---

[4]The idea for these exercises comes from Bowers's and Kalton's book [6], Chapter 4.

Exercise 4.5.13 (Elementary warm up and hint for the next exercise). Let $\{T_n\}$ be a uniformly bounded sequence of bounded operators from a Banach space X to a Banach space Y. Let

$$M = \{x \in X : \lim_n T_n x \text{ exists }\}.$$

Prove that M is a closed subspace of X.

Exercise 4.5.14. As you may recall from the first course (see [14, Chapter 4]), *Fejér's theorem* states that $\sigma_N(f)$ converges uniformly to f for every $f \in C_{per}([0,1])$. This then implies *Weierstrass's theorem on trigonometric approximation*, which states that functions of the form $\sum_{n=-N}^{N} c_n e^{2\pi i n x}$ are dense in $C_{per}([0,1])$. In this exercise, we will prove Fejér's theorem, using Weierstrass's theorem and just simple functional analytic considerations.

1. Show that for every trigonometric polynomial $p(x) = \sum_{n=-N}^{N} c_n e^{2\pi i n x}$, the Cesàro means $\sigma_N(p)$ converge uniformly to p.

2. Voila! Deduce Fejér's theorem (ok, there is something to do here).

Exercise 4.5.15. Recall that for all $f \in L^2$, the partial sums $S_N(f)$ converge to f in the L^2 norm. Now, let us consider L^p, for $p \in [1, \infty)$.

1. Prove that there exists $f \in L^1$ for which $S_N(f)$ does not converge to f in the L^1 norm.

2. Prove that $\sigma_N : f \mapsto \sigma_N(f)$ defines a bounded operator on L^p. (**Hint:** the trick to estimate $\|\sigma_N(f)\|_p$ is to integrate $f * K_N$ against a function $g \in L^q$, change the order of integration and apply Hölder's inequality.)

3. Prove that for all $f \in L^p$, $\sigma_N(f)$ converges to f in the L^p norm.

Chapter 5

Further aspects of duality: weak convergence and the adjoint

5.1 The notion of weak convergence

Let X be the Banach space $C([0,1])$ of continuous functions on the interval $[0,1]$ with the sup norm. We construct a sequence of continuous piecewise linear functions $\{f_n\}$ in X as follows:

1. $f_n(t)$ is equal to 0 for $t = 0$, $t = 2/(n+1)$ and $t = 1$;

2. $f_n(t) = 1$ for $t = 1/(n+1)$;

3. f_n is linear in the remaining intervals.

The sequence tends to zero pointwise, but the norm of X does not detect this because $\|f_n\| = 1$ for all n. The sequence also tends to 0 in the L^1 norm, but the L^1 norm is not in the game. Can the Banach space structure of X itself detect the convergence of f_n to 0?

It turns out that the answer is yes. Recall that by the Riesz representation theorem, the dual of $C([0,1])$ is $M([0,1])$, the space of regular complex Borel measures on $[0,1]$. Now let $\mu \in M([0,1])$. Since $f_n \to 0$ everywhere, it in particular converges to 0 almost everywhere with respect to μ. Since $|f_n| \leq 1$ for all n, the dominated convergence theorem tells us that[1]

$$\langle f_n, \mu \rangle = \int f_n d\mu \to 0.$$

Thus, the dual of X "sees" that the sequence $\{f_n\}$ converges to zero in some sense. Our little discussion motivates the following definition.

Definition 5.1.1. Let X be a normed space. A sequence $\{x_n\}$ is said to **converge weakly** to $x \in X$ if $\langle \phi, x_n \rangle \to \langle \phi, x \rangle$ for all $\phi \in X^*$. We then write $x_n \xrightarrow{w} x$ (in some texts the notation $x_n \rightharpoonup x$ is used).

---

[1]In this chapter, we will use the "pairing" notation $\langle \phi, x \rangle = \langle x, \phi \rangle := \phi(x)$ to depict the action of a functional $\phi \in X^*$ on an element $x \in X$ (see after Definition 1.3.5). This notation is suggestive of the fact that x also acts on ϕ.

DOI: 10.1201/9781003297864-5

We see that the sequence $\{f_n\}$ discussed above converges weakly to 0. It is clear that if $x_n \to x$ in the norm[2] then $x_n \to x$ weakly. The converse is false, as the above example shows. Here is another important example.

Example 5.1.2. Consider an orthonormal sequence $\{e_n\}$ in a Hilbert space H. For any $g \in H$, Bessel's inequality implies that $\sum |\langle g, e_n \rangle|^2 < \infty$, therefore $\langle g, e_n \rangle \to 0$. This means that $e_n \to 0$ weakly. On the other hand, $\|e_n - e_m\|^2 = 2$ for all m, n, so the sequence has no convergent subsequence and does not converge in the norm.

Definition 5.1.3. Let X be a normed space and let X^* be its dual. A sequence $\{\phi_n\} \subseteq X^*$ is said to **weak-∗ converge** to $\phi \in X^*$ if $\langle \phi_n, x \rangle \to \langle \phi, x \rangle$ for all $x \in X$. We write $\phi_n \xrightarrow{w*} \phi$.

Note that weak-∗ convergence is weaker than weak convergence. Of course, in a reflexive space weak and weak-∗ convergence are the same.

Exercise 5.1.4. Give an example of a sequence in a Banach space that is weak-∗ convergent but not weakly convergent.

Exercise 5.1.5. Show that the limit of a weak or weak-∗ convergent sequence is unique.

Proposition 5.1.6. *Every weakly convergent or weak-∗ convergent sequence is bounded.*

Proof. If $\{x_n\}$ is weakly convergent, then the isometrically embedded sequence in $\{\hat{x}_n\}$ in X^{**} has the property that for every $\phi \in X^*$, the sequence $\langle \hat{x}_n, \phi \rangle = \langle \phi, x_n \rangle$ is convergent, and so, in particular, it is a bounded sequence. By the uniform boundedness principle, $\{\hat{x}_n\}$ is bounded in $\hat{X} \subseteq X^{**}$, thus $\sup_n \|x_n\| = \sup_n \|\hat{x}_n\| < \infty$. The boundedness of weak-∗ convergent sequences is an even more direct application of uniform boundedness. $\qquad\square$

It will take the reader some time to fully appreciate how the notions of weak and weak-∗ convergence interact with the structure of a normed space and the operators on it. The following proposition should be considered as an appetizer.

Proposition 5.1.7. *A linear operator T between Banach spaces is bounded if and only if T is weakly sequentially continuous in the sense that $x_n \xrightarrow{w} x$ implies $Tx_n \xrightarrow{w} Tx$.*

Proof. Assume that T is bounded. Let $\{x_n\}$ be weakly convergent to x. Then for all $\psi \in Y^*$ we have that $\psi \circ T \in X^*$, so

$$\langle \psi, Tx_n \rangle = \psi(T(x_n)) = \langle \psi \circ T, x_n \rangle \to \langle \psi \circ T, x \rangle = \langle \psi, Tx \rangle.$$

It follows that $Tx_n \xrightarrow{w} Tx$.

---

[2]Until this point "convergence in the norm" was usually referred to simply as "convergence", since there was no other kind of convergence; we can keep using this terminology so long as we make sure we know what we are talking about. Sometimes convergence in norm is called **strong convergence**, to minimize confusion.

Assume that T is weakly sequentially continuous. Supposing T is not bounded, we shall derive a contradiction as follows. On the one hand, if T is not bounded then there exists a sequence $x_n \to 0$ such that $\|Tx_n\| \to \infty$. But on the other hand $x_n \xrightarrow{w} 0$, so weak sequential continuity of T implies $Tx_n \xrightarrow{w} 0$; Proposition 5.1.6 then gives $\sup_n \|Tx_n\| < \infty$. Thus T must be bounded if it is weakly sequentially continuous. $\qquad\square$

One of the rewards for introducing weaker notions of convergence is an improved likelihood of observing compactness phenomena. Recall that the closed unit ball of an infinite dimensional normed space is never compact [44, Theorem 9.2.3].

Theorem 5.1.8. *Let X be a separable normed space and X^* its dual. Then the closed unit ball of X^* is weak-$*$ sequentially compact in the sense that every sequence $\{\phi_n\} \subset X_1^*$ has a weak-$*$ convergent subsequence. In particular, the closed unit ball of a separable and reflexive space is weakly sequentially compact.*

Proof. Let $\{\phi_n\}_n \subset X_1^*$, and let $\{x_n\}_n$ be a dense sequence in X. Consider the sequence of scalars $\{\phi_n(x_1)\}_n$. This is a bounded sequence in $\mathbb{C}$, so we may extract a convergent subsequence $\{\phi_{1,n}\}_n$ such that the sequence $\phi_{1,n}(x_1)$ converges. We inductively extract for all $k = 2, 3, \dots$ a subsequence $\{\phi_{k,n}\}_n$ such that $\phi_{k,n}(x_i)$ converges for all $i = 1, \dots, k$. Now letting $\psi_n = \phi_{n,n}$, we have a subsequence of the original sequence such that the limit $\lim_n \psi_n(x_k)$ exists for all k. Since $\{x_k\}$ is dense in X and since $\{\psi_n\}$ is contained in the unit ball, we have that $\psi(x) := \lim_n \psi_n(x)$ exists for all $x \in X$. Now it is easy to verify that $\psi \in X^*$ and that $\psi_n \xrightarrow{w*} \psi$, as required. $\qquad\square$

5.2   Application: formal trigonometric series

In this section, we provide a typical yet technically very simple application of weak-$*$ sequential compactness in analysis. As you follow the argument, take note of how the compactness provides the existence of a certain sought after limiting object.

Recall the notation of Section 4.4 and Exercises 4.5.12–4.5.15. For a function $f \in L^p := L^p[0,1]$ (where $p \in [1, \infty]$) one can define Fourier coefficients $\hat{f}(n) = \int_0^1 f(t)e^{-2\pi i n t}dt$, and then one considers the associated Fourier series

$$f \sim \sum_{n \in \mathbb{Z}} \hat{f}(n)e^{2\pi i n t}.$$

Much of harmonic analysis deals with the question of whether this series converges and in what sense. But there is a reverse question one can ask. Suppose that we are given a sequence of scalars $\{c_n\}_{n\in\mathbb{Z}}$. Then we may consider the *formal trigonometric series*

$$\sum_n c_n e^{2\pi int} \tag{5.1}$$

and ask when is this series the Fourier series associated with a function in L^p, that is, when is there a function $f \in L^p$ such that $c_n = \hat{f}(n)$ for all n. For example, we know by the basic theory of Hilbert spaces and of Fourier series in L^2, that (5.1) is the Fourier series of a function $f \in L^2$ if and only $\sum |c_n|^2 < \infty$. The following theorem characterizes when a formal trigonometric series is the Fourier series of an L^p function for $p \in (1, \infty]$.

Theorem 5.2.1. *Let* (5.1) *be a formal trigonometric series. Consider the associated Fejér sums*

$$F_N(t) := \sum_{n=-N}^{N} \left(1 - \frac{|n|}{N+1}\right) c_n e^{2\pi int}.$$

Then for all $p \in (1, \infty]$ we have that (5.1) *is the Fourier series of a function in L^p (for $1 < p \leq \infty$) if and only if $\sup_N \|F_N\|_p < \infty$.*

Proof. If $\{c_n\}_{n\in\mathbb{Z}}$ is the sequence of Fourier coefficients of a function $f \in L^p$ then $F_N = \sigma_N(f) = f * K_N$, and one shows that this sequence is bounded in L^p in the course of the solution of Exercise 4.5.15 (see the exercise for a hint).

Conversely, assume that F_N is a bounded sequence in L^p. By weak-∗ sequential compactness of the closed unit ball of $L^p = (L^q)^*$, there is a weak-∗ convergent subsequence $F_{N_k} \xrightarrow{w*} f \in L^p$. Now, the Fourier coefficients of F_N are given by $\hat{F}_N(n) = \langle e_n, F_N \rangle$, where $e_n(t) = e^{2\pi int} \in L^q$, and likewise for f. It follows that $\hat{F}_{N_k}(n) \xrightarrow{k\to\infty} \hat{f}(n)$. On the other hand, clearly $\hat{F}_{N_k}(n) \xrightarrow{k\to\infty} c_n$. We conclude that $c_n = \hat{f}(n)$, as required. $\qquad\square$

Exercise 5.2.2. Prove that the sequence F_N is bounded in the L^1 norm if and only if the formal trigonometric series (5.1) is the Fourier series of a regular complex Borel measure.

5.3 The adjoint of a linear map

5.3.1 The adjoint map

The idea in the proof of Proposition 5.1.7 motivates the following definition.

Definition 5.3.1. Let $T \in B(X, Y)$. The **adjoint** of T is the operator $T^* : Y^* \to X^*$ given by

$$T^*\psi = \psi \circ T.$$

Warning: some authors use the term **transpose** for the adjoint, and refrain from using the terminology introduced above so as not to confuse this adjoint with the Hilbert spaces adjoint (which can then be referred to as the ***conjugate transpose***). One can keep the concepts distinct by emphasizing, when needed, whether one is talking about a "Hilbert space adjoint" or a "Banach space adjoint".

Theorem 5.3.2. *Let X and Y be normed spaces, and let $T \in B(X,Y)$. Then $T^* \in B(Y^*, X^*)$ and $\|T^*\| = \|T\|$. For all $x \in X, \psi \in Y^*$,*

$$\langle \psi, Tx \rangle = \langle T^*\psi, x \rangle. \tag{5.2}$$

Moreover, T^* is the unique map from Y^* to X^* with this property.

Proof. It is easy to see that T^* is a bounded linear map of norm less than or equal to $\|T\|$ (indeed, $\|T^*\psi\| = \|\psi \circ T\| \leq \|\psi\|\|T\|$) from Y^* into X^*. To see equality of norms,

$$\|T^*\| = \sup_{\psi \in Y_1^*} \|T^*\psi\|$$
$$(*) = \sup\{|\langle T^*\psi, x \rangle| : \psi \in Y_1^*, x \in X_1\}$$
$$= \sup\{|\langle \psi, Tx \rangle| : \psi \in Y_1^*, x \in X_1\}$$
$$(**) = \sup_{x \in X_1} \|Tx\| = \|T\|,$$

where $(*)$ follows from the definition of the norm of the functional $T^*\psi$, and $(**)$ follows from the Hahn–Banach theorem (specifically, Corollary 2.1.8). Uniqueness is trivial. $\qquad\square$

Example 5.3.3. Show that if $X = Y = \mathbb{C}^n$ with the standard inner product, and if the linear operator $T = T_A$ is given by multiplication by an $n \times n$ matrix A

$$T(x) = Ax,$$

then $T^* = T_{A^T}$ is given by multiplication by the *transposed* matrix A^T, i.e. $T^*(y) = A^T y$; as we warned above, A^* is not given by multiplication by the conjugate transposed matrix A^*.

Exercise 5.3.4. Let X be a normed space and let $f \in X^*$. What is the adjoint map f^*?

Exercise 5.3.5. It is well known, that the existence of an adjoint operator in the setting of Hilbert spaces depends on the completeness of the space; for example, there exists a bounded operator T on an inner product space G such that there is no $S \colon G \to G$ for which $\langle Tf, g \rangle = \langle f, Sg \rangle$ for all $f, g \in G$ (give an example). The proof of Theorem 5.3.2, on the other hand, does not depend on completeness. How can this be?

5.3.2 Annihilators and the adjoint

Definition 5.3.6. Let X be a normed space and let $S \subseteq X$. The **annihilator of** S, denoted $S^\perp$, is the set

$$S^\perp = \{\phi \in X^* : \langle \phi, x \rangle = 0 \text{ for all } x \in S\}.$$

Definition 5.3.7. Let X be a normed space and let $T \subseteq X^*$. The **pre-annihilator of** T, denoted $T_\perp$, is the set

$$T_\perp = \{x \in X : \langle \phi, x \rangle = 0 \text{ for all } \phi \in T\}.$$

With the above notation, we can state an important corollary of the Hahn–Banach theorem (Corollary 2.1.9) as follows: *A subspace M contains x in its closure if and only if $x \in (M^\perp)_\perp$.* In other words,

$$\overline{M} = (M^\perp)_\perp. \tag{5.3}$$

The reader is encouraged to pause and consider the analogous reverse situation: if N is a subspace of X^*, does it hold that $\overline{N} = (N_\perp)^\perp$?

Proposition 5.3.8. *Let X and Y be normed spaces, and let $T \in B(X,Y)$. Then $\ker(T^*) = \operatorname{Im}(T)^\perp$ and $\ker(T) = \operatorname{Im}(T^*)_\perp$.*

Proof. Thanks to the notation, the proof is exactly the same as in the Hilbert space case (note, though, that both assertions require proof, and do not follow one from the other by replacing $T \leftrightarrow T^*$ because their roles are not symmetrical). For example,

$$\begin{aligned}
x \in \ker(T) &\iff Tx = 0 \\
&\iff \langle \psi, Tx \rangle = 0 \text{ for all } \psi \in Y^* \\
&\iff \langle T^*\psi, x \rangle = 0 \text{ for all } \psi \in Y^* \\
&\iff x \in \operatorname{Im}(T^*)_\perp.
\end{aligned}$$

$\square$

Corollary 5.3.9. *For $T \in B(X,Y)$, $\ker(T^*)_\perp = \overline{\operatorname{Im}(T)}$.*

Proof. This follows from the proposition and (5.3). $\square$

Remark 5.3.10. In general, it is not true that $\ker(T)^\perp = \overline{\operatorname{Im}(T^*)}$; see Exercise 5.6.4. Knowing this, you can conclude that $\overline{N} = (N_\perp)^\perp$ cannot always hold. We will understand this better later on (see Corollary 6.4.7).

Theorem 5.3.11. *Let X be a normed space and $M \subseteq X$ a closed subspace. Then*

1. *M^* is isometrically isomorphic to $X^*/M^\perp$.*

2. *$(X/M)^*$ is isometrically isomorphic to $M^\perp$.*

Proof. Consider[3] the inclusion map $i\colon M \hookrightarrow X$. The adjoint map $\rho = i^*\colon X^* \to M^*$ is simply the restriction map

$$\rho(\phi) = \phi\big|_M$$

and the kernel of ρ is $M^\perp$. By Theorem 4.3.1, we have an induced bijective map $\dot\rho\colon X^*/M^\perp \to M^*$. We need to show that $\dot\rho$ is isometric. Let $\phi \in X^*$. Then

$$\|\phi + M^\perp\| = \inf\left\{\|\phi + \eta\| : \eta \in M^\perp\right\} = \inf\left\{\|\psi\| : \psi \text{ extends } \phi\big|_M\right\}.$$

All of the functionals appearing in the infimum extend $\phi\big|_M = \dot\rho(\phi + M^\perp)$ therefore

$$\|\dot\rho(\phi + M^\perp)\| \le \|\phi + M^\perp\|.$$

On the other hand, by Hahn–Banach, one of these extensions has the same norm as $\phi\big|_M$, therefore $\|\dot\rho(\phi + M^\perp)\| = \|\phi + M^\perp\|$. That proves the first assertion.

For the second assertion, consider the adjoint map of $\pi\colon X \to X/M$, that is the map $\pi^*\colon (X/M)^* \to X^*$ defined by

$$\pi^*(\psi) = \psi \circ \pi.$$

It is obvious that $\operatorname{Im}(\pi^*) \subseteq M^\perp$, since π vanishes on M. We need to show that π^* is a surjective isometry onto $M^\perp$. To see that π^* is isometric, we compute

$$\|\pi^*(\psi)\| = \sup\left\{|\langle \pi^*(\psi), x\rangle| : \|x\| < 1\right\} = \sup\left\{|\langle \psi, \pi(x)\rangle| : \|x\| < 1\right\}.$$

By Theorem 1.4.4(3), π maps the open unit ball of X onto the open unit ball of X/M, so the right hand side is equal to

$$\sup\{|\langle \psi, y\rangle| : \|y\| < 1\} = \|\psi\|.$$

It remains to show that the range of π^* is equal to $M^\perp$. Let $\phi \in M^\perp$. Then $\ker(\phi) \supseteq M$. It follows that if $x_1 - x_2 \in M$, then $\langle \phi, x_1\rangle = \langle \phi, x_2\rangle$, so we may define a functional f on X/M by

$$f(\dot x) = \langle \phi, x\rangle.$$

We need to show that f is bounded. By definition $\phi = f \circ \pi$. The kernel of f is equal to $\pi(\ker(\phi))$, and is closed by 1.4.4(6). By Proposition 2.1.14, f is bounded. Thus, $f \in (X/M)^*$ is such that $\phi = \pi^*(f)$, so the range of π^* is $M^\perp$. That completes the proof. $\qquad\square$

---

[3]Whenever we need to prove that one object A is isomorphic to a certain quotient B/C, the natural thing to seek is a surjective map $B \to A$ with kernel C. In our case, we seek a map $X^* \to M^*$. But the natural way to get a map between dual spaces is as the adjoint of a map, so in our case we are led to begin by considering a map $M \to X$. These kinds of considerations guide the proof. It might be instructive for the reader to forget this proof and try to write down directly an isomorphism in the direction $M^* \to X^*/M^\perp$.

5.4 Invertibility and the adjoint

Proposition 5.3.8 implies the following interesting corollary:

Corollary 5.4.1. *Let X, Y be normed spaces, and let $T \in B(X, Y)$. T^* is injective if and only if $\operatorname{Im} T$ is dense. Moreover, if T^* has a dense range, then T is injective.*

If T is injective, then $\operatorname{Im} T^*$ might not be dense (see Exercise 5.6.4). We note that in reflexive spaces T is injective or has dense range, respectively, if and only if T^* has dense range or is injective, respectively. This is in line with what we recall from finite dimensional spaces, in which T is injective or surjective, respectively, if and only if T^* is surjective or is injective, respectively.

We now continue to develop the above theme of relating between invertibility properties of T and its adjoint. We shall require the following result, the proof of which is contained in the last paragraph in the proof of the open mapping theorem (Theorem 4.1.8).

Exercise 5.4.2. Let X and Y be Banach spaces and let $T \in B(X, Y)$. Let B_X denote the open unit ball in X. If $\overline{T(B_X)}$ contains an open ball centered at 0 then $T(B_X)$ contains an open ball centered at 0, and consequently T is surjective.

The notation introduced in the above exercise is convenient, and we will continue to use it in this section: henceforth B_X and B_Y denote the open unit balls in X and Y respectively.

Theorem 5.4.3. *Let X and Y be two Banach spaces, and let $T \in B(X, Y)$. T is surjective if and only if there exists $r > 0$ such that $\|T^*\psi\| \geq r\|\psi\|$ for all $\psi \in Y^*$.*

Proof. Suppose that $\|T^*\psi\| \geq r\|\psi\|$ for all $\psi \in Y^*$. To prove that T is surjective, it suffices to prove that $T(B_X)$ contains an open ball centered at 0. By the above exercise, we may content ourselves with showing the closure of $T(B_X)$ contains an open ball; we will show that

$$rB_Y \subseteq \overline{T(B_X)}.$$

To that end, we fix $y_0 \notin \overline{T(B_X)}$, and we shall show that $\|y_0\| > r$. By the Hahn–Banach theorem there exists $\psi \in Y^*$ that separates y_0 from $\overline{T(B_X)}$, that is

$$\operatorname{Re}\langle y, \psi \rangle \leq 1 < \operatorname{Re}\langle y_0, \psi \rangle \tag{5.4}$$

for all $y \in \overline{T(B_X)}$. Since $\overline{T(B_X)}$ is *circular* (i.e., invariant under multiplication by scalars of modulus 1), we find that $|\langle y, \psi \rangle| \leq 1$ for all $y \in \overline{T(B_X)}$. It follows that for all $x \in B_X$,

$$|\langle x, T^*\psi \rangle| = |\langle Tx, \psi \rangle| \leq 1,$$

whence $\|T^*\psi\| \le 1$. So, using the assumption $\|T^*\psi\| \ge r\|\psi\|$ we find that $r\|\psi\| \le 1$. Thus,

$$\|y_0\| \ge r\|y_0\|\|\psi\| \ge r|\langle y_0, \psi\rangle| > r,$$

where for the last strict inequality we used (5.4). That shows that $\|y_0\| > r$, as required.

Conversely, if $T\colon X \to Y$ is surjective and bounded, then the open mapping theorem implies that T is open, that is, there is some $r > 0$ such that $T(B_X) \supseteq rB_Y$. This allows us to compute for all $\psi \in Y^*$,

$$\begin{aligned}
\|T^*\psi\| &= \sup\{|\langle x, T^*\psi\rangle| : x \in B_X\}\\
&= \sup\{|\langle Tx, \psi\rangle| : x \in B_X\}\\
&\ge \sup\{|\langle y, \psi\rangle| : y \in rB_Y\} = r\|\psi\|,
\end{aligned}$$

as required.   □

An operator $T\colon X \to Y$ is said to be **bounded below** if there exists $r > 0$ such that $\|Tx\| \ge r\|x\|$ for all $x \in X$.

Exercise 5.4.4. Let X and Y be two Banach spaces, and let $T \in B(X,Y)$. Prove that T is bounded below if and only if T is injective and $\operatorname{Im}(T)$ is closed.

A corollary of the above theorem and exercise is the following.

Corollary 5.4.5. *Let X, Y be Banach spaces, and let $T \in B(X,Y)$. T is surjective if and only if T^* is injective and has a closed range.*

Exercise 5.4.6. The readers must be wondering: what happens if the roles of T and T^* are reversed? The answer is given away in Exercise 5.6.8; see if you can settle it on your own.

Given an operator $T \in B(X,Y)$, we have by Theorem 5.3.2 a unique operator $T^* \in B(Y^*, X^*)$ that satisfies (5.2). We can repeat this and obtain $T^{**} = (T^*)^* \in B(X^{**}, Y^{**})$, determined by

$$\langle T^{**}f, \psi\rangle = \langle f, T^*\psi\rangle, \quad \text{for all } \psi \in Y^*, f \in X^{**}.$$

Recall that X is identified as a closed subspace $\hat{X} \subseteq X^{**}$ via the embedding $x \mapsto \hat{x}$, where

$$\langle \hat{x}, \phi\rangle = \langle \phi, x\rangle.$$

We will now see that, loosely speaking, $T^{**}\big|_X = T$. Indeed, for every $x \in X$ and $\psi \in Y^*$,

$$\langle T^{**}\hat{x}, \psi\rangle = \langle \hat{x}, T^*\psi\rangle = \langle T^*\psi, x\rangle = \langle \psi, Tx\rangle = \langle \widehat{Tx}, \psi\rangle.$$

Thus if we identify x with $\hat{x}$ for all $x \in X$, we find that $T^{**}\big|_X = T$. If X is reflexive, then $\hat{X} = X^{**}$, so $T^{**} = T$.

These considerations solve one half of Exercise 5.4.6, indeed: if T^* is surjective, then by Corollary 5.4.5, $(T^*)^* = T^{**}$ is injective and has closed range, so in particular it is bounded below (by Exercise 5.4.4). This automatically implies that $T = T^{**}|_X$ is bounded below, so it is injective and has closed range (by Exercise 5.4.4, again). We cannot use this argument to solve the other half of the exercise, since in the non-reflexive case, it is not yet clear that if T is injective and bounded below, then so is T^{**}.

Theorem 5.4.7. *Let X, Y be Banach spaces. A bounded operator $T \in B(X, Y)$ is invertible if and only if T^* is invertible.*

Proof. If T is invertible, then it is straightforward to check that $(T^{-1})^*$ is the inverse of T^*. If T^* is invertible, then T^* is in particular surjective, and the argument in the preceding paragraph shows that T is bounded below and hence is injective. If T^* is invertible then it must also be bounded below, and Theorem 5.4.3 tells us that T is surjective. □

Recall the important notion of spectrum.

Definition 5.4.8. The ***spectrum*** of an operator $T \in B(X)$ is the set

$$\sigma(T) = \{\lambda \in \mathbb{C} : T - \lambda I \text{ is not invertible in } B(X)\}.$$

The spectrum $\sigma(T)$ is defined to be the set of $\lambda \in \mathbb{C}$ such that $T - \lambda I$ does not have a bounded inverse. Note that by the inverse mapping theorem, $\lambda \in \sigma(T)$ if and only $T - \lambda I$ is not bijective. We shall later see that the spectrum is always nonempty (see Theorem 8.3.6).

Corollary 5.4.9. *Let T be a bounded operator on a Banach space X. Then $\sigma(T) = \sigma(T^*)$.*

Proof. By Theorem 5.4.7, the operators $T - \lambda I$ and $T^* - \lambda I = (T - \lambda I)^*$ are either both invertible or both not, hence $\lambda \in \sigma(T) \iff \lambda \in \sigma(T^*)$. □

5.5 Application: invertibility of identity plus compact

Compact operators and their basic spectral theory form an important part of the first course in functional analysis. Here we recall the basic definition and connect it to the present material. See [44, Chapter 9] for the proofs of results that appear here without proof. Recall the following definition and basic facts.

Definition 5.5.1. A linear operator $T \colon X \to Y$ between normed spaces is said to be ***compact*** if the closure of $T(X_1)$ is compact in Y. We let $K(X, Y)$ be the set of all compact operators from X to Y and we write $K(X)$ for $K(X, X)$.

An equivalent definition is as follows: *an operator $T \colon X \to Y$ is compact if and only if every bounded sequence $\{x_n\}_{n=1}^{\infty} \subset X$ has a subsequence $\{x_{n_k}\}_{k=1}^{\infty}$ such that $\{Tx_{n_k}\}_{k=1}^{\infty}$ is a convergent sequence in Y.*

Here is a list of facts about compact operators.

1. $K(X,Y)$ is a closed subspace of $B(X,Y)$.

2. The product of a bounded operator and a compact operator is compact.

3. If $T \in B(X,Y)$ and $\dim \operatorname{Im}(T) < \infty$ then $T \in K(X,Y)$.

4. The identity operator on a normed space X is compact if and only if $\dim X < \infty$ (equivalently, X_1 is compact if and only if $\dim X < \infty$.)

5. Consequently, if X is infinite dimensional and $T \in K(X)$ is compact then $0 \in \sigma(T)$.

6. The spectrum of a compact operator T on an infinite dimensional Banach space X consists of 0 and a sequence which is either finite or converges to 0. Every nonzero element $\lambda \in \sigma(T)$ is an eigenvalue of finite multiplicity, i.e., $\dim \ker(T - \lambda I) < \infty$.

Proposition 5.5.2. *Let X and Y be two Banach spaces, and let $T \in B(X,Y)$. Then T is compact if and only if T^* is compact.*

Proof. Suppose that T is compact, and let $\{f_n\}_{n=1}^{\infty}$ be a sequence in Y_1^*. Our goal is to show that $\{T^* f_n\}_{n=1}^{\infty}$ has a subsequence that is convergent in X^*. We can think of $\{f_n\}$ as a sequence of functions on the compact set $\overline{T(X_1)}$. Then, this sequence of continuous functions is uniformly bounded and equicontinuous, since

$$|f_n(y)| \leq \|f_n\| \|y\| \leq \|T\| \quad \text{and} \quad |f_n(y) - f_n(y')| \leq \|f_n\| \|y - y'\| \leq \|y - y'\|$$

for all $y, y' \in \overline{TX_1}$. By the Arzelà-Ascoli theorem (see [11, Theorem A.4.6]), there is a subsequence $\{f_{n_k}\}$ that converges uniformly on $\overline{TX_1}$. It follows that $\{T^* f_{n_k}\}$ is a convergent sequence in X^*. Indeed,

$$\|T^* f_{n_k} - T^* f_{n_m}\| = \sup\{|f_{n_k}(Tx) - f_{n_m}(Tx)| : x \in X_1\} \xrightarrow{k,m \to \infty} 0$$

because $\{f_{n_k}\}$ converges uniformly on $\overline{TX_1}$. This shows that T^* is compact.

Now, if T^* is compact, then the first part of the proof implies that T^{**} is compact. But then $T = T^{**}\big|_X$ is the restriction of a compact operator to a closed subspace hence is also compact. $\qquad \square$

Let us recall the following theorem from the first course, which is the most important thing to know about the spectral theory of compact operators (see [11, Theorem 9.2.9]).

Theorem 5.5.3 (The Fredholm alternative). *Let X be a Banach space, let $A \in K(X)$, and fix a complex number $\lambda \neq 0$. Then $\dim \ker(\lambda I - A)$ is finite dimensional, and exactly one of the following holds:*

1. *$\ker(\lambda I - A) \neq \{0\}$,*

2. *$\lambda I - A$ is invertible.*

Thus, every nonzero point in $\sigma(A)$ is an eigenvalue of finite multiplicity.

The dichotomy between (1) and (2) is the crux of the "alternative" and is very useful in studying the solvability of linear equations that have the form $(I + A)x = y$ where A is a compact operator. Such equations arise in practice as integral equations, and indirectly in the theory of differential or functional equations, see [44, Chapter 11], or [22, 48]. By the dichotomy, in order to show that there exists a solution for every y, i.e. to show that $I + A$ is surjective, it suffices to prove that $I + A$ is injective. Again, a problem of existence has been traded for a problem of uniqueness — usually a profitable bargain.

A natural question arises: does surjectivity of $I + A$ imply injectivity? We now have enough machinery to answer this question.

Theorem 5.5.4. *Let A be a compact operator on the Banach space X. Then $\mathrm{Im}(I + A) = X$ if and only if $\ker(I + A) = \{0\}$.*

Proof. We know from the Fredholm alternative that $\ker(I + A) = \{0\}$ implies $\mathrm{Im}(I+A) = X$. So suppose that $\mathrm{Im}(I+A) = X$, and consider $I+A^* = (I+A)^*$. By Proposition 5.3.8, $\ker(I + A^*) = \mathrm{Im}(I + A)^\perp = \{0\}$. By Proposition 5.5.2, the operator A^* is also compact, so by the Fredholm alternative applied to $I + A^*$ we have $\mathrm{Im}(I + A^*) = X^*$. By Proposition 5.3.8 again, $\ker(I + A) = \mathrm{Im}(I+A^*)_\perp = (X^*)_\perp$, but by the Hahn–Banach theorem $(X^*)_\perp = \{0\}$. That completes the proof.

The last part of the proof can be replaced with the following argument: from $\ker(I+A^*) = \{0\}$ and $\mathrm{Im}(I+A^*) = X^*$ we have, by the inverse mapping theorem, that $I + A^*$ is invertible, hence $I + A$ is invertible by Theorem 5.4.7. □

5.6   Additional exercises

Exercise 5.6.1. A sequence $\{x_n\}_n$ in a normed space is said to be ***weakly bounded*** if for every $\phi \in X^*$ the sequence of scalars $\{\phi(x_n)\}_n$ is bounded. Prove that a weakly bounded sequence in a Banach space is bounded in the norm. What changes if we consider weakly bounded sequences in a not-necessarily complete normed space?

Exercise 5.6.2. Prove that the closed unit ball in a reflexive space is weakly sequentially compact (that is, remove the separability assumption from Theorem 5.1.8).

Exercise 5.6.3. Prove that a sequence in ℓ^1 converges weakly if and only if it converges in the norm.

Exercise 5.6.4. Consider the operator $T\colon L^1[0,1] \to c_0$ that maps a function to its Fourier coefficients:

$$T(f) = \left(\hat{f}(n)\right)_{n\in\mathbb{Z}} \qquad \text{where} \qquad \hat{f}(n) = \int_0^1 f(t)e^{-2\pi i n t}\,dt.$$

1. Describe the adjoint operator T^*.

2. Show that $\operatorname{Im} T^*$ is not closed, and conclude that T is not surjective (cf. Exercise 4.5.12).

3. Use this example to show that $\ker(T)^\perp \neq \overline{\operatorname{Im}(T^*)}$ in general (see Remark 5.3.10) and also that T being injective does not imply that $\operatorname{Im} T^*$ is dense (cf. Corollary 5.4.1).

Exercise 5.6.5. Let X and Y be Banach spaces, and let $T\colon X \to Y$ be a mapping that "has an adjoint", in the sense that there exists a mapping $S\colon Y^* \to X^*$ so that

$$\langle \psi, T(x)\rangle = \langle S(\psi), x\rangle$$

for all $x \in X$ and $\psi \in Y^*$. Prove that T is a bounded linear map.

Exercise 5.6.6. Prove that X^* is always a complementable subspace of X^{***}. (Hint: consider the natural maps $i: X \to X^{**}$, $j: X^* \to X^{***}$, their adjoints, and the various compositions you can form). Contrast this with Exercise 4.5.8; we see that c_0 cannot be equal to the dual space of any normed space.

Exercise 5.6.7. Try to find a connection between the formalism of weak solutions (see Section 3.3.1) and the notion of the adjoint operator.

Exercise 5.6.8. Prove that Corollary 5.4.5 (equivalently, Theorem 5.4.3) holds with the roles of T and T^* reversed.

Exercise 5.6.9 (The closed range theorem). Prove that $\operatorname{Im} T$ is closed if and only if $\operatorname{Im} T^*$ is closed.

Exercise 5.6.10 (Alternative proof of inverse mapping and open mapping theorems). In Chapter 4, we obtained the inverse mapping theorem (Corollary 4.1.11) as an immediate corollary of the open mapping theorem (Theorem 4.1.8). One can prove the theorems in the reverse order, using different ideas.

1. Adjust the proof of the inverse mapping theorem in Hilbert spaces given in [11] (see Theorem 8.3.4 there) to Banach spaces, so that you have a proof that does not rely on the open mapping theorem or the closed graph theorem (you may want to throw in a reasonable assumption if you get stuck).

2. Obtain the open mapping theorem as a corollary of the inverse mapping theorem by using quotient spaces.

Exercise 5.6.11. Let $\{x_n\}$ be a bounded sequence in a Banach space X, and suppose that there exists $r > 0$

$$\sup_n |\langle \phi, x_n \rangle| \geq r\|\phi\|$$

for all $\phi \in X^*$. Prove that for every $x \in X$ there exists a sequence $\{c_n\} \subset \mathbb{C}^{\mathbb{N}}$ such that $x = \sum_{n=1}^{\infty} c_n x_n$ in the norm. **Bonus question:** Can you come up with nice concrete examples in which this can be applied?

Exercise 5.6.12. Consider the following statement: *Let X and Y be Banach spaces and $T \colon X \to Y$ be a linear map. T is compact if and only if T maps weakly convergent sequences to norm convergent sequences.* Show that the statement is false as stated. Add a (sensible) assumption that corrects it, and prove it.

Chapter 6

Locally convex spaces and weak topologies

6.1 The weak topology

We begin by recalling some basic definitions from general topology.

A *topology* τ on a set T is simply a collection of subsets of T such that $\emptyset, T \in \tau$, and such that τ is closed under finite intersections and arbitrary unions. Sets in τ are called *open* sets. A *base* is sub-collection $\mathcal{B} \subseteq \tau$ such that every open set is equal to the union of elements in $\mathcal{B}$; equivalently, for every $t \in T$ and every $U \in \tau$ containing t, there is $B \in \mathcal{B}$ such that $t \in B \subseteq U$. The sets in $\mathcal{B}$ are called *basic open sets*.

A *local base* at a point t in a topological space T is a family $\mathcal{F}$ of open sets, such that for every open set U containing t, there exists some $V \in \mathcal{F}$, for which $t \in V \subseteq U$.

Given a family of subsets $\mathcal{F}$ of a set T, one can define the *topology generated by* $\mathcal{F}$ as the minimal (under containment) collection of subsets in T that is a topology containing $\mathcal{F}$. This topology is commonly referred to as the *weakest* (or *smallest*) *topology containing* $\mathcal{F}$. The notion is well defined, since it can be got as the intersection of all topologies containing $\mathcal{F}$ (note that the so-called *discrete topology* 2^T on T is a topology containing $\mathcal{F}$, so this intersection is not void).

It turns out that the weakest topology τ containing $\mathcal{F}$ is the topology with the base $\cap_{i=1}^{n} F_i$ where n is any nonnegative integer and $F_i \in \mathcal{F}$. When every point in T belongs to some $F \in \mathcal{F}$, then this topology τ is simply the collection of all unions of all finite intersections of elements in $\mathcal{F}$. The set $\mathcal{F}$ is then called a *subbase* for the topology τ.

Definition 6.1.1. Let X be a normed space. The *weak topology* on X is the weakest topology in which all the functionals in X^* are continuous. In other words, it is the topology generated by the sets

$$V(x; f; \epsilon) = \{y \in X : |f(x - y)| < \epsilon\},$$

where $x \in X$, $f \in X^*$ and $\epsilon > 0$.

DOI: 10.1201/9781003297864-6

We see that the weak topology has a base

$$V(x_1, \ldots, x_k, ; f_1, \ldots, f_k; \epsilon_1, \ldots, e_k) = \cap_{i=1}^k V(x_i; f_i; \epsilon_i).$$

Exercise 6.1.2. Prove that the weak topology has a base given by the sets

$$V(x; f_1, \ldots, f_k; \epsilon) = \{y \in X : |f_i(x - y)| < \epsilon \text{ for all } i = 1, \ldots, k\}. \quad (6.1)$$

If x is fixed, prove that sets of the form (6.1) form a local base at x. Note also that one needs only consider basic sets with $\epsilon = 1$.

Recall that a sequence $\{x_n\}_{n=1}^\infty$ in a topological space is said to converge to a point x, if for every neighborhood U of x, the element x_n is contained in U for all but finitely many integers n. We easily verify that a sequence $\{x_n\}_{n=1}^\infty \subset X$ converges weakly to $x \in X$ in the sense of Definition 5.1.1 precisely when it converges to x in the weak topology of X.

Lemma 6.1.3. *Let f be a linear functional on X, $f_1, \ldots, f_n \in X^*$, and suppose that*

$$V(0; f_1, \ldots, f_n; \epsilon) \subseteq V(0; f; 1).$$

Then there are scalars $a_1, \ldots, a_n$ such that $f = a_1 f_1 + \cdots + a_n f_n$.

Proof. Denote $M = \cap_{i=1}^n \ker(f_i)$. We will first show that $M \subseteq \ker(f)$. Indeed, if $x \in M$, then $nx \in M \subseteq V(0; f_1, \ldots, f_n; \epsilon) \subseteq V(0; f; 1)$ for all n, hence $n|f(x)| < 1$ for all n, whence $f(x) = 0$.

Now we will show that $M \subseteq \ker(f)$ implies that $f = a_1 f_1 + \cdots + a_n f_n$ for some scalars a_i. For definiteness, let the field of scalars be $\mathbb{C}$. Define $T : X \to \mathbb{C}^n$ by $Tx = (f_1(x), \ldots, f_n(x))$. On $\operatorname{Im} T$ we can define a linear functional $\phi(Tx) = f(x)$; this is well defined because $Tx = 0$ implies $f(x) = 0$. Now, ϕ extends from $\operatorname{Im} T$ to $\mathbb{C}^n$. Knowing what functionals on $\mathbb{C}^n$ look like, we conclude that there exist scalars $a_1, \ldots, a_n$ such that

$$f(x) = \phi(Tx) = \sum_{i=1}^n a_i f_i(x)$$

for all $x \in X$. In other words, $f = a_1 f_1 + \cdots + a_n f_n$. $\qquad\square$

Proposition 6.1.4. *The weak topology on X is a Hausdorff topology in which the vector space operations are continuous. In the weak topology, every point has a local base consisting of convex sets. The weak topology is weaker than the norm topology, and if X is infinite dimensional then it is strictly weaker and non-metrizable.*

Proof. The Hahn–Banach theorem implies Hausdorff-ness — we can *strictly* separate any two distinct points with a bounded linear functional, and that gives us disjoint weak neighborhoods of the points. Continuity of the vector space operations is more instructive as an exercise, so it is left to the reader.

Furthermore, the basic open sets that we wrote down in (6.1) above are plainly convex.

The weak topology is the weakest one in which all norm-continuous linear functionals are continuous; the norm topology is certainly such a topology. This means that the weak topology is weaker than the norm topology. We will now show that the weak topology is strictly weaker than the norm topology when X is infinite dimensional. For this, it suffices to show that the weak topology is not metrizable. Suppose, for the sake of later obtaining contradiction, that the weak topology of X was metrizable. Then the point 0 would have a countable local base, and we may assume that this local base consists of basic open sets of the form given in Exercise $6.1.2$. Let $\{f_n\}_{n=1}^{\infty}$ be a sequence which contains all the $f \in X^*$ that arises in this countable local base at 0. For every $g \in X^*$, the set $V(0; g; 1)$ is a neighborhood of 0, hence it must contain some set of the form $V(0; f_{n_1}, \ldots, f_{n_k}; \epsilon)$. But by Lemma $6.1.3$, $V(0; f_{n_1}, \ldots, f_{n_k}; \epsilon) \subseteq V(0; g; 1)$ implies that g is in the span of $f_{n_1}, \ldots, f_{n_k}$. We have just proved that every $g \in X^*$ is in the algebraic span of $\{f_n\}_{n=1}^{\infty}$, and therefore X^* has a countable Hamel basis. On the other hand, if X is infinite dimensional then X^* is also an infinite dimensional space, and by Proposition $1.3.3$ X^* is complete. But an infinite dimensional Banach space so has no countable Hamel basis (see Exercise $1.6.6$), thus we have arrived at a contradiction. We conclude that the weak topology on X is not metrizable. $\qquad \square$

Exercise 6.1.5. Prove that if X is an infinite dimensional normed space, then X^* is also infinite dimensional.

Exercise 6.1.6. Prove that if X is a finite dimensional normed space, then the weak topology coincides with the norm topology.

Since the weak topology differs from the norm topology, we will use some obvious notational modifications. For example, we might denote the norm closure of a set $Y \subseteq X$ by $\overline{Y}^{\|\cdot\|}$ and the weak closure (meaning the closure in the weak topology) as $\overline{Y}^w$, and so forth. We also use phrases like **weakly closed** or **strongly open** to mean closed in the weak topology or open in the topology induced by the norm, etc.

Theorem 6.1.7. *A convex set in a normed space X is weakly closed if and only if it is strongly closed.*

Proof. Let $Y \subseteq X$ be convex. If Y is weakly closed then it is also closed in the stronger norm topology. Conversely, suppose that Y is closed in the norm topology, and let $\overline{Y}^w$ be its weak closure. Let $z \notin Y$. We shall show that $z \notin \overline{Y}^w$. By Hahn–Banach (specifically, Theorem $2.1.19$), there exists $f \in X^*$ and $c \in \mathbb{R}$ and $\epsilon > 0$ such that

$$\operatorname{Re} f(y) \leq c - \epsilon \leq c + \epsilon \leq \operatorname{Re} f(z) , \quad \text{for all } y \in Y.$$

But then

$$V(z; f; \epsilon) = \{x \in X : |f(x) - f(z)| < \epsilon\}$$

is a weakly open neighborhood of z that is disjoint from Y. We conclude that $z \notin \overline{Y}^w$, and this shows that $Y = \overline{Y}^w$, so Y is weakly closed. $\square$

Corollary 6.1.8. *For every convex set $Y \subseteq X$, the weak and strong closures of Y coincide.*

6.2 Nets and topology

We begin this section with an example that shows that sequences are insufficient for determining all topological phenomena.

Example 6.2.1. Let $\{e_n\}_{n=1}^{\infty}$ be an orthonormal basis in a Hilbert space H, and let $x_n = \sqrt{n}e_n$. The sequence $\{x_n\}_{n=1}^{\infty}$ has no bounded subsequence, hence, by Proposition 5.1.6, it has no weakly convergent subsequence. However, the point 0 is contained in the weak closure of $\{x_n\}_{n=1}^{\infty}$. To see this, recall that by Exercise 6.1.2, a local base at 0 is given by neighborhoods of the form
$$V = \{h \in H : |\langle h, h_i \rangle| < 1, \, i = 1, \ldots, k\},$$
where $h_1, \ldots, h_k \in H$. We shall show that, given such a set V, there is some n such that $x_n \in V$. Indeed, if there was no such n such that $x_n \in V$, then for all n we would have $|\langle x_n, h_i \rangle| = |\langle \sqrt{n}e_n, h_i \rangle| \geq 1$ for some i, so

$$\sum_{n=1}^{\infty} \sum_{i=1}^{k} |\langle e_n, h_i \rangle|^2 \geq \sum_{n=1}^{\infty} \frac{1}{n} = \infty.$$

But

$$\sum_{n=1}^{\infty} \sum_{i=1}^{k} |\langle e_n, h_i \rangle|^2 = \sum_{i=1}^{k} \sum_{n=1}^{\infty} |\langle e_n, h_i \rangle|^2 = \sum_{i=1}^{k} \|h_i\|^2 < \infty,$$

so we have a contradiction. Thus, the sequence $\{x_n\}_{n=1}^{\infty}$ has a member in every element of a local base at 0, and therefore 0 belongs to the weak closure of $\{x_n\}_{n=1}^{\infty}$. On the other hand, as we noted, there is no sequence contained in the set $\{x_n\}_{n=1}^{\infty}$ that converges to 0.

Example 6.2.1 shows that sequences are not powerful enough to completely describe the weak topology. We now review the important notion of nets in a topological space — a notion that for many purposes serves as a handy substitute for sequences. A good reference on the material on nets in topological spaces (and also for other things) is [38].

A **directed set** is a pair $(\Lambda, \leq)$ consisting of a set Λ and a reflexive, transitive relation $\leq$ such that for every $\alpha, \beta \in \Lambda$, there exists $\gamma \in \Lambda$ such that $\alpha \leq \gamma$ and $\beta \leq \gamma$. A **net**, formally speaking, is a function $i : \Lambda \to T$ from a directed set $(\Lambda, \leq)$ into another set T. It is common to use the notation

$\{t_\alpha\}_{\alpha \in \Lambda}$ to denote the net $i : \Lambda \ni \alpha \mapsto t_\alpha := i(\alpha) \in T$. In other words, nets are just like sequences, with the difference that they are indexed by arbitrary directed sets, whereas sequences are always indexed by $\mathbb{N}$.

Definition 6.2.2. Let T be a topological space. A net $\{t_\alpha\}_{\alpha \in \Lambda} \subseteq T$ is said to **converge** to $t \in T$ if for every open set $U \ni t$, there exists $\alpha_0 \in \Lambda$ such that $t_\alpha \in U$ for all $\alpha \geq \alpha_0$. In this case, we write

$$\lim t_\alpha = t \quad \text{or} \quad t_\alpha \to t.$$

The definition of a convergent net does not exclude the possibility that a net converges to more than one limit. We shall be mostly be interested in Hausdorff spaces, in which every convergent net has a unique limit.

Exercise 6.2.3. Let T be a topological space. Prove that T is Hausdorff if and only if every convergent net has a unique limit.

Example 6.2.4. The set $\mathbb{N}$ of natural numbers with the usual partial order $\leq$ is a directed set. A net in T indexed by $\mathbb{N}$ (or, more formally, a net from $\mathbb{N}$ into T) is nothing but a sequence. For a sequence, the notion of convergence defined above coincides with the familiar notion of convergence of sequence.

The following example captures the essence of convergent nets and also suggests why we require nets to be indexed by general directed sets rather than just the natural numbers.

Example 6.2.5. Let T be a topological space, let $t_0 \in T$, and let Λ be the set of all open sets containing t_0, directed by reverse containment (that is $U \leq V$ if and only if $U \supseteq V$). Note that Λ is a directed set. Let $\{t_U\}_{U \in \Lambda}$ be a net in T such that $t_U \in U$ for every $U \in \Lambda$. Then $\{t_U\}_{U \in \Lambda}$ is net, and this net converges to t_0.

Example 6.2.6. Let H be an infinite dimensional Hilbert space, and let $\{e_i\}_{i \in I}$ be an orthonormal basis for H. Let Λ be the set of all finite subsets of I, directed by containment (that is $E \leq F$ if and only if $E \subseteq F$). For every $F \in \Lambda$, let $h_F \in H$ be given by

$$h_F = \sum_{i \in F} \langle h, e_i \rangle e_i.$$

Then $\lim_F h_F = h$ by the basic theory of Hilbert spaces. Indeed, we know by [14, Section 3] that for all $\epsilon > 0$ there exists a finite $F_0 \subset I$ such that $\| \sum_{i \in F} \langle h, e_i \rangle e_i - h \| < \epsilon$ for all finite $F \subseteq I$ that contains F_0. But this shows that every ball around h contains h_F for all $F \geq F_0$, which shows that $\lim h_F = h$ by definition (see also Exercise 6.7.2).

The following two exercises show that the convergent nets of a topological space contain all information about the topology.

Exercise 6.2.7. Let T be a set with a topology τ, and let $S \subseteq T$. Prove that a point $t \in T$ belongs to the closure $\overline{S}$ of S if and only if there is a net in S which converges to t.

Exercise 6.2.8. Let T be a set with two topologies defined on it, τ and σ. Prove that the following three conditions are equivalent.

1. $\tau = \sigma$.

2. A set $S \subseteq T$ is closed in the τ-topology if and only if it is closed in the σ-topology.

3. For a net $\{t_\alpha\}_{\alpha \in \Lambda}$ and $t \in T$, we have that $\lim t_\alpha = t$ in the τ-topology if and only if $\lim t_\alpha = t$ in the σ-topology.

With the previous exercise in mind, the following one is not surprising.

Exercise 6.2.9. Let S and T be two topological spaces, and let $f : S \to T$ be a function. Then f is continuous if and only if for every convergent net $s_\alpha \to s$, it holds that $f(s_\alpha) \to f(s)$.

Remark 6.2.10. The reader surely recalls a similar characterization of continuous functions between metric spaces, as well as a characterization of closed spaces as in Exercise 6.2.7, with sequences instead of nets. The reason that sequences suffice in metric spaces to describe all the topological properties is that metric spaces are first countable (every point has a countable local base). In general, sequences suffice for most purposes in first countable spaces, but first countable spaces that are not metrizable do not occur very naturally.

We conclude this section with a brief description of the net analogue of the notion of a subsequence. Suppose that T is a set, Λ is a directed set, and $i : \Lambda \to T$ is a net in T. A **subnet** is a net $i' : \Lambda' \to T$, which is given as $i' = i \circ h$ where $h : \Lambda' \to \Lambda$ is such that for every $\alpha \in \Lambda$, there exists $\beta_0 \in \Lambda'$, such that $h(\beta) \geq \alpha$ for all $\beta \geq \beta_0$[1]. If we write a net $i : \Lambda \to T$ as $\{t_\alpha\}_{\alpha \in \Lambda}$ (that is, $t_\alpha = i(\alpha)$), then the subnet determined by h is written as $\{t_{h(\beta)}\}_{\beta \in \Lambda'}$, and sometimes even as $\{t_\beta\}_{\beta \in \Lambda'}$, especially when $\Lambda' \subseteq \Lambda$.

Exercise 6.2.11. Every subnet of a convergent net is convergent.

One has to be careful with subnets:

Example 6.2.12. Continue with the notation from Example 6.2.6. If we let $\Lambda' = \{\{i\} : i \in I\} \subseteq \Lambda$, then this set is still partially ordered, but it is not directed anymore, so $\{h_F\}_{F \in \Lambda'}$ is not a subnet of $\{h_F\}_{F \in \Lambda}$, because it fails to be a net. If I is countable then we can change the ordering on Λ' so that $\{h_F\}_{F \in \Lambda'}$ becomes a convergent sequence, but it still fails to be a subnet because Λ' is not cofinal in Λ.

---

[1] The image of h is then said to be *cofinal* in Λ.

Now suppose that H is separable, and that $I = \mathbb{N}$. Let $\mathbb{N}_e = \{0, 2, 4, \ldots\}$ be the even natural numbers, and let Γ be the set of all finite subsets of $\mathbb{N}_e$ directed by containment. Then $\{h_F\}_{F \in \Gamma}$ is a convergent net, but it is not a subnet. Indeed, we see that it cannot be a subnet, because every subnet of a convergent net must converge to the same limit (Exercise 6.2.11), while for certain choices of h (e.g., $h = e_1$) h_F no longer converges to h. What condition isn't satisfied?

Exercise 6.2.13. A topological space T is compact if and only if every net in T has a convergent subnet.

6.3   Locally convex spaces and weak topologies

6.3.1   Topological vector spaces and locally convex spaces

Vector spaces with a topology that satisfies the conditions in Proposition 6.1.4 arise frequently enough in analysis to justify a definition.

Definition 6.3.1. Let X be a vector space which is also a Hausdorff topological space. X is said to be a ***topological vector space*** if the maps

$$(\lambda, x) \mapsto \lambda x \quad \text{and} \quad (x, y) \mapsto x + y, \quad \text{where } x, y \in X \text{ and } \lambda \text{ is a scalar,}$$

are continuous.

We restrict our attention to topological vector spaces that are Hausdorff, as these are the spaces of most interest in practice. In fact, a somewhat more restricted class of spaces – to be defined presently – is the one most often considered in analysis.

Definition 6.3.2. A topological vector space X is said to be ***locally convex*** if every point has a local base consisting of convex sets.

Equivalently, a locally convex space is a topological vector space that has a base consisting of convex sets. We shall write "locally convex space" instead of the longer "locally convex topological vector space".

Definition 6.3.3. Let X be a vector space, and let $\mathcal{P}$ be a family of semi-norms on X that is *separating* in the sense that for all $x \in X$ there exists $p \in \mathcal{P}$ such that $p(x) > 0$. The ***topology determined by*** $\mathcal{P}$ is the topology generated by the sets

$$V(x, p, \epsilon) := \{y \in X : p(x - y) < \epsilon\}$$

where $x \in X$, $p \in \mathcal{P}$ and $\epsilon > 0$.

Proposition 6.3.4. *Let X be a vector space, and let $\mathcal{P}$ be a separating family of semi-norms on X. Then X endowed with the topology determined by $\mathcal{P}$ is a locally convex space. Conversely, if X is a locally convex space, then there exists a family of separating semi-norms $\mathcal{P}$ on X such that the topology is equal to the topology generated by $\mathcal{P}$.*

Exercise 6.3.5. Prove the proposition. The direct implication is straightforward, and it is actually the important direction, since in practice this is how locally convex topologies arise. The reverse implication is trickier (and less important). Trick number one: prove that every locally convex space has a local base at 0 consisting of absolutely convex open sets (a set E containing 0 is **absolutely convex** if it is convex and if $x \in E$ implies that $\lambda x \in E$ for every scalar with $|\lambda| \leq 1$). Trick number two: given an absolutely convex neighborhood of zero, show that the Minkowski functional associated with the neighborhood is a semi-norm.

Exercise 6.3.6. Let X be a vector space and let $\mathcal{P}$ be a family of semi-norms on X. Prove that a net $\{x_\alpha\}_{\alpha \in \Lambda}$ converges to x in the topology generated by $\mathcal{P}$ if and only if $p(x_\alpha - x) \to 0$ for every $p \in \mathcal{P}$.

Remark 6.3.7. It is common to define the locally convex topology τ on a vector space X generated by a family of semi-norms $\mathcal{P}$ by the following equivalent description: *τ is the topology in which $x_\alpha \to 0$ if and only if $p(x_\alpha) \to 0$ for every $p \in \mathcal{P}$.*

Example 6.3.8. Suppose that $\mathcal{P}$ is the singleton $\{\|\cdot\|\}$, where $\|\cdot\|$ is a norm. Then $(X, \|\cdot\|)$ is a normed space, and the locally convex topology on X generated by $\mathcal{P}$ is obviously just the topology induced by the norm.

Example 6.3.9. If f is a linear functional on a vector space X, then the function $p(x) = |f(x)|$ is a semi-norm on X. Now if X is a normed space, and $\mathcal{P}$ is the collection of all semi-norms of the form

$$p(x) = |f(x)|, \quad f \in X^*,$$

then the locally convex topology generated by $\mathcal{P}$ is the weak topology.

Example 6.3.10. Let H be a Hilbert space, and let $B(H)$ be the algebra of bounded operators on H. Given $h \in H$, we can define a semi-norm p_h on $B(H)$ by

$$p_h(T) = \|Th\|.$$

The topology generated by the family $\{p_h : h \in H\}$ is called the **strong operator topology** on $B(H)$, or SOT. A net $\{T_\alpha\}$ converges to 0 in the SOT if and only if $T_\alpha h \to 0$ (in norm) for all $h \in H$.

Example 6.3.11. Let H be an infinite dimensional Hilbert space, and let $\{e_i\}_{i \in I}$ be an orthonormal basis for H. Let Λ be the set of all finite subsets of I, directed by containment. For every $F \in \Lambda$, let P_F be the map defined by

$$P_F h = \sum_{i \in F} \langle h, e_i \rangle e_i.$$

Then $\{P_F\}_{F \in \Lambda}$ is a net in $B(H)$ and it follows from the discussion in Example 6.2.6 that it converges in the strong operator topology to I_H. Note that this net does not converge in the norm topology to I_H, since $\|P_F - I_H\| = 1$ for every $F \in \Lambda$.

Example 6.3.12. Let $C(\mathbb{R}^k)$ be the space of all continuous functions on $\mathbb{R}^k$. For every compact subset $K \subset \mathbb{R}^k$, let $p_K(f) = \sup_{x \in K} |f(x)|$. This defines a semi-norm on $C(\mathbb{R}^k)$. Then the family of semi-norms

$$\mathcal{P} = \{p_K : K \subset \mathbb{R}^k \text{ is compact}\}$$

generates a locally convex vector topology on $C(\mathbb{R}^k)$. This topology is called *the topology of uniform convergence on compact sets*[2], because a sequence $\{f_n\} \subset C(\mathbb{R}^k)$ converges to f if and only if f_n converges uniformly to f on every compact subset $K \subset \mathbb{R}^k$.

Did you notice that we spoke about sequences rather than nets in the previous example? Nice catch! But it turns out that the locally convex topological vector space discussed in the previous example is metrizable. In fact, if the topology on X is generated by a countable separating family $\mathcal{P} = \{p_n\}_{n=1}^{\infty}$ of semi-norms, then

$$d(x, y) = \sum_{n=1}^{\infty} \frac{1}{2^n} \frac{p_n(x - y)}{1 + p_n(x - y)}$$

defines a translation invariant metric (*translation invariant* means that $d(x + z, y + z) = d(x, y)$ for all $x, y, z \in X$), and the topology induced by d is the same as the topology generated by $\mathcal{P}$.

A topological vector space in which the topology is induced by a translation invariant metric which is also complete is said to be a **Fréchet space**. The space in Example 6.3.12 is a Fréchet space. In Fréchet spaces versions of the "three big theorems" hold see Rudin's book *Functional Analysis* [12]).

6.3.2   Weak topologies

Definition 6.3.13. Let X be a normed space and let F be a subspace of X^* that separates the points[3] of X. The F–*weak topology on* X, denoted

---

[2]This topology is also known as the **compact-open topology**.
[3]F is said to *separate points* if for all $x \neq y$ there exists $f \in F$ such that $f(x) \neq f(y)$.

$\sigma(X, F)$, is the weakest topology on X in which all of the functionals in F are continuous. In other words, it is the topology generated by the sets

$$V(x; f; 1) = \{y \in X : |f(x - y)| < 1\},$$

where $x \in X$ and $f \in F$. Thus, $\sigma(X, F)$ is the locally convex space determined by the family of semi-norms

$$X \ni x \mapsto |f(x)|, \quad f \in F.$$

Another way to describe the $\sigma(X, F)$-topology on X, is to say that a net $\{x_\alpha\}$ in X converges to x in the $\sigma(X, F)$-topology if and only if $\langle x_\alpha, f \rangle \to \langle x, f \rangle$ for all $f \in F$.

Example 6.3.14. The X^*-weak topology $\sigma(X, X^*)$ is just the weak topology on X.

Definition 6.3.15. Let $\hat{X}$ be the image of X in X^{**}. The $\sigma(X^*, \hat{X})$-topology is called the **weak-∗ topology on** X^*. It is sometimes also denoted $\sigma(X^*, X)$.

Of course, the notion of weak-∗ convergence of sequences that we encountered earlier in Definition 5.1.3 coincides with the notion of convergence of sequences with respect to the weak-∗ topology.

Example 6.3.16. A weak-∗ convergent sequence is bounded (Proposition 5.1.6). This does not hold for nets instead of sequences, as can be seen by recalling Example 6.2.1. We saw in that example that if $\{e_n\}$ is an orthonormal basis of a Hilbert space, then $0 \in \overline{\{\sqrt{n}e_n\}}^w = \overline{\{\sqrt{n}e_n\}}^{w*}$. It follows from Exercise 6.2.7 that there is a net in $\{\sqrt{n}e_n\}$ that is weakly (and also weak-∗) convergent to 0, but this net cannot be bounded.

Example 6.3.17. Let H be a Hilbert space, and let $B(H)$ be the bounded operators on H. For every $g, h \in H$, the functional $w_{g,h}(T) = \langle Tg, h \rangle$ is a bounded linear functional on $B(H)$. Let $F \subseteq B(H)^*$ be the span of all functionals of the form $w_{g,h}$ with $g, h \in H$. Then the $\sigma(B(H), F)$-topology on $B(H)$ is called the **weak operator topology** (or WOT). In this topology a net $\{T_\alpha\}$ converges to 0 if and only if $\langle T_\alpha g, h \rangle \to 0$ for all $g, h \in H$; equivalently, $T_\alpha \xrightarrow{WOT} 0$ if and only if the net $T_\alpha g \xrightarrow{w} 0$ for every $g \in H$.

The weak operator topology is weaker than the strong operator topology, and strictly weaker when H is infinite dimensional. Consider the unilateral shift $S : \ell^2 \to \ell^2$, given by

$$S(x_0, x_1, x_2, \ldots) = (0, x_0, x_1, \ldots), \quad (x_k)_{k=0}^\infty \in \ell^2.$$

It is then easy to see that $S^n \xrightarrow{WOT} 0$. However, $\|S^n x\| = \|x\|$ for all $x \in \ell^2$, so S^n does not converge SOT to 0.

Remark 6.3.18. When H is infinite dimensional, the weak operator topology is strictly weaker than the weak topology on $B(H)$. In fact, $B(H)$ turns out to be the dual of a Banach space (in fact, $B(H) = K(H)^{**}$), and the WOT is strictly weaker than the weak-$*$ topology.

Proposition 6.3.19. *Let $F \subseteq X^*$ be a subspace. A linear functional f on X is continuous in the $\sigma(X, F)$-topology if and only if $f \in F$.*

Proof. One direction follows from the definition. If f is continuous in the $\sigma(X, F)$-topology, then $\{x : |f(x)| < 1\} = V(0; f; 1)$ is $\sigma(X, F)$ open, so there are $f_1, \ldots, f_n \in F$ such that

$$V(0; f_1, \ldots, f_n; \epsilon) \subseteq V(0; f; 1).$$

Lemma 6.1.3 now shows that $f = a_1 f_1 + \cdots + a_n f_n \in F$. $\qquad\square$

Remark 6.3.20. One can show that WOT and SOT have the same continuous linear functionals. Since WOT $\neq$ SOT, it follows that SOT cannot be described as a weak topology generated by functionals.

Example 6.3.21. $X = C([0,1])$, $F = \operatorname{span}\{\delta_x\}_{x \in [0,1]}$. The $\sigma(X, F)$-topology is the ***topology of pointwise convergence***, that is, a net $\{f_\alpha\}$ converges to f in the $\sigma(X, F)$-topology if and only if $f_\alpha(t) \to f(t)$ for all $t \in [0,1]$. Every $\sigma(X, F)$ continuous functional has the form $f \mapsto \sum_{i=1}^n a_i \delta_{x_i}$. We see that the Riemann integral is not a continuous functional with respect to the topology of pointwise convergence, since it is not of this form (why?). Note that this argument doesn't show that the identity $\lim_n \int_0^1 f_n = \int_0^1 \lim_n f_n$ is false in general.

6.4   The continuous dual of a locally convex space

Definition 6.4.1. Let X be a topological vector space. The ***continuous dual*** (or simply the **dual**) of X is the space X' of all continuous linear functionals on X.

The term *continuous dual* is to help distinguish from the **algebraic dual**, which consists of all linear functionals and is sometimes denoted (somewhat annoyingly) as X^*.

Example 6.4.2. By Proposition 6.3.19, if X is a normed space, then the continuous dual of a normed space $(X, \|\cdot\|)$ is equal to the space of bounded linear functionals X^* as we defined earlier, and it is also equal to the continuous dual of the topological vector space $(X, \sigma(X, X^*))$. The continuous dual of $(X^*, w*)$ is X.

Exercise 6.4.3. Let X be a locally convex space with topology generated by a separating family of semi-norms $\mathcal{P}$. Prove that a linear functional $f \colon X \to \mathbb{C}$ belongs to X' if and only if there exist $p_1, \ldots, p_n \in \mathcal{P}$ and positive constant $c > 0$ such that $|f(x)| \le c \max\{p_1(x), \ldots, p_n(x)\}$ for all $x \in X$.

It turns out that there exist topological vector spaces X that have no non-zero continuous linear functionals, that is, $X' = \{0\}$ (see Exercise 6.7.7). Fortunately, this cannot happen in a locally convex space.

Theorem 6.4.4 (Hahn–Banach separation theorem, locally convex). *Let X be a locally convex space. Let K be a convex and closed subset of X and let $x \notin K$. Then there exists $f \in X'$ on X which strictly separates K and x.*

Exercise 6.4.5. Prove the above theorem. (**Hint:** It suffices to treat the real case. If $x \notin K$ then there exists a convex neighborhood U of 0 such that $x + U \cap K = \emptyset$, so a version of Theorem 2.1.19 for locally convex spaces would do the job. Adapt the proof of that theorem together with the results leading up to it to the current setting.)

If the locally convex space in the theorem is the $\sigma(X, F)$-topology on a normed space X, where $F \subseteq X^*$, then Proposition 6.3.19 shows that $f \in F$.

Corollary 6.4.6 (Hahn–Banach separation theorem, weak subspace/point). *Let M be a subspace in a normed space X, and let F be a subspace of X^* that separates the points of X. A point x is in the $\sigma(X, F)$-closure of M if and only if $f(x) = 0$ for every $f \in F$ which annihilates M.*

We are now in a position to complete the picture of the orthogonality relations (Proposition 5.3.8 and Corollary 5.3.9).

Corollary 6.4.7. *Let X be a normed space and $F \subseteq X^*$ a linear subspace. Then $(F_\perp)^\perp = \overline{F}^{w*}$. In particular, if $T \in B(X, Y)$, then $(\ker T)^\perp = \overline{\operatorname{Im} T^*}^{w*}$.*

If X is a locally convex Hausdorff topological vector space with continuous dual X', then one can define on X the $\sigma(X, X')$-topology, the weakest topology in which all functionals in X' are continuous. This topology is in general weaker than the original topology on X (as we know from the case of normed spaces), but by the Hahn–Banach theorem a convex set is closed in X if and only if it is $\sigma(X, X')$-closed (cf. Exercise 6.7.8). Now, the continuous dual X' is also a locally convex topological vector space when endowed with the weak topology defined by X, which we may denote by $\sigma(X', X)$. When no additional structure is present then this is the topology one assumes on X'.

Exercise 6.4.8. Prove that the continuous dual of X' is equal to X endowed with the $\sigma(X, X')$-topology.

6.5 Alaoglu's theorem

Recall that the closed unit ball X_1 of a normed space X is compact if and only if X is finite dimensional (see [44, Theorem 9.2.3]). On the other hand, we showed in Theorem 5.1.8 that the closed unit ball of the dual space of a separable space is sequentially weak-$*$ compact. Since the closed unit ball of the dual of a separable space is metrizable (see Exercise 6.7.9), we conclude that the closed unit ball of the dual of a separable space is weak-$*$ compact. The following fundamental result shows that this is true in general: the unit ball of any dual space X^* is weak-$*$ compact.

Theorem 6.5.1 (Alaoglu's theorem). *Let X be a normed space. The closed unit ball X_1^* of X^* is compact in the weak-$*$ topology.*

Proof. For every $x \in X$, let $D_x = \{z \in \mathbb{C} : |z| \leq \|x\|\}$. Define Z to be the product space

$$Z = \Pi_{x \in X} D_x := \{(\lambda_x)_{x \in X} : \lambda_x \in D_x\}.$$

By Tychonoff's theorem, Z is compact in the product topology. Now, Z can be considered as a space of complex valued functions f on X satisfying $f(x) \in D_x$ for all $x \in X$, and so X_1^* can be naturally identified with a subspace[4] of Z. Note that the topology on X_1^* induced by the product topology is precisely the weak-$*$ topology. If we show that X_1^* is closed, we will be done, because a closed subset of a compact space is compact.

Suppose that $f_\alpha \to z$ in the product topology. By definition of Z, z is a complex valued function on X such that $|z(x)| \leq \|x\|$ for all x. If we show that z is linear we will be done, because then z will be bounded of norm 1. But if $x, y \in X, a \in \mathbb{C}$, then

$$z(ax + y) = \lim_\alpha f_\alpha(ax + y) = \lim_\alpha a f_\alpha(x) + f_\alpha(y) = az(x) + z(y).$$

Here we used "arithmetic of limits" for nets, which the reader should justify. ∎

Remark 6.5.2. From the theorem, it follows that the closed unit ball in a reflexive Banach space is weakly compact. In fact, the closed unit ball of a Banach space is weakly compact if and only if the space is reflexive; see Exercise 6.7.11. It is instructive to use the same argument as above to try (and fail, hopefully) to prove that the unit ball of X is weakly compact in general.

Corollary 6.5.3. *Every Banach space X is isometric to a closed subspace of $C(K)$ for some compact Hausdorff topological space K.*

Proof. Let $K = X_1^*$, and take the inclusion to be $x \mapsto \hat{x}\big|_{X_1^*}$. ∎

---

[4]Z is not a vector space and X_1^* not a linear subspace, so the reader must understand that "subspace" here means "topological subspace", i.e., a subset of a topological space inheriting the induced topology.

Corollary 6.5.4. *Every Banach space X is isometric to a closed subspace of $B(H)$ for some Hilbert space H.*

Proof. By the previous corollary, it suffices to prove this for $X = C(K)$. We take $H = \ell^2(K)$, and $f \mapsto M_f$, where the diagonal multiplication operator M_f is given by
$$M_f(x_k)_{k \in K} = (f(k)x_k)_{k \in K}.$$
It remains to verify that $f \mapsto M_f$ is an isometric linear map from $C(K)$ into $B(H)$. We leave this to the reader. □

6.6 Application: Haar measure on a compact abelian group

Let G be a compact abelian group. By this we mean that G is at once both an abelian group and a compact Hausdorff topological space, and that the group operations are continuous, meaning that $g \mapsto g^{-1}$ is continuous on G and $(g, h) \mapsto g + h$ is continuous as a map from $G \times G$ to G. It is known that there exists a regular Borel measure μ on G, called the *Haar measure*, which is nonnegative, satisfies $\mu(G) = 1$, and is *translation invariant*, in the sense that

$$\mu(g + E) = \mu(E),$$

for all $g \in G$ and for every Borel set $E \subseteq G$. In fact, the Haar measure is known to exist in greater generality (G does not have to be abelian, and if one allows μ to be infinite then G can also be merely locally compact; see [19]). The Haar measure is an indispensable tool in representation theory and in ergodic theory. In this section, we will use Alaoglu's theorem to give a slick proof of the existence of the Haar measure in the abelian compact case. We will need one more ingredient, which is the Kakutani–Markov fixed point theorem, to which we devote the next section.

6.6.1 The Kakutani–Markov fixed point theorem

Theorem 6.6.1 (Kakutani–Markov fixed point theorem). *Let X be a locally convex space and let $\mathcal{F}$ be a commuting family of continuous linear maps on X. Suppose that K is a nonempty, compact and convex subset of X such that $T(K) \subseteq K$ for all $T \in \mathcal{F}$. Then $\mathcal{F}$ has a common fixed point in K, i.e., an element $k \in K$ such that $Tk = k$ for all $T \in \mathcal{F}$.*

Proof. Let us first prove the theorem in the case where $\mathcal{F} = \{T\}$. Choose some $k_0 \in K$. Construct the averages

$$k_N = \frac{1}{N+1} \sum_{n=0}^{N} T^n k_0.$$

Since K is convex and $T(K) \subseteq K$, we have that $k_N \in K$ for all K. Since K is compact, the sequence $\{k_N\}_{N=1}^{\infty}$ contains a subnet $\{k_{N(\alpha)}\}$ that converges to some $k \in K$. We claim that $Tk = k$.

Since X is locally convex and Hausdorff, if $Tk - k \neq 0$ then there exist disjoint open convex neighborhoods $V \ni Tk - k$ and $U \ni 0$ such that $U \cap V = \emptyset$ and therefore $\overline{U} \cap V = \emptyset$. We see that if $Tk \neq k$ then there is an open convex neighborhood U of the origin such that $Tk - k \notin \overline{U}$. Thus, to show that $Tk = k$, it suffices to show that $Tk - k$ belongs to the closure of every open convex neighborhood of 0.

Let U be an open convex neighborhood of the origin. For every $x \in X$, there exists[5] some n_x such that $\frac{1}{n_x}x \in U$. It follows that $X = \cup_n nU$. Since K is compact, the set $K - K$ is compact, because it is the image of the compact set $K \times K$ under the continuous map $X \times X \ni (x, y) \mapsto x - y$. By using compactness of $K - K$ and convexity of U, we see that there is some n_0 such that $K - K \subseteq n_0 U$, and it follows that $\frac{1}{n}x \in U$ for all $n \geq n_0$ and all $x \in K - K$.

For all N, we compute

$$Tk_N - k_N = \frac{1}{N+1}\sum_{n=1}^{N+1} T^n k_0 - \frac{1}{N+1}\sum_{n=0}^{N} T^n k_0 = \frac{1}{N+1}(T^{N+1}k_0 - k_0).$$

Now, $T^{N+1}k_0 - k_0 \in K - K$ for all N, and so by the preceding paragraph we know that for all $N \geq n_0$,

$$\frac{1}{N+1}(T^{N+1}k_0 - k_0) \in U.$$

We conclude that $Tk - k = \lim_\alpha Tk_{N(\alpha)} - k_{N(\alpha)}$ is in $\overline{U}$, as required. That completes the proof of the existence of a fixed point for the case where $\mathcal{F}$ is a singleton.

Now let $\mathcal{F}$ be arbitrary, and for every finite $F \subseteq \mathcal{F}$, define

$$A_F = \{k \in K : Tk = k \text{ for all } T \in F\}.$$

For every F, the set A_F is evidently convex and closed, hence compact. We will show that the family $\{A_F : F \subseteq \mathcal{F} \text{ finite}\}$ has the finite intersection property, meaning that the intersection over every finitely many A_Fs is nonempty. It will follow from compactness that there is some $x \in \cap_F A_F$, which will be the sought after fixed point. Now, $A_{F_1} \cap A_{F_2} = A_{F_1 \cup F_2}$, so we only have to show that every A_F is nonempty. This is done by induction on the cardinality $|F|$ of F. If $|F| = 1$, then $A_F \neq \emptyset$ by the first part of the proof. Suppose that A_F is not empty, and let $T \in \mathcal{F}$. Then for every $y \in A_F$, we have for all $S \in F$

$$STy = TSy = Ty.$$

---

[5]To see this, note that the path $[0,1] \ni t \mapsto \gamma(t) = tx$ is a continuous path in X, so $\gamma^{-1}(U)$ is a neighborhood of 0.

Therefore $T(A_F) \subseteq A_F$, so by the first half of the proof there is some $z \in A_F$ fixed under T. In other words, $z \in A_{F \cup \{T\}}$, so this set is not empty. $\quad\square$

Exercise 6.6.2. The theorem holds also in the non-locally convex setting, try to adapt the argument to that case. Alternatively, find a slicker proof of the part that $Tk = k$ in the locally convex setting by using the Hahn–Banach theorem.

We will use the above theorem in the situation where our space is the dual of a Banach space, $\mathcal{F}$ a commuting family of weak-$*$ continuous linear maps, and K is a nonempty, weak-$*$–compact and convex subset such that $T(K) \subseteq K$ for all $T \in \mathcal{F}$. To be able to use the fixed point theorem, we need to be able to recognize weak-$*$ continuous linear maps. There is, in fact, a rich supply.

Proposition 6.6.3. *Let X, Y be Banach spaces. A linear map $B : Y^* \to X^*$ is weak-$*$ to weak-$*$ continuous if and only if there exists $A \in B(X, Y)$ such that $B = A^*$.*

Proof. We shall prove one implication, which is the one we require below. The other implication is left as an exercise (see Exercise 6.7.13).

Suppose that $B = A^*$ for $A \in B(X, Y)$. Let $\psi_\alpha \xrightarrow{w*} \psi \in Y^*$. Then for all $x \in X$,

$$\langle x, B\psi_\alpha \rangle = \langle Ax, \psi_\alpha \rangle \to \langle Ax, \psi \rangle = \langle x, B\psi \rangle,$$

where we used the fact that ψ_α converges weak-$*$ to ψ in Y^*. This shows that $B\psi_\alpha$ converges weak-$*$ to $B\psi$. Thus, B is weak-$*$ to weak-$*$ continuous. $\quad\square$

6.6.2 The existence of Haar measure for abelian compact groups

We can now prove the existence of Haar measure for abelian compact groups.

Theorem 6.6.4. *Let G be a compact abelian group. Then there exists a Haar measure for G. That is, there is a regular Borel probability measure μ on G that is translation invariant.*

Proof. For every $g \in G$, let $L_g : C(G) \to C(G)$ be the translation operator given by $(L_g f)(h) = f(h - g)$. In fact, it is useful to extend L_g to an operator $\tilde{L}_g$ acting on the space $B(G)$ of all bounded Borel measurable functions on G defined by the same formula: $\tilde{L}_g f(h) = f(h - g)$.

Let $M(G)$ be the space of all regular complex Borel measures on G. For every $g \in G$, we define an operator $T_g : M(G) \to M(G)$ by

$$T_g \mu(E) = \mu(g + E) , \quad \text{for every Borel set } E \subseteq G.$$

It is not hard to check that for every $\mu \in M(G)$, $T_g\mu$ is also a regular Borel measure, and that T_g preserves the total variation of a measure. Moreover, if K denotes the set of all probability measures in $M(G)$, then $T_g(K) \subseteq K$.

Now if $E \subseteq G$ is a Borel set, then

$$\int \chi_E d(T_g\mu) = \mu(g+E) = \int \chi_{g+E} d\mu = \int \tilde{L}_g \chi_E d\mu.$$

It then follows that for every $f \in B(G)$,

$$\int f d(T_g\mu) = \int \tilde{L}_g f d\mu,$$

so for $f \in C(G)$,

$$\langle T_g\mu, f \rangle = \int f d(T_g\mu) = \int L_g f d\mu = \langle \mu, L_g f \rangle.$$

In other words: $T_g = L_g^*$. By Proposition 6.6.3, $\mathcal{F} = \{T_g\}$ is a family of weak-∗ continuous maps sending K into K. Since G is abelian, it follows that $\mathcal{F}$ is commuting family

By Alaoglu's theorem, $(C(G)^*)_1$ is weak-∗ compact. It is easy to see that K is a weak-∗ closed subset of $(C(G)^*)_1$, so K is weak-∗ compact. K is evidently convex, so by the Kakutani–Markov fixed point theorem, $\mathcal{F}$ has a fixed point $\mu \in K$. By definition of K, μ is a regular Borel probability measure on G. This is the required translation invariant measure. □

Remark 6.6.5. Standard measure theoretic arguments show that the Haar measure is unique.

Exercise 6.6.6. It may seem as if the same argument would give the existence of a translation invariant regular Borel probability measure on a locally compact group. Explain why the theorem fails for non-compact spaces. What part of the argument breaks down? Make sure you understand why that part doesn't break down in the compact case.

6.7   Additional exercises

Exercise 6.7.1. Prove that a finite dimensional vector space (over $\mathbb{R}$ or $\mathbb{C}$) can be endowed with one **and only one** topology which makes it into a Hausdorff topological vector space.

Exercise 6.7.2 (Series summation over arbitrary index sets). Let Λ be an arbitrary index set, and let $\{x_i\}_{i \in \Lambda}$ be a family of elements in a normed space

X. We say that the series converges to a limit $x \in X$ if for every $\epsilon > 0$ there exists a finite subset $F_0 \in \Lambda$ such that for all finite $F \subseteq \Lambda$

$$F_0 \subseteq F \implies \left\| \sum_{i \in F} x_i - x \right\| < \epsilon.$$

1. Consider the directed set $\Gamma = \{F \subseteq \Lambda : F \text{ is finite }\}$ directed by containment. Define a net $\{y_F\}_{F \in \Gamma}$ by $y_F = \sum_{i \in F} x_i$. Explain why to say that the net $\{y_F\}_F$ converges to x is the same thing as saying that the series $\sum_{i \in \Lambda} x_i$ converges to x.

2. Prove that $\sum_{i \in \Lambda} x_i$ converges if and only if (i) then there is a countable subset $\Lambda_0 \subseteq \Lambda$ such that $x_i = 0$ for all $i \in \Lambda \setminus \Lambda_0$, and (ii) for every enumeration $\Lambda_0 = \{i_0, i_1, i_2, \dots\}$ it holds that $\lim_{N \to \infty} \sum_{k=0}^{N} x_{i_k} = x$.

3. Explain why in the case $X = \mathbb{C}$ it holds that the series $\sum_{i \in \Lambda} x_i$ converges (in the above sense) if and only if it is absolutely convergent (in the sense that $\sum_i \|x_i\| < \infty$), and provide examples showing why neither implication holds in the case where X is a general normed space.

Exercise 6.7.3. In which topology is $C([0,1])$ dense in $L^\infty[0,1]$:

1. The norm topology.

2. The weak topology.

3. The weak-$*$ topology (where $L^\infty = (L^1)^*$).

Exercise 6.7.4 (Best approximation in normed spaces, revisited). Let K be a nonempty closed and convex set in a reflexive space X. Prove that for every $x \in X$ there exists $k \in K$ such that $\|k - x\| = d(x, K)$.

Exercise 6.7.5. Let $p \in [1, \infty]$. True or false: a sequence $\{a^{(n)}\}_{n=0}^\infty = \{(a_k^{(n)})_{k=1}^\infty\}_{n=0}^\infty \subseteq \ell^p$ converges weakly to 0 if and only if $\{a^{(n)}\}_{n=0}^\infty$ is bounded and for all k, $\lim_n a_k^{(n)} = 0$.

Exercise 6.7.6. Prove the Hahn–Banach separation theorem for locally convex spaces as stated in Theorem 6.4.4 (see Exercise 6.4.5 for a hint).

Exercise 6.7.7. Fix $p \in (0,1)$, and let $L^p = L^p[0,1]$ be the space of all Lebesgue measurable functions $f : [0,1] \to \mathbb{R}$ such that $\int_0^1 |f|^p < \infty$. For $f, g \in L^p$, define $d(f,g) = \int |f - g|^p$. Show the following.

1. (L^p, d) is a complete metric space, and the induced topology makes L^p into a topological vector space.

2. The "ball" of radius $r > 0$ around zero is not convex.

3. L^p is not locally convex.

4. L^p has no continuous linear functionals, that is $(L^p)' = \{0\}$.

Exercise 6.7.8. Let X be a topological vector space. Recall that a set $H \subset X$ is said to be a **half space** if $H = \{x \in X : Ref(x) \leq c\}$, where f is a linear functional on X and $c \in \mathbb{R}$.

1. For $f \neq 0$, $\ker f$ is closed in the given topology, if and only if it is not dense, if and only if f is continuous.

2. Show that a half space $\{x \in X : \operatorname{Re} f(x) \leq c\}$ is closed if and only if only if f is continuous.

3. Now assume that X is a locally convex space. Prove that a set $K \subseteq X$ is convex and closed if and only if K is the intersection of a family of closed half spaces.

Exercise 6.7.9. Suppose that F is an infinite dimensional subspace of the dual of a Banach space X which separates the points of X, and suppose that $\{f_n\}_{n=1}^{\infty}$ is a sequence in F_1 such that $F = \operatorname{span}\{f_n : n = 1, 2, \ldots\}$.

1. Prove that $d(x, y) = \sum_{n=1}^{\infty} \frac{1}{2^n} |\langle x - y, f_n \rangle|$ defines a metric on X_1, which induces on X_1 the same topology as both the $\sigma(X, F)$-topology as well as the $\sigma(X, \overline{F})$-topology.

2. True or false: $\sigma(X, \overline{F}) = \sigma(X, F)$.

Exercise 6.7.10. Give a one line proof that X is weak-$*$ dense in X^{**}. Next, prove the deeper fact, that the closed unit ball X_1 is weak-$*$ dense in the closed unit ball $(X^{**})_1$.

Exercise 6.7.11. Prove that a Banach space X is reflexive if and only if X_1 is weakly compact.

Exercise 6.7.12. Let $T\colon X \to Y$ be a linear map between Banach spaces.

1. Prove that T is bounded if and only if it is weak-to-weak continuous.

2. If F, G are two subspaces (that separate points) in X^* and Y^*, respectively, show that T is $\sigma(X, F)$–to–$\sigma(Y, G)$ continuous if and only if $g \circ T \in F$ for all $g \in G$.

Exercise 6.7.13. Let X, Y be Banach spaces. Prove that a linear map $B : Y^* \to X^*$ is weak$*$-to-weak$*$ continuous if and only if there exists $A \in B(X, Y)$ such that $B = A^*$. (Such an A is sometimes called a **preadjoint** of B, and denoted $A = B_*$.)

Exercise 6.7.14. Recall from Exercise 2.4.8, that a group G is said to be **amenable** if there exists a finitely additive probability measure on G that is translation invariant in the sense that $\mu(gA) = \mu(A)$ for all $g \in G$ and $A \subseteq G$, where $gA = \{ga : a \in A\}$. Prove that every abelian group is amenable.
   (**Hint:** Letting $\ell^{\infty}(G)$ denote the space of all bounded functions $G \mapsto \mathbb{C}$, it suffices to prove the existence of a positive, translation invariant linear functional on $\ell^{\infty}(G)$ that maps the constant function 1 to 1.)

Exercise 6.7.15 (Polar duality). Let X be a locally convex space over $\mathbb{R}$ and X' its continuous dual (see Section 6.4). For a nonempty $A \subseteq X$ we define the ***polar***[6] of A to be

$$A^{\bullet} = \{f \in X' : \langle f, x \rangle \leq 1 \text{ for all } x \in A\}.$$

1. Show that $A^{\bullet}$ is a closed and convex set containing 0.

2. Let A be the closed unit ball in $\mathbb{R}^n$ equipped with the p-norm. What is $A^{\bullet}$?

3. Prove the *bipolar theorem*: if A is a closed and convex set containing 0, then $A^{\bullet\bullet} = A$.

4. What is $A^{\bullet\bullet}$ for a general set A?

---

[6]In the literature one can find variants on this theme, with $\langle f, x \rangle \leq 1$ replaced by $|\langle f, x \rangle| \leq 1$ or $\langle f, x \rangle \geq 0$.

Chapter 7

The Krein–Milman theorem and applications

7.1   Convex sets and convex hulls

Throughout this section, X is a locally convex space.[1]

Definition 7.1.1. Let $A \subseteq X$ be a set. The **convex hull** of A, denoted $\operatorname{co}(A)$, is the smallest convex set containing A. The **closed convex hull** of A, denoted $\overline{\operatorname{co}}(A)$, is the smallest closed convex set containing A.

Since arbitrary intersections of convex sets are convex, one sees that the convex hull of a set A exists by taking the intersection of all convex sets containing A. Similar remarks hold for the closed convex hull.

Exercise 7.1.2. Let $A \subseteq X$. Then

$$\operatorname{co}(A) = \left\{ \sum_{i=1}^{n} a_i x_i : n \in \mathbb{N}, a_1, \ldots, a_n \geq 0, x_1, \ldots, x_n \in A, \sum a_i = 1 \right\}.$$

Lemma 7.1.3. *The closure of a convex set is convex.*

Proof. One can give a one line proof with nets (please go ahead and write such a proof), but let us give a proof using neighborhoods just to get a taste of how it goes.

Let K be a convex set. Let $x, y \in \overline{K}$ and $t \in (0,1)$. Put $z = tx + (1-t)y$. To establish $z \in \overline{K}$, we must show that $(z + V) \cap K \neq \emptyset$ for every convex neighborhood V of 0. Now, because x, y are in the closure of K, then there are points $x_0 \in x + V$ and $y_0 \in y + V$ which are in K. Put $z_0 = tx_0 + (1-t)y_0$. Then $z_0 \in K$, and also

$$z_0 = tx_0 + (1-t)y_0$$
$$\in t(x + V) + (1-t)(y + V) = z + tV + (1-t)V \subseteq z + V,$$

because V is convex. Thus $(z + V) \cap K \neq \emptyset$ for any such V, so $z \in \overline{K}$. □

---

[1] Readers or instructors who feel that this is too abstract can safely focus on the case that X is a normed space endowed either with its norm topology or with a weak topology $\sigma(X, F)$, where $F \subseteq X^*$ separates the points of X.

DOI: 10.1201/9781003297864-7

Corollary 7.1.4. $\overline{co}(A) = \overline{co(A)}$.

Proof. By the lemma, the right hand side is convex and closed, so it must contain the left hand side. On the other hand $co(A)$ is contained in the closed $\overline{co}(A)$, so the closure $\overline{co(A)}$ is contained in $\overline{co}(A)$. □

Definition 7.1.5. Let $K \subseteq X$ be convex. A *face* of K is a nonempty convex subset $F \subseteq K$ such that for all $t \in (0, 1)$ and all $x, y \in K$,

$$tx + (1-t)y \in F \implies x, y \in F.$$

If $x_0 \in K$, and if the singleton $\{x_0\}$ is a face of K, then x_0 is said to be an *extreme point* of K. The set of extreme points of K is denoted $\text{Ext}(K)$.

Example 7.1.6. Consider the extreme points and faces of the closed and open unit balls in $\mathbb{R}^2$ with respect to the p norms, $p = 1, 2, \infty$. We see here that $\text{Ext}(\overline{B}_1(0)) \subseteq \{x : \|x\| = 1\}$, and it is easy to see that this is what happens in general. For $p = 1, \infty$ the extreme points are a proper subset of the "unit sphere" $\{x : \|x\| = 1\}$, which consists of four extreme points and four one dimensional faces.

Example 7.1.7. The faces of a three-dimensional cube are (i) the six two-dimensional square sides of the cube[2] (ii) the twelve one-dimensional edges between adjacent squares, (iii) the eight zero-dimensional vertices at the corners of the cube, and (iv) the cube itself. Here $\text{Ext}(\text{cube})$ consists of the eight vertices. Note that the convex hull of the extreme points is equal to the cube. This is a good example to refute a common misconception: the convex hull of a set A is not equal to $\{ta + (1-t)b : t \in [0, 1], a, b \in A\}$; stick to the description given in Exercise 7.1.2.

Example 7.1.8. In an inner product space, the extreme points of the closed unit ball are precisely the unit vectors. To see that every unit vector is extreme, invoke the uniform convexity of inner product spaces (see Exercise 1.6.12).

The following example shows that in infinite dimensional spaces, the notion of extreme point might be subtler than what one expects from experience in finite dimensional spaces.

Example 7.1.9. Let $X = L^1[0, 1]$. Then $\text{Ext}(X_1) = \emptyset$, i.e. X_1 has no extreme points! Indeed, for any $\|f\| = 1$, let t_0 be such that $\int_0^{t_0} |f(t)|dt = 1/2$. Then one easily constructs $g \in X_1$ supported on $[0, t_0]$ and $h \in X_1$ supported on $[t_0, 1]$ such that $f = g/2 + h/2$.

Exercise 7.1.10. The intersection of faces is again a face (so long as it is nonempty). If we have convex sets $F \subseteq G \subseteq H$, then if F is a face of G and G is a face of H, then F is a face of H. In particular, $\text{Ext}(G) \subseteq \text{Ext}(H)$ for every two convex sets $G \subseteq H$.

---

[2]These are what most normal people would call the faces of the cube.

Lemma 7.1.11. *Let K be convex and F a face of K. Suppose that $\sum_{i=1}^{n} a_i x_i \in F$, where $x_i \in K$ and $a_1, \ldots, a_n$ are nonnegative and sum to 1. Then $a_i \neq 0 \implies x_i \in F$.*

Proof. We content ourselves with proving the statement for $i = 1$. Suppose $a_1 > 0$. If $a_1 = 1$ then all other as are zero, so $x_1 \in F$. Otherwise, put $t = \sum_{i=2}^{n} a_i$. Now, t is in $(0, 1)$. Define $y = \sum_{i=2}^{n} (a_i / t) x_i$. Then $x_1, y \in K$ and $(1 - t)x_1 + ty = \sum_{i=1}^{n} a_i x_i \in F$. Thus, from the definition of face, it follows that x_1 (and also y) is in F. $\qquad\square$

Lemma 7.1.12. *Let K be a nonempty, compact convex subset of X. Let f be a continuous functional and define $b = \sup\{\operatorname{Re} f(x) : x \in K\}$. Then the set $F = \{x \in K : \operatorname{Re} f(x) = b\}$ is a compact face in K.*

Proof. $\operatorname{Re} f$ is continuous on a compact set, so attains its maximum, thus $F \neq \emptyset$. Moreover, F is a closed subset of a compact Hausdorff space, therefore it is compact. To see that F is a face, note first that F is convex. Next, if $x, y \in K$, $t \in (0, 1)$ and $tx + (1 - t)y \in F$, then

$$t \operatorname{Re} f(x) + (1 - t) \operatorname{Re} f(y) = b$$

while $\operatorname{Re} f(x) \leq b$ and $\operatorname{Re} f(y) \leq b$. It follows that $\operatorname{Re} f(x) = \operatorname{Re} f(y) = b$, and therefore $x, y \in F$. $\qquad\square$

7.2 The Krein–Milman theorem

Theorem 7.2.1 (Krein–Milman theorem). *Let K be a nonempty, compact and convex subset of a locally convex space X. Then $K = \overline{\operatorname{co}}(\operatorname{Ext}(K))$.*

Remark 7.2.2. In particular, the result implies that $\operatorname{Ext}(K)$ is not empty. Reconsidering Example 7.1.9, we see that the unit ball of $L^1[0, 1]$ is not compact with respect to any weak topology, and in particular it is not compact with respect to the weak topology. This also implies that $L^1[0, 1]$ is not the dual of any Banach space, since otherwise the closed unit ball would be weak-$*$ compact by Alaoglu's theorem, and would then have extreme points by the Krein–Milman theorem.

Proof. The first step is to show that $\operatorname{Ext}(K)$ is not empty. Let $\mathcal{F}$ be the family consisting of all compact faces of K. The family $\mathcal{F}$ is not empty because $K \in \mathcal{F}$. Order $\mathcal{F}$ by inclusion. An application of Zorn's lemma (by compactness, the intersection of a totally ordered chain in $\mathcal{F}$ is a nonempty face) shows that $\mathcal{F}$ has a minimal element, which we shall denote as F_0. We shall show that F_0 is a singleton, hence $\operatorname{Ext}(K) \neq \emptyset$.

Assume for contradiction that there are two distinct elements x, y in F_0. By the Hahn–Banach theorem for locally convex spaces (Theorem 6.4.4), there is a continuous functional f such that

$$\operatorname{Re} f(x) \neq \operatorname{Re} f(y). \tag{7.1}$$

Now let $b = \sup\{\operatorname{Re} f(z) : z \in F_0\}$, and define $F = \{z \in F_0 : \operatorname{Re} f(z) = b\}$ as in Lemma 7.1.12. The lemma then implies that F is a compact face of F_0, therefore it is a compact face of K. But (7.1) implies that $F \subsetneq F_0$, which contradicts the minimality of F_0. Therefore, $F_0 = \{x_0\}$, and $\operatorname{Ext}(K) \neq \emptyset$.

We shall now prove that $K = \overline{\operatorname{co}}(\operatorname{Ext}(K))$. Assume that there is some $x_0 \in K$ which is not in $\overline{\operatorname{co}}(\operatorname{Ext}(K))$. Once again, the Hahn–Banach separation theorem implies that there is a continuous functional f and some $a \in \mathbb{R}$ such that

$$\operatorname{Re} f(x_0) > a \geq \operatorname{Re} f(y) \tag{7.2}$$

for all $y \in \overline{\operatorname{co}}(\operatorname{Ext}(K))$. Put $b = \sup\{\operatorname{Re} f(z) : z \in K\}$ and $F = \{z \in K : \operatorname{Re} f(z) = b\}$. By Lemma 7.1.12, F is a compact face of K. By the first half of the proof, F has an extreme point x_1. But then x_1 is an extreme point of K, so by (7.2) we have that $\operatorname{Re} f(x_1) \leq a$. On the other hand,

$$\operatorname{Re} f(x_1) = b \geq \operatorname{Re} f(x_0) > a,$$

and we have a contradiction. It follows that $K = \overline{\operatorname{co}}(\operatorname{Ext}(K))$, as required. $\qquad\square$

7.3   Minimality of $\operatorname{Ext}(K)$

The following is a kind of converse to the Krein–Milman theorem.

Theorem 7.3.1 (Milman's Theorem). *Let K be a nonempty, compact and convex subset of a locally convex space X. If $L \subseteq K$ is a closed set such that $\overline{\operatorname{co}}(L) = K$, then $\operatorname{Ext}(K) \subseteq L$.*

Proof. Let $x \in \operatorname{Ext}(K)$. We need to show that $x \in L$. To motivate the following proof, note that if the assumption was that $\operatorname{co}(L) = K$, then $x \in L$ would follow immediately. Indeed, by the elementary Lemma 7.1.11, if x was equal to a convex combination $x = \sum_i a_i x_i$ of points $x_i \in L$, then x would have to be equal to one of the x_is. The problem is that we don't know that x is actually equal to such a convex combination. The point of the following delicate proof is to hit x on the nose with a convex combination, albeit a convex combination of points that are close to points in L. This, in turn, will imply that x can be approximated by points in L, and by closedness of L we will obtain $x \in L$.

So let us begin the formal proof, assuming that $x \in \operatorname{Ext}(K)$. Let V be an open convex neighborhood of 0, so $x + V$ is a neighborhood of x. By replacing

V with $V \cap (-V)$, we may assume that V is *balanced*, i.e. that $V = -V$. We will be done once we show that for every such V, it holds that $(x+V) \cap L \neq \emptyset$. Indeed, if $(x + V) \cap L \neq \emptyset$ for every such neighborhood V, then $x \in \overline{L} = L$.

Let U be a convex neighborhood $0 \in U$ such that $\overline{U} \subseteq V$. If the topology on X is given either by a norm or by some weak topology $\sigma(X, F)$, then it is clear how to find such a U, indeed: if V is the ball of radius r then U can be taken to be the ball of radius $r/2$; if $V = V(0; f_1, \ldots, f_m; r)$ where $f_1, \ldots, f_m \in F$, then U can be taken to be $V(0; f_1, \ldots, f_m; r/2)$. We leave the justification of the existence of such a U in general as an exercise.

For all $y \in L$, let $U_y = y + U$. Then the open sets U_y cover L, and since L is compact there are $y_1, \ldots, y_m \in L$ such that $L \subseteq \cup_{j=1}^m U_{y_j}$. Denote $K_j = K \cap \overline{U_{y_j}}$ and $S = \mathrm{co}(\cup_{j=1}^m K_j)$. We claim that S is compact, and leave this as an exercise (see Exercise 7.3.3).

We have that $L = L \cap K \subseteq \cup_j K_j$. Whence $K = \overline{\mathrm{co}}(L) \subseteq S$. In particular, $x \in S$, so $x = \sum_{j=1}^m a_j x_j$, where $x_j \in K_j$ and $a_j \geq 0$ for all j, and $\sum_j a_j = 1$. Since x is an extreme point there is some j such that $x = x_j \in \overline{U_{y_j}}$ (recall Lemma 7.1.11). Thus,

$$x - y_j = x_j - y_j \in \overline{U_{y_j}} - y_j = \overline{U} \subseteq V = -V.$$

for $j = 1, \ldots, n$. It follows that $y_j \in x + V$, so $(x + V) \cap L \neq \emptyset$, and the proof is complete. $\qquad\square$

Exercise 7.3.2. Prove that in a topological vector space, for every neighborhood V of the origin, there exists a neighborhood U of the origin such that $\overline{U} \subseteq V$.

Exercise 7.3.3. Give two proofs (one with nets and one without nets) of the following fact: if $K_1, \ldots, K_m$ are compact convex sets in a topological vector space, then $\mathrm{co}(\cup_{j=1}^m K_j)$ is compact (it is important here that K_j is compact *and* convex for all j).

7.4   Application: the Stone–Weierstrass theorem

In this section, we will use the Krein–Milman theorem together with the Hahn–Banach theorem to give a proof of the Stone–Weierstrass theorem. The proof we present does not make use of the Weierstrass polynomial approximation theorem, so we will have here an alternative proof of that classical theorem as well. We prove the real valued version of the Stone–Weierstrass Theorem. The complex version follows readily (see [41, Chapter 1]).

7.4.1 Preliminaries

In this section, we are interested in the space $C_{\mathbb{R}}(K)$ of continuous real valued functions on K, where K is a compact Hausdorff topological space. Recall from Section 3.4 that the dual of the space $C(K)$ of continuous complex valued functions on K is equal to the space $M(K)$ of regular Borel complex measures on K. It follows that $C_{\mathbb{R}}(K)^* = M_{\mathbb{R}}(K)$, the space of regular Borel real valued measures on K. The correspondence between linear functionals and measures is given by

$$F \leftrightarrow \mu \quad \text{where} \quad F(g) = \langle \mu, g \rangle := \int g d\mu \ , \ g \in C_{\mathbb{R}}(K).$$

Below we will identify a measure with the linear functional that it gives rise to. Recall that the norm of a measure $\mu \in M_{\mathbb{R}}(K)$ is defined to be its total variation $\|\mu\| = |\mu|(X)$.

For every real valued measure μ on K, there exists a partition $K = P \cup N$ such that $\mu(E) \geq 0$ for every $E \subseteq P$ and $\mu(E) \leq 0$ for every $E \subseteq N$ (of course, every set appearing is Borel). This partition is unique up to measure zero. One can then define

$$\mu_+(E) = \mu(E \cap P) \quad \text{and} \quad \mu_-(E) = -\mu(E \cap N).$$

Then $\mu = \mu_+ - \mu_-$ and we have that $\|\mu\| = \mu_+(K) + \mu_-(K)$. This decomposition is called the **Jordan-Hahn decomposition** of μ. If $\mu = \mu_+$ ($\mu = \mu_-$) then μ is called a **positive** (**negative**) measure. If μ is a positive Borel measure on K, the **support** of μ is the set

$$\text{supp}(\mu) = \{x \in K : \mu(U) > 0 \text{ for every neighborhood } U \ni x\}.$$

The simplest kind of measure on K is given by the **Dirac measure** δ_k, where $k \in K$, which is defined by

$$\delta_k(E) = \begin{cases} 1 & k \in E \\ 0 & k \notin E \end{cases}.$$

The bounded linear functional corresponding to δ_k is clearly the evaluation functional $f \mapsto f(k)$, in other words, $\delta_k(f) = f(k)$.

Exercise 7.4.1. Show that $\text{supp}(\delta_k) = \{k\}$, and, conversely, if the support of a regular Borel measure μ is single point k, then $\mu = c\delta_k$ for some scalar c.

If $f \in C_{\mathbb{R}}(K)$ and μ is a measure, we define the measure $f\mu$ to be the measure given by

$$f\mu(E) = \int_E f d\mu.$$

Equivalently, $f\mu$ is the measure corresponding to the functional

$$g \mapsto \int g f d\mu \ , \ g \in C_{\mathbb{R}}(K).$$

If $f \geq 0$ and if μ is positive, then the norm of the measure $f\mu$ is given by $\|f\mu\| = \int f\,d\mu$.

Lemma 7.4.2. *A measure μ is an extreme point of $C_{\mathbb{R}}(K)_1^* = M_{\mathbb{R}}(K)_1$ if and only if there is some $k \in K$ and $c \in \{\pm 1\}$ such that $\mu = c\delta_k$.*

Remark 7.4.3. A similar result holds in the complex case, where the only change is that we allow also complex c with $|c| = 1$.

Proof. Suppose that δ_k is equal to a convex combination $\delta_k = a\mu + b\nu$, with $\|\mu\|, \|\nu\| \leq 1$, $a \in (0, 1)$ and $b = 1 - a$. Necessarily then $\|\mu\| = \|\nu\| = 1$, because otherwise, the triangle inequality would show that $\|\delta_k\| < 1$. Applying both sides of $\delta_k = a\mu + b\nu$ to $f \equiv 1$, we find

$$1 = a(\mu_+(K) - \mu_-(K)) + b(\nu_+(K) - \nu_-(K)),$$

and this forces μ and ν to both be positive. Moreover, at least one of the measures, say μ, has k in its support. We now show that $\mu = \delta_k$. If $\mu \neq \delta_k$, then (by regularity) there is a compact set $L \subset K$ that does not contain k, such that $\mu(L) > 0$. Now let $f \in C_{\mathbb{R}}(K)_1$ be a function that takes nonnegative values, satisfies $f(k) = 1$, and vanishes on L. Then

$$1 = \delta_k(f) = a\int_{K \setminus L} f\,d\mu + b\int f\,d\nu \leq a\mu(K \setminus L) + b\nu(K) < 1.$$

This contradiction establishes that $\mu = \delta_k$. It follows that $\nu = \delta_k$ as well, and we conclude that δ_k is an extreme point. In the same way, one sees that $-\delta_k$ is also an extreme point.

To show the converse, consider the set $L = \{c\delta_k : k \in K, |c| = 1\}$. This set is compact, as it is the image in $C_{\mathbb{R}}(K)_1^*$ of the compact set $K \times \{c : |c| = 1\}$ under the weak-$*$ continuous map $(k, c) \mapsto c\delta_k$. By Milman's theorem, to see that $\mathrm{Ext}(C_{\mathbb{R}}(K)) \subseteq L$ it suffices to show that $\overline{\mathrm{co}}(L) = C_{\mathbb{R}}(K)_1^*$. But if not, there exists a weak-$*$ continuous functional $\hat{f} \in \widehat{C_{\mathbb{R}}(K)}$ that strictly separates some $\mu \in C_{\mathbb{R}}(K)_1^*$ from L. This means that there is some $a \in \mathbb{R}$ such that

$$\int f\,d\mu = \langle \mu, f \rangle > a \geq \langle c\delta_k, f \rangle = cf(k)$$

for all $k \in K$ and all c such that $|c| = 1$. It follows that

$$\|f\|_\infty = \sup_{k \in K} |f(k)| \leq a < \left| \int f\,d\mu \right| \leq \|f\|_\infty,$$

a contradiction. We conclude that $\mathrm{Ext}(C_{\mathbb{R}}(K)) \subseteq L$ and the proof is complete. $\qquad\square$

7.4.2 Statement and proof of the Stone–Weierstrass theorem

Theorem 7.4.4 (Stone–Weierstrass theorem). *Let A be a subalgebra of $C_{\mathbb{R}}(K)$ that contains the constant functions and separates the points of K. Then A is dense in $C_{\mathbb{R}}(K)$.*

The following proof is a simplified version of the one given in [42, Section 5.7], where a somewhat more general theorem (Bishop's theorem) is proved.

Proof. Consider the closed unit ball $(A^{\perp})_1$ of the annihilator $A^{\perp}$ of A

$$(A^{\perp})_1 = \left\{ \mu \in M_{\mathbb{R}}(K)_1 : \int f d\mu = 0 \text{ for all } f \in A \right\}.$$

The fact that A is an algebra has the following consequence: if $g \in A$ and $\mu \in A^{\perp}$ then $g\mu \in A^{\perp}$. Indeed, for all $f \in A$ we have

$$\int f(g d\mu) = \int (fg) d\mu = 0.$$

Our goal is to show that $(A^{\perp})_1 = \{0\}$ because the Hahn–Banach theorem would then tell us that $\overline{A} = (A^{\perp})_{\perp} = C_{\mathbb{R}}(K)$.

Suppose, for the sake of reaching a contradiction, that $(A^{\perp})_1 \neq \{0\}$. Clearly, $(A^{\perp})_1$ is a convex and weak-* compact subset of $M_{\mathbb{R}}(K)_1$. By the Krein–Milman theorem, $(A^{\perp})_1$ has an extreme point ν; the assumption $(A^{\perp})_1 \neq \{0\}$ forces $\|\nu\| = 1$.

We will now show that the support of ν must be a single point. Assume that we have two distinct points $x, y \in \text{supp}(\nu)$. Then by the assumption that A is an algebra that contains the constant functions and separates points, one can find $f \in A$ such that $f(x) = 0, f(y) = 1$. Define $g = \frac{f^2}{2\|f\|^2}$. Then $g \in A$ and $0 \leq g \leq 1/2$. There is a neighborhood U of y such that g is positive on U. Thus, since y is in the support of ν, $\|g\nu\| > 0$. Likewise, $\|(1-g)\nu\| > 0$.

We now form the measures $\nu_1 = \frac{g}{\|g\nu\|}\nu$ and $\nu_2 = \frac{(1-g)}{\|(1-g)\nu\|}\nu$. We have noted above that $\nu_1, \nu_2 \in A^{\perp}$, and clearly $\|\nu_i\| \leq 1$. From the fact that g is non-constant, it follows that neither ν_1 nor ν_2 is equal to ν.

Recall that by the Jordan-Hahn decomposition theorem, there is a partition $K = P \cup N$ of K into disjoint sets P, N such that $\nu(E) \geq 0$ if $E \subseteq P$ and $\nu(E) \leq 0$ if $E \subseteq N$. Compute

$$\|(1-g)d\nu\| = \int_P (1-g)d\nu - \int_N (1-g)d\nu = \|\nu\| - \|g\nu\| = 1 - \|g\nu\|.$$

Putting $t = \|g\nu\|$, we find $\nu = t\nu_1 + (1-t)\nu_2$. This is in contradiction to the assumption that ν is an extreme point in $(A^{\perp})_1$. The contradiction shows that the support of ν must be a singleton $\{x\}$.

But if $\text{supp}(\nu) = \{x\}$, then $\nu = c\delta_x$ for $|c| = 1$, by Exercise 7.4.1. Recalling that $\nu \in A^{\perp}$, it follows that

$$cf(x) = \int f c\delta_x = \int f\nu = 0,$$

for all $f \in A$. But this is absurd, since $1 \in A$. The contradiction implies that $A^{\perp} = \{0\}$, and, as a consequence, that $\overline{A} = C_{\mathbb{R}}(K)$.   □

7.5   Application: ridge functions and neural networks revisited

The most immediate and most important consequences of the Stone–Weierstrass theorem are the uniform density of the polynomials and of the trigonometric polynomials in several variables in $C_{\mathbb{R}}(K)$ where K is any compact subset in $\mathbb{R}^k$. It follows easily that the complex trigonometric exponentials are dense in $C(K)$. This, in turn, implies several basic results in harmonic analysis with almost no effort; for example the convergence of Fourier series in L^2 (see [11, Chapter 4]), and the convergence Fejér sums in the uniform and the L^p norms (see Exercises 4.5.13, 4.5.14, 4.5.15).

The Stone–Weierstrass theorem and its consequences also arise in the theory of neural networks. We will present here[3] the ultimate result in the theory, obtained by Leshno, Lin, Pinkus and Schocken [32], following the presentation in [39] (see also [12, 26]). As in Section 3.5, we let $\sigma \colon \mathbb{R} \to \mathbb{R}$ be a fixed function (the *activation function*), and consider the space

$$\mathcal{M}(\sigma) = \mathrm{span}\{\sigma(y \cdot x + b) : y \in \mathbb{R}^k, b \in \mathbb{R}\}$$

which consists of all functions expressible as neural networks with a single hidden layer, k-dimensional input and 1-dimensional output. The question of interest is, for what activation functions is $\mathcal{M}(\sigma)$ dense in some space of functions? For definiteness, we shall study the density of $\mathcal{M}(\sigma)$ in the space $C_{\mathbb{R}}(\mathbb{R}^k)$ of continuous functions with the topology of uniform convergence on compact sets (see Example 6.3.12).

The form of functions generating $\mathcal{M}(\sigma)$ motivates the following definition.

Definition 7.5.1. A continuous function $f \colon \mathbb{R}^k \to \mathbb{R}$ is said to be a ***ridge function*** if it has the form

$$f(x) = g(y \cdot x)$$

for some $y \in \mathbb{R}^k$ and $g \in C_{\mathbb{R}}(\mathbb{R})$. We let $\mathcal{R}$ denote the space spanned by ridge functions.

Proposition 7.5.2. *The space $\mathcal{R}$ is dense in $C_{\mathbb{R}}(\mathbb{R}^k)$.*

---

[3]This is surely not the most profound application of the Krein–Milman theorem; not only is it a second hand application going through the Stone–Weierstrass theorem, but also as an application of the Stone–Weierstrass theorem it is admittedly not particularly remarkable. Yet, it was too cool to be left out. For a more gratifying pure mathematical application of the Krein–Milman theorem see Exercise 7.7.6.

Proof. Let $\mathcal{E}$ be the space spanned by all real exponentials of the form $x \mapsto e^{y \cdot x}$ for some $y \in \mathbb{R}^k$. Then $\mathcal{E} \subset \mathcal{R}$, and it is an algebra that separates points and contains the constants. By the Stone–Weierstrass theorem, for every $f \in C_{\mathbb{R}}(\mathbb{R}^k)$, every compact $K \subset \mathbb{R}^k$ and every $\epsilon > 0$, there is a function $r \in \mathcal{E}$ such that

$$\sup_{x \in K} |r(x) - f(x)| < \epsilon.$$

This shows that $\overline{\mathcal{E}} = C_{\mathbb{R}}(\mathbb{R}^k)$ and consequently $\overline{\mathcal{R}} = C_{\mathbb{R}}(\mathbb{R}^k)$. $\qquad\square$

Theorem 7.5.3. *Let $\sigma \in C(\mathbb{R})$ be a continuous function. The space $\mathcal{M}(\sigma)$ is dense in $C_{\mathbb{R}}(\mathbb{R}^k)$ in the topology of uniform convergence on compact sets, if and only if σ is not a polynomial.*

Proof. If σ is a polynomial of degree m then $\mathcal{M}(\sigma)$ consists only of polynomials of degree at most m and is therefore closed. Thus, the property of *not* being a polynomial is necessary for $\mathcal{M}(\sigma)$ to be dense. Let us turn to sufficiency.

By Proposition 7.5.2, in order to show that $\mathcal{M}(\sigma)$ is dense in $C_{\mathbb{R}}(\mathbb{R})$ it suffices to show that for every $y \in \mathbb{R}^k$ and $g \in C_{\mathbb{R}}(\mathbb{R})$, the ridge function $r(x) = g(y \cdot x)$ is in the closure of $\mathcal{M}(\sigma)$. Consider the space

$$\mathcal{N}(\sigma) = \overline{\mathrm{span}}\{\sigma(ax + b) : a, b \in \mathbb{R}\}.$$

A moment of reflection shows that if $g \in \mathcal{N}(\sigma)$, then $r \in \overline{\mathcal{M}(\sigma)}$. Thus, the theorem is reduced to the case $k = 1$, and our task is now to prove that $\mathcal{N}(\sigma) = C_{\mathbb{R}}(\mathbb{R})$.

Assume that σ is not a polynomial. Let us treat first the special case in which $\sigma \in C^{\infty}$, that is, σ is infinitely differentiable. For any $a, b, h \in \mathbb{R}$, we form the difference quotient

$$\frac{\sigma((a + h)t + b) - \sigma(at + b)}{h} \in \mathcal{N}(\sigma).$$

Letting $h \to 0$ and noting that we have a family of functions in t that converges uniformly on compact sets to the function $t \mapsto t\sigma'(at+b)$, we find that $t\sigma'(at+ b) \in \mathcal{N}(\sigma)$. Repeating this trick, we find that $t^m \sigma^{(m)}(at + b) \in \mathcal{N}(\sigma)$ for all $m = 1, 2, \ldots$. Choosing $a = 0$ and b such that $\sigma^{(m)}(b) \neq 0$ (remember that σ is not a polynomial), we obtain that the functions $1, t, t^2, \ldots$ are all contained in $\mathcal{N}(\sigma)$. By the Stone–Weierstrass theorem the polynomials are dense in $C_{\mathbb{R}}(\mathbb{R})$. Thus, $\mathcal{N}(\sigma) = C_{\mathbb{R}}(\mathbb{R})$, and the proof is complete in the case that $\sigma \in C^{\infty}$.

Now let σ be a continuous function such that $\mathcal{N}(\sigma)$ is not dense. Then there is some m such that $t^m \notin \mathcal{N}(\sigma)$. Let h be a nonnegative infinitely differentiable function with compact support such that $\int h = 1$, and define $h_n(t) = nh(nt)$. Then $\sigma_n := \sigma * h_n \in C^{\infty}$ and $\sigma_n \to \sigma$ in $C_{\mathbb{R}}(\mathbb{R})$[4]. Consideration of the Riemann integral defining the convolution $\sigma * h_n$ shows that $\sigma_n = \sigma * h_n \in \mathcal{N}(\sigma)$ for all n, therefore $\mathcal{N}(\sigma_n) \subseteq \mathcal{N}(\sigma)$. The first part of the proof shows that $t^m \in \mathcal{N}(\sigma_n)$

---

[4]See e.g. [11, Section 12.5] for an argument that can be adapted to this setting.

unless $\sigma_n^{(m)}(b) = 0$ for all $b \in \mathbb{R}$, that is unless σ_n is a polynomial of degree at most $m - 1$. Since $t^m \notin \mathcal{N}(\sigma)$, this means that σ_n is a polynomial of degree at most $m - 1$ for all n. It now follows that $\sigma = \lim_n \sigma_n$ is a polynomial (of degree at most $m - 1$). $\qquad\square$

7.6   The isometric structure of $C(K)$

The family of spaces $C(K)$ gives a rich source of "concrete" examples of Banach spaces. Recall that every Banach space is isometric to a closed subspace of $C(K)$, where K is some compact Hausdorff topological space. Although the spaces consisting of *all* continuous functions on some compact Hausdorff space are far from exhausting the examples of Banach spaces, it is still of interest to try to classify these Banach spaces and understand their structure. We now show how extreme points can be used in this pursuit.

If K and L are two compact Hausdorff spaces, and if $h : K \to L$ is a homeomorphism, then it is clear that the map $C(L) \to C(K)$ given by $f \mapsto f \circ h$ is an isometric isomorphism. Also, if $u \in C(K)$ is such that $|u| = 1$, then $f \mapsto uf$ is an isometric automorphism of $C(K)$. It is quite remarkable that every isometric isomorphism arises as the composition of such maps.

Theorem 7.6.1 (Banach–Stone theorem). *Let K and L be two compact Hausdorff spaces. Suppose that there is an isometric isomorphism $T : C(L) \to C(K)$. Then there exists a homeomorphism $h : K \to L$ and $u \in C(K)$ with $|u| \equiv 1$ such that for all $f \in C(L)$ and all $k \in K$,*

$$(Tf)(k) = u(k)f(h(k)).$$

In particular, $C(K)$ and $C(L)$ are isometric if and only if K and L are homeomorphic.

Proof. Consider the adjoint map $T^* : C(K)^* \to C(L)^*$ given by $T^*(\phi) = \phi \circ T$. By Theorem 5.4.7 we have that T^* is also an isomorphism and Theorem 5.3.2 applied to T and to T^{-1} implies that T^* is isometric. It follows that T^* takes the set of extreme points of $(C(K)^*)_1$ bijectively onto the set of extreme points of $(C(L)^*)_1$. In other words, by Lemma 7.4.2, it takes a delta measure to a unimodular constant times a delta measure.

Thus, for every $k \in K$, there is some $h(k) \in L$ and some $|u(k)| = 1$ such that $T^*(\delta_k) = u(k)\delta_{h(k)}$. Spelling out the duality, we have

$$(Tf)(k) = \langle \delta_k, Tf \rangle = \langle T^*\delta_k, f \rangle = \langle u(k)\delta_{h(k)}, f \rangle = u(k)f(h(k)).$$

It remains to show that u and h are continuous, and that h is bijective. First, note that $u(k) = T1(k)$, that is, $u = T1$, so $u \in C(K)$. Since $|u| = 1$, for every $f \in C(L)$ we have $u^{-1}(k)(Tf)(k) = f(h(k))$, so $f \circ h$ is continuous. Since this holds for every $f \in C(L)$, it follows that h must be continuous. We leave the remaining details to the reader. □

Exercise 7.6.2. Fill in the missing details in the above proof.

7.7 Additional exercises

Exercise 7.7.1. Find all extreme points in the unit ball of $C_0(X)$, where X is a locally compact topological Hausdorff space. (Separate into two cases: X compact, and X not compact).

Exercise 7.7.2. The previous exercise applies to c and c_0 (why?).

1. Deduce that c and c_0 are not isometrically isomorphic.

2. Prove that c_0 and c are isomorphic.

3. Recall that $c^* \cong \ell^1$ and that $c_0^* \cong \ell^1$. We see that the predual of a space is not determined uniquely (at least, not up to isometric isomorphism).

4. Explain why the dualities (ℓ^1, c) and (ℓ^1, c_0) must be given by a different pairing.

Exercise 7.7.3. Find the extreme points in the unit ball of ℓ^p, for each $p \in [1, \infty]$.

Exercise 7.7.4. True or false: For a compact convex set K in a locally convex space, $\mathrm{Ext}(K)$ is closed (if that was too hard or too easy, consider the question for finite dimensional spaces).

Exercise 7.7.5. Let K be a closed, convex and bounded set in a reflexive normed space X, and let $\phi \colon K \to \mathbb{R}$ be a lower semicontinuous and convex function (see Exercise 2.4.15).

1. Prove that ϕ being lower semicontinuous with respect to the norm implies that it is lower semicontinuous with respect to the weak topology.

2. Prove that ϕ attains a minimum in K.

3. If ϕ is continuous, prove that ϕ attains a maximum at one of the extreme points of K.

4. Conclude that if ϕ is a continuous real linear functional then it attains its maximum and its minimum on the set of extreme points of K.

Exercise 7.7.6 (Existence and uniqueness of Haar measure on compact groups). Let G be a compact topological group, and let $P(G)$ denote the set of all regular Borel probability measures on G. Recall that a **left Haar measure** is a probability measure $\mu \in P(G)$ such that $\mu(gE) = \mu(E)$ for every Borel $E \subseteq G$. Similarly, a **right Haar measure** is a $\mu \in P(G)$ such that $\mu(Eg) = \mu(E)$ for every Borel $E \subseteq G$. The goal of this exercise is to prove that on every compact group, there exists a unique left Haar measure, which is also the unique right Haar measure; this measure will simply be called **the Haar measure on G**.

1. Show that if there exists a left Haar measure μ and a right Haar measure ν, then $\mu = \nu$. (**Hint:** For every $f \in C(G)$, integrate the function $(g, h) \mapsto f(gh)$ with respect to $\mu \times \nu$ using Fubini's theorem.)

2. Below we shall prove that a left Haar measure exists. Explain why it must be unique.

3. And now for existence. Take a five minute break.

4. For $g \in G$, define the shift operators L_g and $T_g = L_g^*$ as in the proof of Theorem 6.6.4.

5. Prove that for every $f \in C(G)$, the map $G \ni g \mapsto L_g f$ is continuous. Deduce that for every $\mu \in P(G)$, the map $G \ni g \mapsto T_g \mu$ is weak-$*$ continuous.

6. Prove that $T_g(P(G)) \subseteq P(G)$ for all $g \in G$.

7. Let $K \subseteq P(G)$ be a minimal weak-$*$ compact and convex subset of $P(G)$ that is invariant under all T_g, $g \in G$ (why does K exist?). We need to show that K is a singleton. Assume therefore that $\mu, \nu \in K$ and construct $\lambda = \frac{1}{2}\mu + \frac{1}{2}\nu$. Our goal is to show that $\lambda = \mu = \nu$.

8. Prove that the set $E = \{T_g \lambda : g \in G\}$ is weak-$*$ compact. Explain why $\overline{\operatorname{co}(E)}^{w*} = K$.

9. We have set things in such a way that what remains now for you to do, is to figure out how to use Krein–Milman's theorem and Milman's theorem to finish the proof.

Exercise 7.7.7. Let K be a compact and convex set in a locally convex space X. A point $x_0 \in K$ is said to be **exposed** if there exists a continuous linear functional $f \in X'$ such that

$$\operatorname{Re} f(x_0) > \operatorname{Re} f(x), \quad \text{for all } x \in K \setminus \{x_0\}.$$

Let $\operatorname{Exp}(K)$ denote the set of exposed points in K. Disconnect from the web, and then decide which of the following statements is true, in the simplest setting when $X = \mathbb{R}^n$.

1. $\text{Exp}(K) \subseteq \text{Ext}(K)$? If so, is it dense in it?

2. $\text{Ext}(K) \subseteq \text{Exp}(K)$? If so, is it dense in it?

Exercise 7.7.8. Let X be a normed space and let $Y \subseteq X$ be a subspace. True or false: every extreme point in Y_1^* can be extended to an extreme point in X_1^*.

Exercise 7.7.9. Let $\mathbb{T}$ be the unit circle, and let $\mathbb{D}$ be the open unit disc, in the complex plane. Let $A(\mathbb{D})$ denote the closure of the complex polynomials in $C(\mathbb{T})$.

1. Prove that $\text{Ext}((A(\mathbb{D})^*)_1) = \{\lambda \delta_z : \lambda, z \in \mathbb{T}\}$.

2. For a point $z \in \mathbb{D}$, show "with your hands" that δ_z is not extreme in $(A(\mathbb{D})^*)_1$. In fact, you may consider $z = 0$, this is sufficient (explain why).

Exercise 7.7.10. True or false: there exists a Banach space X such that $C_{\mathbb{R}}([0,1]) \cong X^*$.

Exercise 7.7.11. An $n \times n$ matrix $A = (a_{ij})_{i,j=1}^n$ with nonnegative entries is said to be a **doubly stochastic matrix** if all $\sum_{i=1}^n a_{ij} = 1$ for all j and $\sum_{j=1}^n a_{ij} = 1$ for all i. Let S be the set of all doubly stochastic matrices. Describe $\text{Ext}(S)$.

Exercise 7.7.12. Prove that an operator T on a Hilbert space H is an extreme point of $B(H)_1$ if T is an isometry ($T^*T = I$) and if T is a co-isometry ($TT^* = I$). (In fact, it is true that these are precisely the extreme points of $B(H)_1$, but that's significantly more difficult to prove.)

Exercise 7.7.13 (Open ended question). A **cone**[5] in a real vector space X is set C such that (i) $C + C \subseteq C$ and (ii) $\alpha C \subseteq C$ for every $\alpha \geq 0$ (iii) $C \cap (-C) = \{0\}$.

1. Prove that every cone is convex. Prove that the only extreme point in a cone is 0.

2. An **extreme ray** of a cone C is a face F that is equal to $F = \{\alpha v : \alpha \geq 0\}$ for some $v \in C$.

3. Give an example of a closed cone in a reflexive space that is not generated by (i.e. not the closed convex hull of) its extreme rays.

4. State and prove a theorem that under certain reasonable conditions, a cone is generated by its extreme rays.

---

[5]Sometimes "cone" is used to refer to something a little more general, and what we are defining would be called a "proper convex cone".

Chapter 8

Banach algebras

8.1 Basic notions and first examples

Definition 8.1.1. An ***algebra*** is a vector space A that is at the same time a ring (with the same addition), such that the multiplication by scalars and by elements of the algebra commute, that is, for all $a, b \in A$ and every scalar λ,

$$\lambda(ab) = (\lambda a)b = a(\lambda b).$$

An algebra is said to be a ***normed algebra*** if the underlying vector space has a norm on it. A normed algebra A is said to be a ***Banach algebra*** if it is complete with respect to the norm, and if, additionally,

$$\|ab\| \leq \|a\| \|b\|. \tag{8.1}$$

The crucial consequence of (8.1) is continuity of multiplication.

Exercise 8.1.2. Prove that the multiplication in a Banach algebra is jointly continuous, that is, if $a_n \to a$ and $b_n \to b$, then $a_n b_n \to ab$.

Definition 8.1.3. A map $\varphi\colon A \to B$ between two algebras is said to be a ***homomorphism*** if it is a linear map that respects multiplication in the sense that $\varphi(ab) = \varphi(a)\varphi(b)$ for all $a, b \in A$. In the context of algebras, an ***isomorphism*** is a bijective homomorphism.

The definition of the word "isomorphism" depends on context (recall that we defined the notion of isomorphism for Banach spaces in Definition 1.3.13).

Exercise 8.1.4. Let A be a normed algebra, which is complete with respect to the norm, and in which multiplication is a separately continuous function in each variable (that is, for every $a \in A$, the maps $A \ni b \mapsto ab$ and $A \ni b \mapsto ba$ are both continuous with respect to the norm). Prove that there exists an equivalent norm that makes A into a Banach algebra.

Let us recall some basic algebraic notions. An algebra A is ***commutative*** (or ***abelian***) if $ab = ba$ for all $a, b \in A$. If A is an algebra, a ***subalgebra*** of A is a subspace $B \subseteq A$ such that $ab \in B$ for all $a, b \in B$. A closed subalgebra of a Banach algebra is a Banach algebra. An ***ideal*** in A is a subspace $I \subseteq A$ such

DOI: 10.1201/9781003297864-8

that $ax, xa \in I$ for all $x \in I$ and $a \in A$. Ideals are sometimes referred to as *two sided ideals*; *left* and *right* ideals are defined similarly. An ideal $I \subseteq A$ is *proper* if $I \neq A$, *nontrivial* if it is proper and nonzero, and *maximal* if it is proper and not strictly contained in any proper ideal.

Example 8.1.5. Let X be a normed space. Then the algebra of all bounded operators $B(X)$ with $ab := a \circ b$ is a normed algebra. Inequality (8.1) is easy to check: for all $a, b \in B(X)$

$$\|ab\| = \sup_{x \in X_1} \|abx\| \leq \sup_{x \in X_1} \|a\|\|bx\| \leq \sup_{x \in X_1} \|a\|\|b\|\|x\| \leq \|a\|\|b\|.$$

If X is complete, then $B(X)$ is complete and thus a Banach algebra. Every closed subalgebra $A \subseteq B(X)$ is then also a Banach algebra.

Definition 8.1.6. A *unit* in a Banach algebra A is an element $e \in A$ such that $ae = ea = a$ for all $a \in A$. The unit, if it exists, is unique. We will call a Banach algebra *unital* if it has a unit e such that $\|e\| = 1$. The unit of a unital Banach algebra A is often denoted 1_A or 1. A homomorphism $\varphi \colon A \to B$ between two unital algebras is said to be *unital* if $\varphi(1_A) = 1_B$.

When A is a unital algebra it is convenient to identify $\mathbb{C}1_A$ with a copy of the complex numbers embedded in A and then write λ instead of $\lambda 1_A$.

Exercise 8.1.7. Let A be a Banach algebra. Prove that the Banach space $\tilde{A} = A \oplus_1 \mathbb{C}$ becomes a unital Banach algebra if one endows it with the multiplication
$$(a, \lambda) \cdot (b, \mu) = (ab + \mu a + \lambda b, \lambda \mu).$$
(Recall that the norm on $A \oplus_1 \mathbb{C}$ is given by $\|(a, \lambda)\| = \|a\| + |\lambda|$). Thus every Banach algebra can be embedded as an ideal of codimension one in a unital Banach algebra. The algebra $\tilde{A}$ is sometimes referred to as *the unitization of A*, even though there are other unitizations in use (as we shall see below).

Exercise 8.1.8. Prove that a Banach algebra with unit can be given an equivalent norm $\|\cdot\|'$ under which it is a unital Banach algebra (i.e., $\|e\|' = 1$).

Example 8.1.9. Let X be a locally compact Hausdorff topological space. Then the algebra $C_b(X)$ of bounded continuous functions is a commutative unital Banach algebra, when the multiplication is the usual pointwise multiplication: $fg(x) = f(x)g(x)$. The algebra $C_0(X)$ is a closed subalgebra of $C_b(X)$ thus it is also a Banach algebra; it is unital if and only if X is compact. When X is not compact, $C_b(X)$ is a unital algebra containing $C_0(X)$, but it is different from $\widetilde{C_0(X)}$. The algebra $C_0(X) + \mathbb{C}1$ is another unital Banach algebra containing $C_0(X)$, which is also different from $\widetilde{C_0(X)}$. The algebra $C_0(X) + \mathbb{C}1$ can be identified with $C(K)$ for some compact Hausdorff space K. Can you see what K must be?

Example 8.1.10. Consider the previous case with $X = \mathbb{N}$. Then $C_b(\mathbb{N}) = \ell^\infty$ and $C_0(\mathbb{N}) = c_0$. What is $C_0(\mathbb{N}) + \mathbb{C}1$?

Example 8.1.11. On $\ell^1(\mathbb{Z})$ we define the **convolution product** as follows

$$f * g(n) = \sum_{k \in \mathbb{Z}} f(k)g(n-k) \ , \ \text{for all } n \in \mathbb{Z}, f, g \in \ell^1(\mathbb{Z}).$$

The above sum can be written more symmetrically as $\sum_{k+m=n} f(k)g(m)$, so $g * f = f * g$. To see that $f * g$ is in $\ell^1(\mathbb{Z})$, we estimate

$$
\begin{aligned}
\sum_n |f * g(n)| = \sum_n \left| \sum_{k \in \mathbb{Z}} f(k)g(n-k) \right| \\
\leq \sum_{n,k} |f(k)g(n-k)| \\
\leq \sum_k |f(k)| \sum_n |g(n-k)| \\
= \|g\|_1 \sum_k |f(k)| = \|g\|_1 \|f\|_1.
\end{aligned}
$$

Thus $f * g \in \ell^1(\mathbb{Z})$ and $\|f * g\|_1 \leq \|f\|_1 \|g\|_1$. Another routine calculation shows that the convolution is an associative multiplication, and all the other algebraic requirements are pretty clear. Thus, $\ell^1(\mathbb{Z})$ is a Banach algebra when it is endowed with the convolution product. This is a unital algebra; the sequence $e_0 = (\ldots, 0, 0, 1, 0, 0, \ldots)$ is a unit and $\|e_0\| = 1$.

Exercise 8.1.12. Is $\ell^1(\mathbb{Z})$ a Banach algebra with respect to pointwise multiplication?

Example 8.1.13. Example 8.1.11 is a special case of the following general class of examples of Banach algebras. If G is a locally compact group and μ is a left Haar measure, then $L^1(G)$ becomes a Banach algebra if one equips it with the convolution product:

$$f * g(x) = \int f(y)g(y^{-1}x)d\mu(y) \ , \ \text{for all } f, g \in L^1(G).$$

One can check that $f * g(x)$ is defined for almost every $x \in G$, and that this product makes $L^1(G)$ into a Banach algebra. Another important special case is given by $\mathbb{R}$ with the Lebesgue measure. The convolution then takes the familiar form

$$f * g(x) = \int_{-\infty}^{\infty} f(y)g(x-y)dy.$$

Exercise 8.1.14. Prove that $L^1(\mathbb{R})$ has no unit.

Example 8.1.15 (The disc algebra). Let $A(\mathbb{D})$ be the closure of analytic polynomials in $C(\mathbb{T})$. Note that by *analytic polynomial* we mean a function of the form $p(z) = \sum_{n=0}^{N} \hat{p}(n)z^n$ consisting of nonnegative powers of the complex variable z; we do **not** mean a polynomial in the real variables $x = \mathrm{Re}\, z$ and $y = \mathrm{Im}\, z$. Since $z \in \mathbb{T}$, it is sometimes convenient to think of p as $p(e^{i\theta}) = \sum_{n=0}^{N} \hat{p}(n)e^{in\theta}$, and we see that $\hat{p}(n)$ is the nth Fourier coefficient of p, as the notation suggests. Thus, by definition, the algebra $A(\mathbb{D})$ consists of the closure in $C(\mathbb{T})$ of the complex trigonometric polynomials whose negative Fourier coefficients are 0. Consider the following algebras.

1. $A_1 = $ all functions f in $C(\mathbb{T})$ whose Fourier coefficients satisfy $\hat{f}(n) = 0$ for all $n < 0$.

2. $A_2 = $ the closure of analytic polynomials in $C(\overline{\mathbb{D}})$.

3. $A_3 = $ all functions f in $C(\overline{\mathbb{D}})$ such that $f\big|_{\mathbb{D}}$ is analytic.

We claim that $A_1 = A(\mathbb{D})$, $A_2 = A_3$, and that $f \mapsto f\big|_{\mathbb{T}}$ is an isometric isomorphism of $A_2 = A_3$ onto $A(\mathbb{D})$. These algebras are called **the disc algebra**.

Every analytic polynomial is in A_1. Now, $f \mapsto \hat{f}(n)$ is a continuous functional on $C(\mathbb{T})$, therefore every uniform limit of analytic polynomials is also in A_1. Thus $A(\mathbb{D}) \subseteq A_1$. Conversely, if $f \in A_1$, then Fejér's theorem shows that $f = \lim_N \sigma_N(f)$ is the norm limit of analytic polynomials (see Exercise 4.5.14); whence $A_1 = A(\mathbb{D})$.

The polynomials are certainly in A_3, and uniform limits of analytic functions are analytic, so $A_2 \subseteq A_3$. Conversely, if $f \in A_3$, then define $f_r(z) = f(rz)$ for all $r \in (0, 1)$. The function f_r extends to an analytic function on the open disc of radius $1/r$ around the origin, and so its Taylor series converges to f_r uniformly on $\overline{\mathbb{D}}$. Therefore $f_r \in A_2$. But uniform continuity of f on $\overline{\mathbb{D}}$ implies that $f_r \xrightarrow{r \to 1} f$ uniformly on $\overline{\mathbb{D}}$, so f is in A_2.

Finally, it is clear that the map $f \mapsto f\big|_{\mathbb{T}}$ is a contractive homomorphism from $C(\overline{\mathbb{D}})$ into $C(\mathbb{T})$[1]. By the maximum modulus principle, if f is analytic in $\mathbb{D}$ and continuous in $\overline{\mathbb{D}}$, then

$$\max_{z \in \mathbb{T}} |f(z)| = \max_{z \in \overline{\mathbb{D}}} |f(z)|,$$

in other words, the map $f \mapsto f\big|_{\mathbb{T}}$ is an isometry from A_3 into $A(\mathbb{D})$. The range of this isometry contains all analytic polynomials; being closed, it must be $A(\mathbb{D})$.

8.2 Some motivation: ODEs in Banach spaces

Studying function algebras from a mix of analytic and algebraic points of view is a natural thing to do, and we have already seen how it can be useful

---

[1]In fact, by the Tietze extension theorem (see Theorem 35.1 in [35]), the restriction map is onto.

for approximation theory (see Sections 7.4 and 7.5). Another impetus for the development of the theory of Banach algebras is to facilitate the solution of various problems involving operators. In this section, we make some motivational discussion of how the need to consider Banach algebras may arise.

Let X be a Banach space. In X we can consider paths $\{x(t) : t \in I\}$, where $I \subseteq \mathbb{R}$ is an (open) interval and $x : I \to X$ is a continuous map. The path $t \mapsto x(t)$ is said to be ***differentiable*** if

$$\frac{dx}{dt}(t) := \lim_{s \to t} \frac{x(t) - x(s)}{t - s}$$

exists for every $t \in I$. Once we have such a notion, we can consider the following first order differential equation

$$\frac{dx}{dt}(t) = Ax(t), \tag{8.2}$$

where A is a linear operator. Such an equation arises in an undergraduate course in ordinary differential equations when $X = \mathbb{R}^n$ and (8.2) is a compact way of writing down a system of first order linear differential equations with constant coefficients. More generally, many equations describing the evolution of a system in time can be described in this way, for example the heat equation or Schrödinger's equation — in this case, X will be some function space, and A will be a differential operator.

Let us try to solve equation (8.2) "numerically", assuming that A is bounded. Already in the simplest case where $X = \mathbb{R}$, we know that we cannot determine the solution uniquely unless we have some initial condition, so let us assume that $x(0) = x_0$ is given. We discretize $\mathbb{R}$; that is we choose a small step size $h > 0$ and consider the values of x on $\{0, h, 2h, \ldots\}$. An approximation to the differential equation is given by

$$\frac{x(nh + h) - x(nh)}{h} = Ax(nh), \quad n = 0, 1, 2, \ldots.$$

Hopefully, if h is very small, a solution to this problem will give a good approximation to the solution of the differential equation. Since we know $x(0) = x_0$, this determines

$$x(h) = x_0 + hAx_0 = (1 + hA)x_0.$$

Then we can find

$$x(2h) = x(h) + hAx(h) = (1 + hA)x(h) = (1 + hA)^2 x_0.$$

Inductively,

$$x(nh) = (1 + hA)^n x_0.$$

Let us now fix $t > 0$. The approximate solution to (8.2), given by solving the discretized equation with step size $h = t/n$, is

$$x(t) \approx \left(1 + \frac{tA}{n}\right)^n x_0.$$

It is natural to hope that this will converge to a solution as one takes the limit $n \to \infty$. If $A \in \mathbb{R}$, then $\lim_{n \to \infty} (1 + \frac{tA}{n})^n = e^{tA}$. We therefore get the idea $x(t) = e^{tA}x_0$, where the exponent of an operator has yet to be defined. One way to define e^A is simply $\lim_{n \to \infty} (1 + \frac{A}{n})^n$ and this does work in the case at hand (we dropped the t, since A will now be an arbitrary operator). But the simplest way to define the exponential is by plugging A into the power series of $\exp(z)$

$$\exp(A) = e^A = \sum_{n=0}^{\infty} \frac{1}{n!} A^n.$$

Up to this point, our discussion was cruising along heuristically and the arguments only served as a means to motivate a guess for what the solution should be. Now that we have an idea, we work rigorously to show that it works. Since

$$\sum_{n=0}^{\infty} \left\| \frac{1}{n!} A^n \right\| \leq \sum_{n=0}^{\infty} \frac{\|A^n\|}{n!} \leq \sum_{n=0}^{\infty} \frac{\|A\|^n}{n!} = e^{\|A\|},$$

we see that the series defining $\exp(A)$ converges absolutely, so it converges in $B(X)$. Note that this series converges for every $A \in B(X)$, and so we can define e^{tA} for every $t \in \mathbb{R}$. Thanks to absolute convergence, one may manipulate series freely to obtain

$$\exp((t+s)A) = \exp(tA)\exp(sA)$$

for every $s, t \in \mathbb{R}$, just by repeating the proof from calculus. But now we can check that for every $x_0 \in X$, the path defined by

$$x(t) = \exp(tA)x_0$$

satisfies

$$\lim_{h \to 0} \frac{x(t+h) - x(t)}{h} = \lim_{h \to 0} \frac{1}{h}(\exp((t+h)A) - \exp(tA))x_0$$

$$= \lim_{h \to 0} \frac{1}{h}(\exp(hA) - 1)\exp(tA)x_0 = Ax(t),$$

granted that we can show that $\lim_{h \to 0} \frac{1}{h}(\exp(hA) - 1) = A$. Indeed,

$$\frac{1}{h}(\exp(hA) - 1) = \frac{1}{h}\left(1 + \frac{hA}{1!} + \frac{h^2 A^2}{2!} + \cdots - 1\right)$$

$$= A + h\left(\frac{A^2}{2!} + \frac{A^3 h}{3!} + \cdots\right) \xrightarrow{h \to 0} A,$$

because the series $\frac{A^2}{2!} + \frac{A^3 h}{3!} + \ldots$ also converges absolutely.

The above discussion shows that manipulating algebraically and analytically the elements of the algebra generated by an operator is useful both for explorations of solutions as well as for rigorously justifying methods of solving problems. This kind of analysis is what gave birth to the theory of Banach algebras.

8.3   Invertibility and the spectrum

Coming to the notions of spectrum and invertibility, we will from now on consider Banach algebras over the complex numbers only. The utility of this assumption will become evident within a few pages. The reader is encouraged to keep track of what works and what doesn't in the case of Banach spaces over the reals.

8.3.1   Invertible elements

Definition 8.3.1. Let A be a unital Banach algebra. An element $a \in A$ is said to be ***invertible*** if there exists $b \in A$ such that $ab = ba = 1$. In this case, b is said to be the ***inverse*** of a, and is denoted by a^{-1}. An element b is said to be a ***right inverse*** (respectively, a ***left inverse***) for a if $ab = 1$ (respectively, $ba = 1$); in this case a is said to be ***right invertible*** (respectively, ***left invertible***). The group of invertible elements in A is denoted by $G(A)$ or by A^{-1}.

Here are some easy-to-see facts: an element is invertible if and only if it is right and left invertible; the inverse of an element, if it exists, is unique; the inverse of an invertible is invertible; the product of two invertibles is invertible and conversely, if the product ab is invertible, and either a or b is invertible, then the other one is invertible, too.

Proposition 8.3.2. *Let A be a unital Banach algebra. If $a \in A$ is such that $\|a\| < 1$, then $1 - a \in G(A)$. Moreover, the inverse of $1 - a$ is given by*

$$(1 - a)^{-1} = \sum_{n=0}^{\infty} a^n.$$

Proof. By (8.1) we have that $\|a^n\| \leq \|a\|^n$, so the series $\sum \|a^n\|$ is majorized by a geometric series $\sum \|a\|^n$. The series $\sum a^n$ therefore converges absolutely, thus it converges to some element $b = \sum a^n \in A$. By continuity of the product,

$$\sum_{n=0}^{N} a^n (1 - a) = (1 - a) \sum_{n=0}^{N} a^n \to (1 - a)b = b(1 - a).$$

On the other hand, we have a telescopic sum

$$\sum_{n=0}^{N} a^n (1 - a) = 1 - a + a - a^2 + a^2 - \cdots - a^{N+1} = 1 - a^{N+1} \to 1,$$

because $\|a^{N+1}\| \leq \|a\|^{N+1} \xrightarrow{N \to \infty} 0$. Thus $(1 - a)b = b(1 - a) = 1$, so

$$\sum a^n = b = (1 - a)^{-1}.$$

$\square$

Remark 8.3.3. The series $\sum a^n$ is sometimes referred to as the **Neumann series**.

Corollary 8.3.4. *Let A be a unital Banach algebra. Then $G(A)$ is an open set, and the map $a \mapsto a^{-1}$ is continuous on $G(A)$.*

Proof. Suppose that $a \in G(A)$. If $\|h\| < \|a^{-1}\|^{-1}$, then $1 - a^{-1}h \in G(A)$ by the previous proposition and $(1 - a^{-1}h)^{-1} = \sum(a^{-1}h)^n$. It follows that $a - h = a(1 - a^{-1}h)$ is also invertible and

$$(a - h)^{-1} = \left(a(1 - a^{-1}h)\right)^{-1} = \left(\sum(a^{-1}h)^n\right)a^{-1}.$$

This shows that the open unit ball with center at a and radius $\|a^{-1}\|^{-1}$ is contained in $G(A)$. For continuity, we estimate

$$\|a^{-1} - (a - h)^{-1}\| \leq \|1 - \sum_{n=0}^{\infty}(a^{-1}h)^n\|\|a^{-1}\|$$

$$\leq \|a^{-1}\|\sum_{n=1}^{\infty}\|a^{-1}h\|^n$$

$$= \|a^{-1}\|\frac{\|a^{-1}h\|}{1 - \|a^{-1}h\|} \leq \frac{\|a^{-1}\|^2\|h\|}{1 - \|a^{-1}\|\|h\|}.$$

The right hand side clearly tends to zero as $\|h\| \to 0$, and the proof is complete.

$\square$

8.3.2 The spectrum

Definition 8.3.5. Let A be a unital Banach algebra. For every $a \in A$, the **spectrum** of a (or, if we want to be precise, **the spectrum of a in A**) is defined to be the set

$$\sigma_A(a) = \{\lambda \in \mathbb{C} : a - \lambda \notin G(A)\}.$$

The **resolvent set** of a (in A) is the set $\rho_A(a) = \mathbb{C} \setminus \sigma_A(a)$.

When there is no ambiguity regarding the algebra A in which the element a is considered, we drop the subscript A and write $\sigma(a) = \sigma_A(a)$. We will see in the examples below that the spectrum depends, in general, on the algebra in which a is considered, that is if $a \in A \subseteq B$ and A and B share the same unit, then $\sigma_A(a)$ is in general different from $\sigma_B(a)$ (though it is quite clear that $\sigma_B(a) \subseteq \sigma_A(a)$). By a common convention, if T is a bounded operator on some Banach space X, then $\sigma(T)$ denotes the spectrum of T in $B(X)$, that is, $\sigma(T) = \sigma_{B(X)}(T)$, and, by the inverse mapping theorem, this is the set of $\lambda \in \mathbb{C}$ such that $T - \lambda I$ is not bijective.

Theorem 8.3.6. *For every element a in a unital Banach algebra A, its spectrum $\sigma_A(a)$ is a nonempty compact subset of $\mathbb{C}$, which is contained in $\{z \in \mathbb{C} : |z| \leq \|a\|\}$.*

Proof. If $|\lambda| > \|a\|$, then $a - \lambda = \lambda(\lambda^{-1}a - 1) \in G(A)$ by Proposition 8.3.2, and so $\sigma(a) \subseteq \{z \in \mathbb{C} : |z| \le \|a\|\}$, as claimed. Since $G(A)$ is open, $\rho(a)$ is open, so $\sigma(a)$ is closed. It remains to prove that the spectrum is nonempty.

Suppose that $\sigma(a) = \emptyset$. Then $a - \lambda$ is invertible for every $\lambda \in \mathbb{C}$. For every $f \in A^*$ we may therefore define a function $u : \mathbb{C} \to \mathbb{C}$ by

$$u(z) = f((a - z)^{-1}).$$

Claim 1: u is an entire function.

Proof of Claim 1. For $z, w \in \mathbb{C}$, we have the identity

$$(a - z)^{-1} - (a - w)^{-1} = (a - z)^{-1}[(a - w) - (a - z)](a - w)^{-1}$$
$$= (z - w)(a - z)^{-1}(a - w)^{-1}.$$

Then

$$\frac{u(z) - u(w)}{z - w} = f\left((z - w)^{-1}[(a - z)^{-1} - (a - w)^{-1}]\right)$$
$$= f\left((a - z)^{-1}(a - w)^{-1}\right) \xrightarrow{z \to w} f\left((a - w)^{-2}\right),$$

where the limit exists by Corollary 8.3.4, together with continuity of f. We have shown that u is differentiable at every $w \in \mathbb{C}$, so u is entire.

Claim 2: $u(z) \xrightarrow{z \to \infty} 0$ and u is bounded.

Proof of Claim 2. As in the proof of Corollary 8.3.4, we can use the Neumann series to show that for $|z| > \|a\|$

$$\|(a - z)^{-1}\| = |z|^{-1} \|(1 - z^{-1}a)^{-1}\| \le \frac{|z|^{-1}}{1 - \|z^{-1}a\|}.$$

Applying f, we find that

$$|u(z)| = |f((a - z)^{-1})| \le \frac{\|f\||z|^{-1}}{1 - \|z^{-1}a\|} \xrightarrow{z \to \infty} 0.$$

This proves that $u(z) \xrightarrow{z \to \infty} 0$. Since u is continuous, we also get that u is bounded.

From the above two claims together with Liouville's theorem[2], u must be a constant, and this constant must be 0. Thus, we have found that for every $z \in \mathbb{C}$ and every $f \in A^*$,

$$f\left((a - z)^{-1}\right) = 0.$$

By the Hahn–Banach theorem $(a - z)^{-1} = 0$ for every $z \in \mathbb{C}$. Since $(a - z)^{-1}$ is supposed to be an invertible element, this is impossible. This contradiction shows that $\sigma(a) \ne \emptyset$. $\qquad\square$

---

[2]See Theorem 10.23 in [13].

Remark 8.3.7. If we were working over the reals, then it would not be true that the spectrum is nonempty. This is one of the most significant differences between operator theory over the real numbers and operator theory over the complex numbers. Recall that in finite dimensional linear algebra, the fact that every operator on a complex space has an eigenvalue follows from the fundamental theorem of algebra. In infinite dimensional spaces, the proof that $\sigma(A)$ is nonempty also uses complex function theory in a nontrivial way.

Corollary 8.3.8 (Gelfand–Mazur theorem). *Let A be a unital Banach algebra which is also a division algebra (that is, every nonzero element has an inverse). Then A is isometrically isomorphic to $\mathbb{C}$.*

Proof. Let $\theta : \mathbb{C} \to A$ be the isometric inclusion of $\mathbb{C}$ in A given by $\theta(\lambda) = \lambda 1$. We need to show that θ is onto. Let $a \in A$. By Theorem 8.3.6, there exists some $\lambda \in \mathbb{C}$ such that $a - \lambda 1$ is not invertible. It follows that $a - \lambda 1 = 0$, or $a = \lambda 1 = \theta(\lambda)$, and the proof is complete. $\square$

Remark 8.3.9. The analogous problem over the real numbers has a similar but slightly more complicated solution. Mazur proved that every real Banach division algebra A is one of the following three possibilities: either $A = \mathbb{R}$, or $A = \mathbb{C}$, or $A = \mathbb{H}$ (the quaternions).

Remark 8.3.10. The Gelfand–Mazur theorem implies that there is no proper field extension of $\mathbb{C}$ that can be given a norm that makes it into a Banach algebra. Consider, for example, the field $\mathbb{C}(z) = \{p/q : p, q \in \mathbb{C}[z], q \neq 0\}$ of all rational functions, or the field of all meromorphic functions (the field of fractions formed from holomorphic functions) on $\mathbb{C}$. It is interesting that there is something in the algebraic structure of these fields that encodes the fact that there is no way to endow them with a complete, submultiplicative norm.

Definition 8.3.11. Let $A \in B(X)$. If $\lambda \in \mathbb{C}$ is a scalar for which there exists a nonzero $x \in X$ such that $Ax = \lambda x$, then λ is said to be an ***eigenvalue of*** A, and x is said to be an ***eigenvector*** corresponding to λ. The space spanned by all eigenvectors corresponding to an eigenvalue λ is called the ***eigenspace*** corresponding to λ. The ***point spectrum*** of A, denoted by $\sigma_p(A)$ is the set of all eigenvalues of A.

In other words, λ is an eigenvector if and only if $A - \lambda I$ is not injective. If X is finite dimensional, $\sigma(A) = \sigma_p(A)$. However, if X is infinite dimensional, then the spectrum of an operator can contain points that are not eigenvalues.

Proposition 8.3.12 (The spectral mapping theorem). *Let A be a unital Banach algebra, let $a \in A$, and let $p \in \mathbb{C}[z]$ be a polynomial. Then*

$$\sigma\left(p(a)\right) = p\left(\sigma(a)\right) = \{p(\lambda) : \lambda \in \sigma(a)\}.$$

Proof. Let $\lambda \in \mathbb{C}$, and write $p(z) - \lambda = c(z - \lambda_1) \cdots (z - \lambda_n)$ as the product of linear factors, where $\lambda_1, \ldots, \lambda_n$ are precisely the solutions of the equation

$p(z) = \lambda$. It is plain to see that this factorization holds also after plugging a into both sides:

$$p(a) - \lambda = c(a - \lambda_1) \cdots (a - \lambda_n).$$

We claim that the product on the right hand side is invertible if and only if all the factors are invertible. Indeed, clearly a product of invertible elements is invertible. Conversely, if the product is invertible with inverse b, then $bc(a - \lambda_1) \cdots (a - \lambda_n) = 1$, so $a - \lambda_n$ is left invertible. As all the factors commute, a symmetric argument shows that $a - \lambda_n$ is also right invertible hence it is invertible (the other factors are treated in the same manner). Said differently, the right hand side is *not* invertible if and only if at least one of the factors is not invertible, or $\lambda \in \sigma(p(a))$ if and only if $\lambda_i \in \sigma(a)$ for some i. But $\lambda_1, \ldots, \lambda_n$ are precisely the roots of the equation $p(z) = \lambda$, so we see that $\lambda \in \sigma(p(a))$ if and only if λ is the image under p of a point in $\sigma(a)$, in other words $\lambda \in \sigma(p(a))$ if and only if $\lambda \in p(\sigma(a))$.   □

8.3.3 Examples

Example 8.3.13. If $A \in M_n(\mathbb{C})$, then $\sigma(A)$ is the set of eigenvalues of A.

Example 8.3.14. Let $S : \ell^p \to \ell^p$ be the unilateral shift

$$S(x_0, x_1, x_2, \ldots) = (0, x_0, x_1, x_2, \ldots), \quad (x_0, x_1, x_2, \ldots) \in \ell^p.$$

We wish to compute $\sigma(S)$, that is, we will find the spectrum of S in the Banach algebra $B(\ell^p)$. First, it is easy to see that $\sigma_p(S) = \emptyset$. As for the remaining part of $\sigma(S)$, it turns out that it is easier to compute $\sigma(S^*)$. One checks easily[3] that

$$S^*(x_0, x_1, x_2, \ldots) = (x_1, x_2, \ldots) \quad , \quad (x_0, x_1, x_2, \ldots) \in \ell^q.$$

Some straightforward calculations show that $\sigma_p(S^*) = \mathbb{D}$ when the conjugate exponent $q = \frac{p}{p-1} < \infty$ and $\sigma_p(S^*) = \overline{\mathbb{D}}$ when $q = \infty$. On the other hand $\sigma(S^*) \subseteq \{z : |z| \leq \|S^*\|\} = \overline{\mathbb{D}}$. By compactness, $\sigma(S^*) = \overline{\mathbb{D}}$. Finally, by Corollary 5.4.9, $\sigma(S) = \overline{\mathbb{D}}$.

Example 8.3.15. Consider the algebra $C(K)$ of continuous functions on a compact Hausdorff space K. When is a function f invertible in $C(K)$? Precisely when there exists a function $g \in C(K)$ such that $fg = 1$ everywhere. This happens precisely when $f(k) \neq 0$ for all $k \in K$, because then one can take $g = \frac{1}{f}$. It follows that for every $f \in C(K)$, the spectrum is simply the image of K under f, that is

$$\sigma_{C(K)}(f) = f(K).$$

Example 8.3.16. Consider the disc algebra $A(\mathbb{D})$. We saw in Example 8.1.15 that one may think of $A(\mathbb{D})$ either as a subalgebra of $C(\mathbb{T})$ or as a subalgebra

---

[3]Why is this OK when $p = \infty$?

of $C(\overline{\mathbb{D}})$. So every $f \in A(\mathbb{D})$ can be considered as an element in one of the following algebras: $A(\mathbb{D})$, $C(\mathbb{T})$ or $C(\overline{\mathbb{D}})$. How does the spectrum of f depend on the algebra relative to which the spectrum is defined? We saw above that $\sigma_{C(\mathbb{T})}(f) = f(\mathbb{T})$ and $\sigma_{C(\overline{\mathbb{D}})}(f) = f(\overline{\mathbb{D}})$. What is $\sigma_{A(\mathbb{D})}(f)$?

Now, if f is invertible in $A(\mathbb{D})$, then it is invertible in both $C(\overline{\mathbb{D}})$ and $C(\mathbb{T})$. It follows that $\sigma_{A(\mathbb{D})}(f) \supseteq \sigma_{C(\mathbb{T})}(f) = f(\mathbb{T})$ and that $\sigma_{A(\mathbb{D})}(f) \supseteq \sigma_{C(\overline{\mathbb{D}})}(f) = f(\overline{\mathbb{D}})$. On the other hand, if $\lambda \notin f(\overline{\mathbb{D}})$, then $f - \lambda$ does not vanish on $\overline{\mathbb{D}}$, and so $(f - \lambda)^{-1}$, its inverse in $C(\overline{\mathbb{D}})$, is analytic and so it is an element of $A(\mathbb{D})$. Therefore, $\sigma_{A(\mathbb{D})}(f) = f(\overline{\mathbb{D}})$.

8.4   The spectral radius

Definition 8.4.1. Let A be a unital Banach algebra and $a \in A$. The **spectral radius** of a is defined to be

$$r(a) = \sup \{|\lambda| : \lambda \in \sigma_A(a)\}.$$

A striking property of the spectral radius, which follows from the spectral radius formula that we prove below, is that it is independent of the algebra A in which we consider a.

Example 8.4.2. We return to Example 8.3.16. Consider $f \in A(\mathbb{D})$, when the latter is considered as a unital subalgebra of $C(\mathbb{T})$. Then $\sigma_{A(\mathbb{D})}(f) = f(\overline{\mathbb{D}})$ and so the spectral radius, when computed in $A(\mathbb{D})$, is equal to $\sup_{z \in \overline{\mathbb{D}}} |f(z)|$. On the other hand $\sigma_{C(\mathbb{T})}(f) = f(\mathbb{T})$ and so the spectral radius, when computed in $C(\mathbb{T})$, is equal to $\sup_{z \in \mathbb{T}} |f(z)|$. By the maximum modulus principle, $\sup_{z \in \overline{\mathbb{D}}} |f(z)| = \sup_{z \in \mathbb{T}} |f(z)|$, and we are able to verify directly in this case that, although the spectrum depends on the algebra, the spectral radius does not.

From Theorem 8.3.6, we have that $r(a) \leq \|a\|$. By Proposition 8.3.12 we find, in addition, that $r(a)^n = r(a^n) \leq \|a^n\|$, and so

$$r(a) \leq \inf_{n \geq 1} \|a^n\|^{1/n}.$$

Theorem 8.4.3 (The spectral radius formula). *For every element a in a unital Banach algebra,*

$$r(a) = \lim_{n \to \infty} \|a^n\|^{1/n}.$$

Proof. We have already noted above that

$$r(a) \leq \inf_{n \geq 1} \|a^n\|^{1/n} \leq \liminf_{n \to \infty} \|a^n\|^{1/n}.$$

It remains to prove that

$$r(a) \geq \limsup_{n \to \infty} \|a^n\|^{1/n}. \tag{8.3}$$

We claim that if $|z| < r(a)^{-1}$ (interpreted as $|z| < \infty$, when $r(a) = 0$), then the sequence $\{z^n a^n\}_{n=1}^{\infty}$ is bounded. Assuming this for the moment, we find that there is some $M > 0$ such that $|z|^n \|a^n\| \leq M$, or $\|a^n\|^{1/n} \leq M^{1/n}/|z|$, for all n. This implies that

$$\limsup_n \|a^n\|^{1/n} \leq \frac{1}{|z|},$$

for every $0 < |z| < r(a)^{-1}$, and letting $|z| \nearrow r(a)^{-1}$, we obtain (8.3).

To complete the proof, we now prove that $\{z^n a^n\}_{n=1}^{\infty}$ is bounded for every $z \in \mathbb{C}$ such that $0 < |z| < r(a)^{-1}$. Fix $f \in A^*$. Consider the function

$$u(z) = f\left((1 - za)^{-1}\right),$$

which is well defined in the open disc $\{|z| < r(a)^{-1}\}$. For $0 < |z| < r(a)^{-1}$ we can write

$$u(z) = \frac{1}{z} f\left((z^{-1} - a)^{-1}\right).$$

We know from the proof of Theorem 8.3.6 that $w \mapsto f\left((w - a)^{-1}\right)$ is analytic on the complement of $\sigma(a)$, thus u is analytic in the punctured disc $U = \{0 < |z| < r(a)^{-1}\}$. On the other hand, if $|z| < \|a\|^{-1}$ then by Proposition 8.3.2 we have $(1 - za)^{-1} = \sum_{n=0}^{\infty}(za)^n$ so

$$u(z) = f\left(\sum_{n=0}^{\infty}(za)^n\right) = \sum_{n=0}^{\infty} f(z^n a^n) = \sum_{n=0}^{\infty} f(a^n) z^n$$

in the disc $\{z : |z| < \|a\|^{-1}\}$ and u is analytic at the origin. By the uniqueness of Taylor series from complex analysis, this power series converges in the (perhaps larger) disc $\{|z| < r(a)^{-1}\}$. Therefore, if $|z| < r(a)^{-1}$ then the sequence $\{f(z^n a^n)\}_{n=0}^{\infty}$ is bounded. Fixing z, we see that for every $f \in A^*$, $\{f(z^n a^n)\}_{n=0}^{\infty}$ is bounded, in other words, the sequence $\{z^n a^n\}_{n=0}^{\infty}$ is *weakly bounded*. It follows from the uniform boundedness principle that the sequence $\{z^n a^n\}_{n=0}^{\infty}$ is norm bounded (see Exercise 5.6.1), as claimed. □

Definition 8.4.4. An element a in a unital Banach algebra is said to be *quasinilpotent* if $\lim_n \|a^n\|^{1/n} = 0$.

By the spectral radius formula, a is quasinilpotent if and only if $\sigma(a) = \{0\}$.

Example 8.4.5. If a is *nilpotent* (in the sense that $a^n = 0$ for some n) then a is quasinilpotent. In particular, if $a^n = 0$ for some n, then $\sigma(a) = \{0\}$. This also follows from the spectral mapping theorem (Proposition 8.3.12). In finite dimensional spaces the converse is also true.

Exercise 8.4.6. If $a \in M_n(\mathbb{C})$ is quasinilpotent, then a is nilpotent.

Example 8.4.7. In $C(K)$, a function f has $\sigma(f) = f(K)$, thus $r(f) = \|f\|_\infty$. So the only quasinilpotent element is 0.

Example 8.4.8. The Volterra operator $V : C([0,1]) \to C([0,1])$ defined by

$$(Vf)(x) = \int_0^x f(t)dt$$

is quasinilpotent, but it is not nilpotent (see Exercise 8.7.6).

8.5 Application: Lomonosov's invariant subspace theorem

In this section, we give a proof of Lomonosov's invariant subspace theorem based on the spectral radius formula. Our presentation is inspired by that in Rudin's book (see [42, Theorem 10.35], and also the surrounding comments regarding the history of this proof).

Let X be a complex Banach space, and let $T \in B(X)$. A linear subspace $Y \subseteq X$ is said to be ***invariant*** for T if $Tx \in Y$ for all $x \in Y$. For many years, one of the most important open problems in operator theory was the *invariant subspace problem*, which asks whether every bounded operator on an infinite dimensional and separable Banach space has a nontrivial closed invariant subspace (the problem is not interesting for finite dimensional or non-separable spaces, and it is also not interesting to ask about non-closed invariant subspaces). In the 1970s, Enflo found a counter example in a complicated Banach space, and later, in the 1980s, Read found an example of an operator on ℓ^1 with no invariant subspace. The problem in the Hilbert space setting remains open to this day[4].

There are, however, positive results. In the Hilbert space setting the problem has an affirmative answer for normal operators. In general Banach spaces, it was shown by Aronszajn and Smith that every compact operator has a nontrivial invariant subspace. It was later proved by Bernstein and Robinson that every operator T on a Hilbert space which is *polynomially compact* (in the sense that there exists a polynomial p such that $p(T)$ is compact) has a nontrivial invariant subspace. Both of these results involving compact operators are subsumed by the following theorem of Lomonosov. Before stating the theorem, let us introduce the following definition.

---

[4]On [31, p. 282] Lax writes that it is an open question whether every operator on a separable space has a nontrivial invariant subspace "and it is an open question whether this question is interesting". Surely that's a bit of an exaggeration, but it does seem to be the case that the invariant subspace problem is not as interesting to the research community as it used to be. We leave it to the reader to ponder this.

Definition 8.5.1. Let X be a Banach space and let $T \in B(X)$. A subspace $Y \subseteq X$ is said to be **hyper-invariant** for T if $SY \subseteq Y$ for every $S \in B(X)$ that commutes with T. A hyper-invariant subspace Y is said to be **nontrivial** if it is different from 0 and X.

Theorem 8.5.2 (Lomonosov's theorem). *Let X be an infinite dimensional complex Banach space, and let $T \in B(X)$ be a nonzero compact operator. Then T has a closed, nontrivial hyper-invariant subspace.*

Proof. Let
$$M = \{T\}' := \{S \in B(X) : TS = ST\}.$$

M is clearly a closed and unital subalgebra of $B(X)$. It follows from the spectral theory of compact operators that if $\sigma(T)$ contains a nonzero point λ, then λ is a nonzero eigenvalue and the corresponding eigenspace V_λ is finite dimensional (see the review below Definition 5.5.1), and hence a nontrivial closed invariant subspace. Elementary algebra shows that V_λ is invariant under any $S \in M$. Indeed, if $v \in V_\lambda$ and $S \in M$, then $TSv = STv = S\lambda v = \lambda Sv$, so $Sv \in V_\lambda$. We rephrase this in the following way: *if T has no nontrivial hyper-invariant subspace, then the spectral radius $r(T)$ must be zero.*

Now we shall assume that T has no nontrivial hyper-invariant subspace and this will lead us to conclude that $\lim_n \|T^n\|^{1/n} > 0$. By the spectral radius formula, we then have $r(T) > 0$, in contradiction with the previous paragraph.

Choose some $x_0 \in X$ such that $Tx_0 \neq 0$ (this is possible because T is nonzero). Let $\delta > 0$ and let B be an open ball centered at x_0, such that $\|x\| > \delta$ and $\|Tx\| > \delta$ for all $x \in B$. Define $K = \overline{T(B)}$. Since T is a compact operator, K is compact. Note that our choice of B also guarantees that $0 \notin K$.

For every $x \in X$, the space
$$\overline{\operatorname{span}\{Sx : S \in M\}} = \overline{\{Sx : S \in M\}}$$

is a closed hyper-invariant subspace. If we assume that there are no proper hyper-invariant subspaces, we must have $X = \overline{\{Sx : S \in M\}}$ for every $x \neq 0$. Therefore, for every $k \in K$, there exists an operator $S \in M$ such that $Sk \in B$. By continuity, there is a neighborhood $U \ni k$ such that $S(U) \subset B$. Since K is compact, we may find $k_1, \ldots, k_n \in K$ and corresponding operators $S_1, \ldots, S_n \in M$ and $U_1, \ldots, U_n$, such that $K \subseteq \cup_{i=1}^n U_i$ and $S_i(U_i) \subseteq B$ for all $i = 1, \ldots, n$.

Now, $Tx_0 \in K$. There is therefore some $i_1 \in \{1, \ldots, n\}$ such that $Tx_0 \in U_{i_1}$. By construction, $S_{i_1} Tx_0 \in B$. Therefore, $TS_{i_1} Tx_0 \in K$, and as a consequence $TS_{i_1} Tx_0 \in U_{i_2}$ for some i_2. Consequently, $S_{i_2} TS_{i_1} Tx_0 \in B$. It follows then that $TS_{i_2} TS_{i_1} Tx_0$ is back in K. We can continue this forever, and we find that for every N, there are $i_1, \ldots, i_N \in \{1, \ldots, n\}$ such that
$$S_{i_N} T \ldots S_{i_2} TS_{i_1} Tx_0 = S_{i_N} \ldots S_{i_2} S_{i_1} T^N x_0 \in B.$$

Now if we put $\mu = \max\{\|S_1\|, \ldots, \|S_n\|\}$, then
$$\delta < \|S_{i_N} \ldots S_{i_2} S_{i_1} T^N x_0\| \leq \mu^N \|T^N\| \|x_0\|.$$

We conclude that

$$r(T) = \lim_{N \to \infty} \|T^N\|^{1/N} \geq \lim_{N \to \infty} \left(\frac{\delta}{\|x_0\|} \right)^{1/N} \frac{1}{\mu} = \frac{1}{\mu} > 0.$$

We have already remarked that $r(T) > 0$ implies the existence of a nontrivial hyper-invariant subspace, but we have just derived this inequality from the assumption that there are no nontrivial hyper-invariant subspaces. It follows that T must have a nontrivial hyper-invariant subspace. □

Corollary 8.5.3. *Every bounded operator on a complex Banach space that commutes with a nonzero compact operator has a closed nontrivial invariant subspace. In particular, every compact operator has one.*

8.6 Vector valued holomorphic functions and the functional calculus

The beautiful and profound idea underlying the proofs of the nonemptiness of the spectrum and of the spectral radius formula is to take full advantage of the fact that our Banach spaces are over the field of complex numbers by harnessing the powerful theory of holomorphic functions. Both theorems solve problems that seemingly have nothing to do with holomorphic functions, yet each proof introduces some holomorphic function that works magic.

The interplay between function theory and Banach algebras is rich and deep, and is more than a trick pulled out of the sleeve in the course of a slick proof. In this section, we will develop a few of the tools which clarify one aspect of this connection: the possibility of applying a holomorphic function f to an element $a \in A$ in a Banach algebra to obtain another element $f(a) \in A$. This notion plays an important role in further developments of Banach algebras and operator theory.

8.6.1 Functional calculus generalities

Let us begin by explaining the cryptic phrase in the title of this section. A *functional calculus* is a homomorphism from some algebra of functions into some Banach algebra A, that enjoys several natural properties. For example, let a be an element of a unital Banach algebra A. Then we have a natural functional calculus from the algebra $\mathbb{C}[z]$ of polynomials in one variable into A, given by "evaluating a polynomial at a". That is, if $p(z) = \sum_{k=0}^{n} c_k z^k$, we define $p(a)$ by simply plugging $z = a$ into the defining formula:

$$p(a) = \sum_{k=0}^{n} c_k a^k. \tag{8.4}$$

One can check that the map

$$\mathbb{C}[z] \ni p \mapsto p(a)$$

is a unital (i.e., unit preserving) homomorphism from $\mathbb{C}[z]$ into A.

We might wish to extend this elementary functional calculus to a larger class of functions, for example, the rational functions. Clearly, this will not work unconditionally, because if a is not invertible, then there will be no sensible way to evaluate $\frac{1}{z}$ at a. We easily fix this problem by limiting our attention to the algebra $\mathrm{Rat}(\sigma(a))$ of all rational functions whose poles lie in the complement of the spectrum $\sigma(a)$ of a. If $p, q \in \mathbb{C}[z]$ are two polynomials with no common root, then $p/q \in \mathrm{Rat}(\sigma(a))$ if and only if q has no zeros in $\sigma(a)$. In this case $0 \notin \sigma(q(a)) = q(\sigma(a))$, and so we can define $(p/q)(a) = p(a)q(a)^{-1}$. We leave to the reader to show that

$$p/q \mapsto p(a)q(a)^{-1}$$

is a well defined homomorphism from $\mathrm{Rat}(\sigma(a))$ into A. It gives a natural and useful notion of "evaluating a rational function at an element $a \in A$".

One can go further. Let $\mathrm{Hol}(D_R)$ be the algebra of all functions holomorphic in a disc $D_R = \{z \in \mathbb{C} : |z| < R\}$. Every $a \in A$ with spectral radius smaller than R gives rise to a holomorphic functional calculus $\mathrm{Hol}(D_R) \to A$ as follows. If $f \in \mathrm{Hol}(D_R)$ then it has a power series $f(z) = \sum_{n=0}^{\infty} c_n z^n$ that converges in D_R, and by the Cauchy-Hadamard formula $\limsup_{n \to \infty} |c_n|^{1/n} \leq R^{-1}$. On the other hand, if $a \in A$ and $r(a) < R$, then by the spectral radius formula $\lim_{n \to \infty} \|a^n\|^{1/n} = r(a) < R$. It follows that $\limsup \|c_n a^n\|^{1/n} < 1$, so there is some $t < 1$ such that $\|c_n a^n\| \leq t^n$ for all sufficiently large n. Thus, the series $\sum c_n a^n$ converges absolutely and we can define

$$f(a) = \sum_{n=0}^{\infty} c_n a^n.$$

Showing that $f \mapsto f(a)$ is an algebra homomorphism or that it is continuous in the appropriate sense is a bit tricky. We shall not attempt it, since there is a more general and better way to go.

Let $\mathrm{Hol}(\sigma(a))$ denote the algebra of all functions that are holomorphic in some open neighborhood of $\sigma(a)$. Note that $\mathrm{Hol}(\sigma(a))$ contains both $\mathrm{Rat}(\sigma(a))$ as well as $\mathrm{Hol}(D_R)$ for all $R > r(a)$. Then one can define a functional calculus

$$\mathrm{Hol}(\sigma(a)) \ni f \mapsto f(a)$$

that extends the polynomial, the rational and the power series functional calculi defined above and is continuous in an appropriate sense. We will define this functional calculus later in this chapter after some preparations, but to motivate these preparations let us write down the formula that defines it.

Recall that the decisive tool in complex function theory was not power series but the complex integral, and in particular Cauchy's theorem and integral formula. If $f \in \mathrm{Hol}(\sigma(a))$, then there is an open set U containing $\sigma(a)$ such that $f \in \mathrm{Hol}(U)$. Assume for simplicity that U is simply connected, and let Γ be a smooth closed loop inside U, which is disjoint from $\sigma(a)$ and winds around every point in $\sigma(a)$ once. By Cauchy's integral formula[5],

$$f(z_0) = \frac{1}{2\pi i} \int_{\Gamma} f(z)(z - z_0)^{-1} dz$$

for all $z_0 \in \sigma(a)$. What happens if we plug $z_0 = a$ into this formula? It turns out that defining

$$f(a) = \frac{1}{2\pi i} \int_{\Gamma} f(z)(z - a)^{-1} dz \tag{8.5}$$

gives rise to the ultimate version of the holomorphic functional calculus. The remainder of this chapter is devoted to making sense of this integral and to studying the properties of the mapping $f \mapsto f(a)$.

8.6.2 Integration of Banach space valued functions

Let X be a Banach space, fixed throughout this section.

Definition 8.6.1. Let $f \colon [a, b] \to X$ be a continuous function. We define the *integral* of f on the interval $[a, b]$ to be

$$\int_a^b f(t)dt = \lim \sum_{j=0}^{n-1} f(t_j)(t_{j+1} - t_j),$$

where the limit[6] is taken as $n \to \infty$ and $\max\{t_{j+1} - t_j : j = 0, 1, \ldots, n-1\} \to 0$ over all partitions $a = t_0 < t_1 < \ldots < t_n = b$ of $[a, b]$.

Since a continuous function from a compact metric space into another metric space is uniformly continuous, one can show, in a similar manner to how one proceeds for scalar valued Riemann integrals, that the above limit exists and defines a unique element in X.

Exercise 8.6.2. Prove that the above limit exists. Prove that the above Riemann-type integral enjoys the familiar properties of the scalar valued Riemann integral from calculus:

1. **Linearity:** $\int_a^b (f(t) + \lambda g(t))dt = \int_a^b f(t)dt + \lambda \int_a^b g(t)dt$.

2. **Additivity:** $\int_a^c f(t)dt = \int_a^b f(t)dt + \int_b^c f(t)dt$.

---

[5] See [13, Theorem 10.35]

[6] This limit can also be rigorously formulated as a limit of a norm convergent net in X.

3. **Triangle inequality:** $\left\| \int_a^b f(t)dt \right\| \leq \int_a^b \|f(t)\|dt.$

Finally, show that the integral also has the property, that for all $\rho \in X^*$,

$$\rho\left(\int_a^b f(t)dt\right) = \int_a^b \rho(f(t))dt. \tag{8.6}$$

We will now define line integrals of vector valued functions. We shall use the simplest version that is sufficient for developing the functional calculus. For the required background in the theory of functions in a complex variable, consult [13]. In the following definition, a function $\gamma\colon [a,b] \to \mathbb{C}$ is said to be *piecewise* C^1 if it is continuous, if it is differentiable at all but finitely many points in $[a,b]$, and if its derivative is piecewise continuous.

Definition 8.6.3. A set $\Gamma \subset \mathbb{C}$ is said to be a *simple piecewise* C^1 *curve* if there is a piecewise C^1 function $\gamma\colon [a,b] \to \mathbb{C}$, injective with the exception of perhaps the endpoints, such that $\Gamma = \gamma([a,b])$. Γ is said to be *closed* if $\gamma(a) = \gamma(b)$.

Exercise 8.6.4. Let Γ be a simple piecewise C^1 curve determined by a piecewise C^1 function $\gamma\colon [a,b] \to \mathbb{C}$, and let $f\colon \Gamma \to X$ be continuous. Show that

$$\int_a^b f(\gamma(t))\gamma'(t)dt = \lim \sum_{j=0}^{n-1} f(\gamma(t_j))(\gamma(t_{j+1}) - \gamma(t_j)),$$

where the limit is taken over the partitions of $[a,b]$.

In order to define integration over piecewise C^1 curves, we shall need to be able to specify in which direction the integration is carried out. If Γ is a curve determined by $\gamma\colon [a,b] \to \mathbb{C}$, then Γ inherits from γ an *orientation*, which, roughly speaking, is the direction in which the point $\gamma(t)$ traverses Γ as t moves from a to b. We say that Γ is an *oriented curve* to indicate that we take Γ together with a choice of orientation.

Definition 8.6.5. Let Γ be an oriented simple piecewise C^1 curve determined by a piecewise C^1 function $\gamma\colon [a,b] \to \mathbb{C}$, and let $f\colon \Gamma \to X$ be continuous. The *line integral* of f on Γ is defined to be

$$\int_\Gamma f(z)dz = \int_a^b f(\gamma(t))\gamma'(t)dt.$$

Note that the right hand side of the above equation, which is used to define the left hand side, is an integral of the type defined earlier in Definition 8.6.1. The orientation of Γ is implicit in the notation. Exercise 8.6.4 shows that the line integral does not depend on the choice of parameterization γ so long as the orientation is preserved.

The definition of line integral extends easily to more general sets. First, note that if two simple non-closed C^1-curves are concatenated at their endpoints, then one obtains a curve that is piecewise C^1. Further, one can integrate against unions of curves.

Definition 8.6.6. Let $\Gamma = \Gamma_1 \cup \ldots \cup \Gamma_n$ be a union of oriented simple piecewise C^1 curves such that $\Gamma_i \cap \Gamma_j$ is finite for all $i \neq j$. We define

$$\int_\Gamma f(z)dz = \sum_{k=1}^n \int_{\Gamma_k} f(z)dz.$$

Definition 8.6.7. We shall say that the boundary ∂V of a bounded open set $V \subseteq \mathbb{C}$ is **piecewise** C^1 if ∂V is equal to the union $\Gamma_1 \cup \cdots \cup \Gamma_n$ of finitely many simple closed piecewise C^1 curves such that $\Gamma_i \cap \Gamma_j$ is finite for all $i \neq j$. If V is a bounded open set such that $\partial V = \Gamma_1 \cup \cdots \cup \Gamma_n$ is piecewise C^1, and if $f : \partial V \to \mathbb{C}$ is continuous, then we define

$$\int_{\partial V} f(z)dz = \sum_{k=1}^n \int_{\Gamma_k} f(z)dz,$$

where the orientation on each Γ_i is chosen in the so-called *standard* manner, which means that if you imagine yourself moving inside the complex plane along the boundary of V, then the set V is always to your immediate left.

Exercise 8.6.8. Suppose that $f_n : \Gamma \to X$ are continuous functions from a piecewise C^1 curve into X that converge uniformly to a function $f : \Gamma \to X$, that is

$$\lim_{n \to \infty} \sup_{z \in \Gamma} \|f_n(z) - f(z)\| = 0.$$

Prove that

$$\lim_{n \to \infty} \int_\Gamma f_n(z)dz = \int_\Gamma f(z)dz.$$

8.6.3 Holomorphic functions with values in a Banach space

In this section, X will continue to denote a Banach space.

Definition 8.6.9. Let U be an open set in $\mathbb{C}$ and let $f : U \to X$.

1. f is said to be **weakly holomorphic** if for all $\rho \in X^*$ the composition $\rho \circ f : U \to \mathbb{C}$ is holomorphic.

2. f is said to be **strongly holomorphic**, or simply **holomorphic**, if for every $z_0 \in U$ the following limit exists in the norm

$$\lim_{z \to z_0} \frac{f(z) - f(z_0)}{z - z_0}.$$

When the limit exists it is called the **derivative** of f at z_0 and is denoted by $f'(z_0)$.

3. f is said to be **analytic** if for every $z_0 \in U$ there exists $r > 0$ such that in the open disc $D_r(z_0)$ of radius r around z_0 the function f is given by a uniformly and absolutely convergent series

$$f(z) = \sum_{n=0}^{\infty} (z - z_0)^n x_n, \quad z \in D_r(z_0)$$

where $x_n \in X$ for $n = 0, 1, 2, \ldots$.

It is easy to show that an analytic function is strongly holomorphic and clear that a strongly holomorphic function is weakly holomorphic. We shall see below that the converse implications also hold, and thereafter we shall simply refer to **holomorphic** functions. In the literature, the words *holomorphic* and *analytic* are often used interchangeably. Strictly speaking, these terms are not synonymous but rather equivalent, by the following theorem.

Theorem 8.6.10. *Let U be an open set in $\mathbb{C}$ and let $f : U \to X$ be a weakly holomorphic function. Then the following hold:*

1. *f is continuous.*

2. *If V is bounded and open such that $\overline{V} \subset U$ and ∂V is piecewise C^1, then $\int_{\partial V} f(z)dz = 0$ and $f(z_0) = \frac{1}{2\pi i} \int_{\partial V} f(z)(z - z_0)^{-1}dz$ for all $z_0 \in V$.*

3. *f is analytic and hence is strongly holomorphic.*

Proof. For (1), we assume without loss of generality that $0 \in U$ and prove that f is continuous at 0. Fix $\rho \in X^*$, and consider the function

$$g(z) = \frac{\rho(f(z)) - \rho(f(0))}{z}.$$

The fact that f is weakly holomorphic implies that $\rho \circ f$, and hence g, is holomorphic in a neighborhood of 0. In particular, g is bounded in some small disc D_r of radius r centered at the origin. Since this is true for every $\rho \in X^*$, we conclude that the set

$$\{z^{-1}(f(z) - f(0)) : 0 < |z| < r\}$$

is weakly bounded, hence by the uniform boundedness principle, it is bounded in norm. It follows that $\|f(z) - f(0)\| \leq M|z|$ for some constant M and all $z \in D_r$, which readily implies that f is continuous at 0.

From the continuity of f it follows that the integrals in assertion (2) exist. To prove (2), we use property (8.6) from Exercise 8.6.2 to see that

$$\rho \left(\int_{\partial V} f(z)dz \right) = \int_{\partial V} \rho \circ f(z)dz = 0,$$

for all $\rho \in X^*$, where the second equality follows from Cauchy's theorem because f is weakly holomorphic. By Hahn–Banach we conclude that $\int_{\partial V} f(z)dz = 0$. The second assertion in (2) is proved in a similar manner.

To prove (3), we again assume that $0 \in U$ and show that f has a power series that converges absolutely and uniformly in a disc D_r centered at 0; as noted above, this will imply that f is differentiable at 0. Let r be such that $\overline{D_{2r}} \subset U$. By (2),

$$f(z) = \int_{\partial D_{2r}} f(w)(w - z)^{-1} dw$$

for $z \in D_{2r}$. We write $(w - z)^{-1} = w^{-1}(1 - z/w)^{-1} = w^{-1} \sum_{n=0}^{\infty} \left(\frac{z}{w}\right)^n$ for $z \in D_{2r}$ and $w \in \partial D_{2r}$. This gives

$$f(z) = \sum_{n=0}^{\infty} \left(\int_{\partial D_{2r}} \frac{f(w)}{w^{n+1}} dw \right) z^n.$$

It is easy to show that this series converges absolutely and uniformly in D_r. $\square$

Example 8.6.11. In the course of the proof of Theorem 8.3.6 we showed that for every $f \in A^*$, the function $u(z) = f((a - z)^{-1})$ is holomorphic in $\mathbb{C} \setminus \sigma(a)$. In other words, we showed that the A-valued function $z \mapsto (a - z)^{-1}$ is weakly holomorphic. By the above theorem it follows that $z \mapsto (a - z)^{-1}$ is strongly holomorphic. Note however that it is not hard to verify this directly.

8.6.4 The holomorphic functional calculus

For a compact set $K \subset \mathbb{C}$, we let $\mathrm{Hol}(K)$ denote the algebra of functions holomorphic in an open neighborhood of K. Thus, $f \in \mathrm{Hol}(K)$ if there is an open $U \supset K$ such that f is holomorphic in U. For brevity, we shall write $\mathrm{Hol}(a)$ for $\mathrm{Hol}(\sigma(a))$. A bit of care is needed in verifying that $\mathrm{Hol}(K)$ is an algebra: if $f, g \in \mathrm{Hol}(K)$, then there are sets $U, V \supset K$ such that $f \in \mathrm{Hol}(U)$ and $g \in \mathrm{Hol}(V)$. The sum $f + g$ and the product fg are then defined and holomorphic on the open set $U \cap V \supset K$, so $f + g$ and fg are in $\mathrm{Hol}(K)$.

For the definition of the functional calculus, we shall require the following fact: if U is an open set that contains a compact set K, then there exists a bounded open set V such that $K \subset V \subset \overline{V} \subset U$ and such that ∂V is piecewise C^1. Roughly, such a set V can be found by covering K with finitely many discs $\{D_i\}_{i=1}^n$ of a radius that is sufficiently small such that $\overline{\cup_{i=1}^n D_i} \subset U$, obtaining an open set $V := \cup_{i=1}^n D_i$ with a piecewise C^1 boundary.

Definition 8.6.12 (The holomorphic functional calculus). Let A be a unital Banach algebra. Given $a \in A$ and $f \in \mathrm{Hol}(a)$ we define $f(a) \in A$ as follows. Let $U \supset \sigma(a)$ be an open set in which f is holomorphic, and let V be an open bounded set such that $\sigma(a) \subset V \subset \overline{V} \subset U$ and such that ∂V is piecewise C^1. Define

$$f(a) = \frac{1}{2\pi i} \int_{\partial V} f(z)(z - a)^{-1} dz. \tag{8.7}$$

The mapping $f \mapsto f(a)$ is called **the holomorphic functional calculus** for a or simply the **functional calculus**.

Theorem 8.6.13. *Let A be a unital Banach algebra and let $a \in A$. The functional calculus $f \mapsto f(a)$ has the following properties:*

1. **Well definedness:** $f(a)$ *is well defined, and in particular does not depend on the choice of U and V.*

2. **Homomorphism:** *The mapping $f \mapsto f(a)$ is a homomorphism from* $\mathrm{Hol}(a)$ *into A.*

3. **Naturality:** *If f is a polynomial, a rational function with poles off $\sigma(a)$, or a function with a power series representation that converges in a neighborhood of $\sigma(a)$, then the definition of $f(a)$ coincides with the polynomial, rational and power series functional calculi defined in Section 8.6.1.*

4. **Continuity:** *If f_n, f are holomorphic in $U \supset \sigma(a)$, and if $f_n \to f$ uniformly on compact subsets of U, then $f_n(a) \to f(a)$.*

5. **Spectral mapping theorem:** $\sigma(f(a)) = f(\sigma(a))$ *for all $f \in \mathrm{Hol}(a)$.*

Proof. **Well definedness.** First, we fix U, and show that $f(a)$ is independent of V. Let V, V' be as in the definition of the functional calculus. Let W be a bounded open set containing $\overline{V \cup V'}$ such that $\overline{W} \subset U$ and such that ∂W is piecewise C^1. The boundary of $Y = W \backslash \overline{V}$ is equal to $\partial W \cup \partial V$. Using Example 8.6.11, it is easy to check that $z \mapsto f(z)(z-a)^{-1}$ is weakly holomorphic in a neighborhood of $\overline{Y}$. By Theorem 8.6.10(2),

$$\int_{\partial Y} f(z)(z-a)^{-1}dz = 0.$$

But by the convention of the direction of integration, $\int_{\partial Y} = \int_{\partial W} - \int_{\partial V}$, therefore

$$\int_{\partial W} f(z)(z-a)^{-1}dz = \int_{\partial V} f(z)(z-a)^{-1}dz.$$

Similarly,

$$\int_{\partial W} f(z)(z-a)^{-1}dz = \int_{\partial V'} f(z)(z-a)^{-1}dz.$$

This shows that the definition of $f(a)$ does not depend on the choice of V. Now let U and U' be two sets playing the role of U in the definition. Let $f_U(a)$ and $f_{U'}(a)$ be the elements obtained by the functional calculus with U and U', respectively. Choose a bounded $V \supset \sigma(a)$ with a piecewise C^1 boundary such that $\overline{V} \subset U \cap U'$. The boundary of V can be used to compute both $f_U(a)$ and $f_{U'}(a)$. Thus $f_U(a)$ and $f_{U'}(a)$ are obtained as (8.7), hence $f_U(a) = f_{U'}(a)$.

Homomorphism. It is clear that $f + g \mapsto f(a) + g(a)$ since the A valued integral is linear. To see that $fg \mapsto f(a)g(a)$, let $f, g \in \mathrm{Hol}(a)$. Suppose that both functions are holomorphic in an open set $U \supset \sigma(a)$. Let V, W be bounded open sets with piecewise C^1 boundary such that

$$\sigma(a) \subset V \subset \overline{V} \subset W \subset \overline{W} \subset U.$$

Now,

$$f(a)g(a) = \frac{-1}{(2\pi)^{-2}} \int_{w \in \partial W} f(w)(w-a)^{-1} dw \int_{v \in \partial V} g(v)(v-a)^{-1} dv, \quad (8.8)$$

while, on the other hand,

$$fg(a) = \frac{1}{2\pi i} \int_{\partial V} f(v)g(v)(v-a)^{-1} dv.$$

For every $v \in \partial V$ we write $f(v) = \frac{1}{2\pi i} \int_{\partial W} f(w)(w-v)^{-1} dw$. Thus we get

$$fg(a) = \frac{-1}{(2\pi)^{-2}} \int_{\partial V} \left(\int_{\partial W} f(w)(w-v)^{-1} dw \right) g(v)(v-a)^{-1} dv. \quad (8.9)$$

We want to show that the right hand sides of (8.8) and (8.9) are equal. We add and subtract a term in (8.8) and rewrite $f(a)g(a)$ as a sum $I_1 + I_2$, where

$$I_1 = \frac{-1}{(2\pi)^{-2}} \int_{\partial V} \int_{\partial W} \frac{f(w)}{w-v} g(v)(v-a)^{-1} dw dv,$$

and

$$I_2 = \frac{-1}{(2\pi)^{-2}} \int_{\partial V} \int_{\partial W} f(w)[(w-a)^{-1} - (w-v)^{-1}]g(v)(v-a)^{-1} dw dv.$$

The first integral I_1 is equal to (8.9), so to finish we must show that the second integral I_2 vanishes. Using the identity

$$(w-a)^{-1} - (w-v)^{-1} = (v-a)(v-w)^{-1}(w-a)^{-1},$$

and applying a Fubini type theorem for Banach space valued functions, we rewrite $-(2\pi)^2 I_2$ as

$$\int_{\partial V} \int_{\partial W} f(w)(v-a)(v-w)^{-1}(w-a)^{-1}g(v)(v-a)^{-1} dw dv$$

$$= \int_{\partial W} f(w)(w-a)^{-1} \left(\int_{\partial V} \frac{g(v)}{v-w} dv \right) dw.$$

For every fixed $w \in \partial W$, the inner integral is the integral of a function analytic in a neighborhood of $\overline{V}$, since $\overline{V} \subseteq W$. Hence it is zero by Cauchy's theorem, and so is I_2, as we wanted to show. This establishes the equality $f(a)g(a) = fg(a)$, and completes the proof of the multiplicativity of the functional calculus.

Continuity. Crashing through with the triangle inequality for A-valued integrals, it is easy to see that if $\overline{V} \subset U$ and $f_n \to f$ uniformly on ∂V, then

$$2\pi i f_n(a) = \int_{\partial V} f_n(z)(z-a)^{-1}dz \to \int_{\partial V} f(z)(z-a)^{-1}dz = 2\pi i f(a).$$

Naturality. Let ζ be the entire function $\zeta(z) = z$. Because the functional calculus is a continuous homomorphism, it suffices to prove that $\zeta(a) = a$ (explain why). For the same price, we shall show that $\zeta^n(a) = a^n$ for all $n \in \mathbb{N}$. Since ζ is entire, we may define

$$\zeta^n(a) = \frac{1}{2\pi i} \int_C z^n(z-a)^{-1}dz$$

where C is a circle with a radius bigger than the norm of a. But then we have

$$\frac{1}{2\pi i} \int_C z^n(z-a)^{-1}dz = \frac{1}{2\pi i} \int_C z^{n-1}(1-a/z)^{-1}dz$$

$$= \frac{1}{2\pi i} \sum_{k=0}^{\infty} \int_C a^k z^{n-k-1}dz = a^n,$$

because $\int_C a^k z^{n-k-1}dz$ is equal to $a^k \int_C z^{n-k-1}dz$, which is equal to $a^n 2\pi i$ if $k = n$ and to 0 otherwise.

Spectral mapping theorem. The reader will be able to prove this after acquiring additional tools; see Exercise 9.6.9. $\qquad \square$

    The holomorphic functional calculus is sometimes attributed to F. Riesz; the above presentation was influenced by that in Riesz and Sz.-Nagy's book on functional analysis [11]. A somewhat more recent treatment of the functional calculus with an eye towards applications to operator theory can be found in [23]. A few applications appear in the exercises.

8.7 Additional exercises

Exercise 8.7.1. Prove that every Banach algebra A is isometric to a closed subalgebra of $B(X)$ for some Banach algebra X.

Exercise 8.7.2. True or false: an element a in a Banach algebra with unit is invertible if and only if it is not contained in any maximal ideal.

Exercise 8.7.3. Read Section 8.2.

1. For an element in a unital Banach algebra A define $\exp(a) = \sum_{n=0}^{\infty} \frac{1}{n!}a^n$. Prove that if $a, b \in A$ commute (i.e., $ab = ba$) then $\exp(a+b) = \exp(a)\exp(b)$.

2. Let a be an element in a Banach algebra. Prove that

$$\lim_{n\to\infty} \left(I + \frac{1}{n}a\right)^n = \exp(a).$$

Exercise 8.7.4. Let T be the operator on $L^2[0,1]$ given by $Tf(x) = xf(x)$. Find the spectrum of T. Prove that T has no eigenvalues.

Exercise 8.7.5. Let $a = (a_n)_{n=1}^\infty \in \ell^\infty(\mathbb{N})$. Let $\{e_n\}_{n=1}^\infty$ be an orthonormal basis of a Hilbert space H.

1. Show that there exists a bounded operator $A \in B(H)$ satisfying $Ae_n = a_n e_{n+1}$ for all $n = 1, 2, \ldots$. Such an operator is called a **unilateral weighted shift**.

2. Compute $\|A\|$.

3. Show that for every $\lambda \in \mathbb{T}$, there is a unitary U on H such that $UAU^{-1} = \lambda A$.

4. Prove that the spectrum of A is the union of the origin with a family of concentric circles about the origin.

5. Compute $\sigma(A)$ for the case where $a = (1, 1, 1, \ldots,)$.

6. Compute $\sigma(A)$ for the case where $a = (1, 1/2, 1/3, \ldots, 1/n, \ldots)$. (**Hint:** use the spectral radius theorem.)

Exercise 8.7.6 (The Volterra operator). Let $V : C([0,1]) \to C([0,1])$ be given by

$$(Vf)(x) = \int_0^x f(t)dt.$$

This operator is called the **Volterra operator**.

1. Prove that V is compact.

2. Prove directly that V has no eigenvalues.

3. Find (explicitly) a closed nontrivial invariant subspace for V.

4. Prove that $\|V\| = 1$ and that $\|V^n\|^{1/n} \xrightarrow{n\to\infty} 0$.

5. Find the spectrum of V, and deduce again that V has no eigenvalues.

Exercise 8.7.7. Let $A = C^{(n)}[0,1]$ be endowed with the norm $\|f\| = \sum_{k=0}^n \sup_{t\in[0,1]} |f^{(k)}(t)|$.

1. For what n is A a unital commutative Banach algebra with a normalized unit? For those n, what are the maximal ideals?

2. For which n can A be re-normed (by an equivalent norm) to be a Banach algebra?

Exercise 8.7.8. Let A be a unital Banach algebra.

1. Let $a, b \in A$. Show that $1 - ab$ is invertible if and only if $1 - ba$ is invertible. (**Hint:** Consider the formal Neumann series $\sum(ab)^n$ and $\sum(ba)^n$. How are they (formally) related?)

2. Deduce that $\sigma(ab) \setminus \{0\} = \sigma(ba) \setminus \{0\}$ for all a, b in A.

3. Give an example where $\sigma(ab) \neq \sigma(ba)$.

Exercise 8.7.9. Let X be a Banach space, and let $T \in B(X)$. Recall that the **point spectrum** of T is defined as $\sigma_p(T) = \{\lambda : \lambda I - T \text{ is not injective }\}$. The **continuous spectrum** of T is defined to be

$$\sigma_c(T) = \left\{\lambda : \lambda I - T \text{ is injective, and } \operatorname{Im}(\lambda I - T) \neq \overline{\operatorname{Im}(\lambda I - T)} = X\right\}.$$

The **residual spectrum** of T is defined to be

$$\sigma_r(T) = \left\{\lambda : \lambda I - T \text{ is injective, and } \overline{\operatorname{Im}(\lambda I - T)} \neq X\right\}.$$

1. Explain why $\sigma(T) = \sigma_p(T) \cup \sigma_c(T) \cup \sigma_r(T)$, and why this requires X to be Banach.

2. In previous exercises where you were asked to find the spectrum, find, if you can, the point, continuous, and residual spectra.

3. Prove that $\sigma_r(T) \subseteq \sigma_p(T^*) \subseteq \sigma_p(T) \cup \sigma_r(T)$.

Exercise 8.7.10. Let X be a Banach space, $T \in B(X)$ and $f : [0, \infty) \to X$ a continuous function. Prove that the non-homogeneous differential equation

$$\begin{cases} \frac{dx}{dt}(t) = Ax(t) + f(t), & t \geq 0 \\ x(0) = x_0 \end{cases}$$

has a unique solution, that is given by

$$x(t) = \exp(tA)x_0 + \int_0^t \exp\left[(t-s)A\right] f(s)ds.$$

Exercise 8.7.11. Let $\log z$ denote the principal branch of the logarithm defined on the slit plane $\mathbb{C} \setminus (-\infty, 0]$. Let a be an element in a unital Banach algebra A such that $\sigma_A(a) \subseteq \mathbb{C} \setminus (-\infty, 0]$.

1. Prove that $\exp(\log(a)) = a$.

2. Show that if $\|1 - a\| < 1$ then there exists $b \in A$ such that $a = \exp(b)$.

3. Deduce that the connected component of the identity in $G(A)$ is equal to $\exp(A) = \{\exp(b) : b \in A\}$.

Exercise 8.7.12. Suppose that the spectrum of an element a in a unital Banach space A is contained in the open left half plane. Prove that

$$(z - a)^{-1} = \int_0^\infty e^{-zt} \exp(ta)dt$$

for all z with $\operatorname{Re} z \geq 0$, and that

$$e^{ta} = \lim_{R \to \infty} \frac{1}{2\pi i} \int_{-iR}^{iR} e^{zt}(z - a)^{-1}dz$$

for all $t > 0$.

Exercise 8.7.13. Let A be a unital Banach algebra. Recall that an **idempotent** is an element $e \in A$ such that $e^2 = e$; an idempotent e is **nontrivial** if $e \neq 0, 1$.

1. Show that A contains a nontrivial idempotent if and only if A contains an element a such that $\sigma(a)$ is disconnected.

2. Show that if a bounded operator T on a Banach space X has a disconnected spectrum, then T has a nontrivial invariant subspace.

3. Use the above to prove that an operator on a complex n-dimensional vector space that has n distinct eigenvalues is diagonalizable.

Exercise 8.7.14. Let T be a bounded operator on a Banach space X. A compact set $L \subset \mathbb{C}$ is said to be a K-*spectral set* for T if $\sigma(T) \subseteq L$ and if

$$\|r(T)\| \leq K \sup_{z \in L} |r(z)| \cdot$$

for every r in the algebra $\operatorname{Rat}(L)$ of rational functions with no poles in L.

1. Prove that for every compact set L that contains an open neighborhood of $\sigma(T)$ there exists K such that L is a K-spectral set for T.

2. Conclude that if T is a contraction (that is, $\|T\| \leq 1$) then for all $\epsilon > 0$, the closed disc of radius $1 + \epsilon$ is a K-spectral set for T, for some K.

3. Is the spectrum $\sigma(T)$ a K-spectral set for T?

4. (Difficult. Requires a different approach than the above.) Prove that if T is a contraction on a Hilbert space then the closed unit disc is a K-spectral set for T, for $K = 1$.

Exercise 8.7.15. Use the above exercise together with Runge's theorem to give an alternative proof that the functional calculus is a homomorphism.

Chapter 9

Commutative Banach algebras

This chapter is dedicated to the theory of commutative Banach algebras, also known as *Gelfand theory* (see [20]). We begin, however, with a short section containing basic results that hold for not-necessarily-commutative Banach algebras.

9.1 Ideals and quotients

Recall that an ***ideal*** in a Banach algebra A is a subspace $I \subseteq A$ such that $ax, xa \in I$ for all $x \in I$ and $a \in A$. If I is an ideal in A, then $\bar{I}$ (the closure of I) is also an ideal. In case that A is unital and I is proper, then $\bar{I}$ is also proper; indeed, if I is proper then every $b \in I$ is not invertible, hence $\|1 - b\| \geq 1$ by Proposition 8.3.2. The closure of a proper ideal in a nonunital Banach algebra need not be a proper ideal.

Proposition 9.1.1. *Let A be a Banach algebra, and let I be a closed ideal in A. On the quotient space A/I the following product*

$$(a + I)(b + I) = ab + I, \quad a, b \in I,$$

is well defined, and it makes A/I into a Banach algebra. If A is unital, then so is A/I.

Proof. We know by Theorem 1.4.4 that A/I is a Banach space with respect to the norm

$$\|a + I\| = \inf_{b \in I} \|a - b\|.$$

It is easy to see (and probably similar to something that the reader encountered in a course on ring theory) that multiplication is well defined. To see that this norm is sub-multiplicative, we calculate:

$$\|a + I\|\|b + I\| \geq \inf_{y,z \in I} \|(a - y)(b - z)\|$$
$$= \inf_{y,z \in I} \|ab - az - yb + yz\|$$
$$\geq \inf_{w \in I} \|ab - w\| = \|(a + I)(b + I)\|.$$

DOI: 10.1201/9781003297864-9

Now assume that A is unital[1]. Then it is clear that $1 + I$ is the unit of A/I. Since $0 \in I$, we have $\|1 + I\| \le \|1 + 0\| = 1$. On the other hand, every $z \in I$ is not invertible, so $\|1 + z\| \ge 1$ for every $z \in I$. Therefore $\|1 + I\| = 1$, as required. $\square$

Proposition 9.1.2. *Let A and B be Banach algebras. Let $\varphi : A \to B$ be a bounded homomorphism, and let us write $K = \ker \varphi$. Let $\pi : A \to A/K$ be the quotient map. Then there exists a unique bounded homomorphism $\overline{\varphi} : A/K \to B$ that satisfies $\overline{\varphi} \circ \pi = \varphi$ and $\|\overline{\varphi}\| = \|\varphi\|$. If φ is surjective then $\overline{\varphi}$ is an isomorphism.*

Proof. This result was stated in the context of bounded linear maps between Banach spaces (see Theorem 4.3.1). It remains just to check that when φ is multiplicative, $\overline{\varphi}$ is too. $\square$

Proposition 9.1.3. *In a unital Banach algebra, every proper ideal is contained in a maximal ideal, and every maximal ideal is closed.*

Proof. Containment of every proper ideal in a maximal ideal is a standard application of Zorn's lemma, using the fact that an ideal is proper if and only if it does not contain 1. Now, if I is a maximal ideal in a unital Banach algebra, then $\overline{I}$ is also a proper ideal, whence $I = \overline{I}$ by maximality. $\square$

9.2   The maximal ideal space and the Gelfand transform

Throughout this section, A will denote a commutative and unital Banach algebra. In such an algebra, maximal ideals are tightly related to the notion of invertibility.

Proposition 9.2.1. *An element $a \in A$ is invertible if and only if it is not contained in any maximal ideal.*

Proof. If $a \in G(A)$ then $1 = a^{-1}a$ belongs to every ideal that contains a, so a is not contained in any proper ideal (this implication is true in any unital algebra). Conversely, if a is not invertible, we consider the set

$$Aa = \{ba : b \in A\}.$$

Using commutativity and the existence of a unit, one checks that Aa is an ideal that contains a. Since $a \notin G(A)$, the unit 1 does not belong to the ideal Aa, so it is proper. By Proposition 9.1.3 there is a maximal ideal I such that $a \in Aa \subseteq I$. $\square$

---

[1] Recall that "unital" means that A has unit 1 and also that $\|1\| = 1$.

In a noncommutative unital Banach algebra it is possible for an element to be non-invertible while still not belonging to any maximal ideal (see Exercise 8.7.2).

We introduce temporarily the following notation:

$$\mathcal{M}(A) = \{I \subset A : I \text{ is a maximal ideal in } A\}.$$

Definition 9.2.2. The **spectrum** of A, denoted $sp(A)$, is defined to be the set of all nonzero homomorphisms from A to $\mathbb{C}$ (that is, all nonzero multiplicative linear functionals on A). The elements of the spectrum are also referred to as **characters**.

If $\varphi \in sp(A)$, then $\varphi(1)$ is an identity for $\varphi(A)$, so the fact that $\varphi \neq 0$ forces $\varphi(1) = 1$. This will be used repeatedly below.

Theorem 9.2.3. *There exists a bijective correspondence between $sp(A)$ and $\mathcal{M}(A)$, given by*

$$\varphi \mapsto \ker \varphi. \tag{9.1}$$

Proof. If $\varphi \in sp(A)$, then $\ker \varphi$ is an ideal in A. Now, $\varphi \neq 0$, so by Lemma 1.4.1, $\ker \varphi$ is a maximal subspace, and therefore $\ker \varphi$ must be a maximal ideal. This shows that the map (9.1) sends $sp(A)$ into $\mathcal{M}(A)$.

If I is a maximal ideal in A, then I is closed, so A/I is a unital Banach algebra. If J is a proper ideal in A/I then $\pi^{-1}(J)$ is a proper ideal in A containing I, so by maximality of I the quotient A/I has no proper ideals but the zero ideal. Therefore by Proposition 9.2.1 every nonzero element in A/I is invertible, so $A/I \cong \mathbb{C}$, by the Gelfand–Mazur theorem (Theorem 8.3.8). I is then clearly the kernel of the multiplicative functional

$$A \xrightarrow{\pi} A/I \xrightarrow{\cong} \mathbb{C}.$$

Thus the map (9.1) is onto $\mathcal{M}(A)$.

Finally, if $\varphi, \psi \in sp(A)$ and $\ker \varphi = \ker \psi$, then for all $a \in A$

$$a - \varphi(a)1 \in \ker \varphi = \ker \psi,$$

and this implies $\psi(a) - \psi(\varphi(a)1) = 0$ or $\psi(a) = \varphi(a)$. □

Because of the above identification, the spectrum of A is commonly referred to as the **maximal ideal space** of A.

Proposition 9.2.4. *If $\varphi \in sp(A)$ then φ is bounded and $\|\varphi\| = 1$.*

Proof. Since $\ker \varphi$ is maximal, it is closed. By Proposition 2.1.14, the closedness of the kernel $\ker \varphi$ of the linear functional φ implies that it is bounded.

Since $\varphi(1) = 1$ it follows that $\|\varphi\| \geq 1$. Now if $\|\varphi\| > 1$, then there exists $a \in A$ such that $\|a\| < 1$ and $|\varphi(a)| > 1$, so that $a^n \to 0$, while $|\varphi(a^n)| \geq 1$ for all n. This contradicts boundedness, so we conclude that $\|\varphi\| = 1$. □

We now wish to endow the spectrum of A with a topology. As a consequence of the previous result, $sp(A) \subseteq A_1^*$. This allows us to endow $sp(A)$ with the weak-$*$ topology on A_1^*. The restriction of the weak-$*$ topology to $sp(A)$ is sometimes called the **Gelfand topology**. Recall that by Alaoglu's theorem (Theorem 6.5.1), A_1^* is compact in the weak-$*$ topology. Clearly, $sp(A)$ is weak-$*$ closed in A_1^*, hence compact. We summarize this as follows.

Proposition 9.2.5. *The spectrum $sp(A)$ of a commutative unital Banach algebra A, when endowed with the Gelfand topology, is a compact Hausdorff space.*

From now on we shall let Δ denote the space $sp(A)$ endowed with the Gelfand topology. Recall that for every $a \in A$, we let $\hat{a} \in A^{**}$ denote the linear functional on A^* given by

$$\hat{a}(f) = f(a).$$

We write $\hat{a}$ also for the restriction $\hat{a}\big|_{\Delta}$, and then $\hat{a}$ can be considered as a continuous complex valued function on Δ.

Definition 9.2.6. The map $\Gamma : A \to C(\Delta)$ given by $\Gamma(a) = \hat{a}$ is called **the Gelfand transform**. Sometimes, the element $\hat{a}$ is referred to as **the Gelfand transform of** a.

Theorem 9.2.7. *The Gelfand transform $\Gamma : A \to C(\Delta)$ is a unital homomorphism with the following properties:*

1. $\Gamma(A)$ *is a unital subalgebra of $C(\Delta)$ that separates points.*

2. $\|\hat{a}\|_\infty \leq \|a\|$ *for all $a \in A$.*

3. a *is invertible in A if and only if $\hat{a}$ never vanishes.*

4. *For every $a \in A$,*

$$\sigma_A(a) = \{\hat{a}(\varphi) : \varphi \in \Delta\} = \sigma_{C(\Delta)}(\hat{a}).$$

5. $r(a) = \|\hat{a}\|_\infty.$

Proof. Since every character is unital, $\Gamma(1) = \hat{1}$ is the constant function 1. Clearly, Γ is linear. It is multiplicative because

$$\Gamma(a)\Gamma(b)(\varphi) = \left(\widehat{ab}\right)(\varphi) = \hat{a}(\varphi)\hat{b}(\varphi) = \varphi(a)\varphi(b) = \varphi(ab) = \Gamma(ab)(\varphi).$$

Therefore Γ is an algebra homomorphism, so $\Gamma(A)$ is an algebra, and, by the definition of equality in $\Delta \subseteq A^*$, the algebra $\Gamma(A)$ separates points. This establishes (1).

For every $\varphi \in \Delta$, $|\hat{a}(\varphi)| = |\varphi(a)| \leq \|\varphi\|\|a\| = \|a\|$, and this implies (2).

If a is invertible then so is $\hat{a}$ because Γ is a unital homomorphism (so $\widehat{aa^{-1}} = \widehat{aa^{-1}} = \hat{1} = 1$). Conversely, if a is not invertible then a is contained in some maximal ideal $\ker\psi$, for some $\psi \in \Delta$. We see that $\hat{a}(\psi) = 0$, so $\hat{a}$ is not invertible. Thus we have (3). From this (4) follows because

$$a - \lambda \text{ is not invertible} \iff \hat{a} - \lambda \text{ is not invertible.}$$

Finally, (5) follows immediately from (4). $\qquad\qquad\qquad\qquad\qquad\qquad\square$

9.3   Examples of Gelfand transforms

Example 9.3.1. Let $a = \left(\begin{smallmatrix}0&1\\0&0\end{smallmatrix}\right)$, and let A be the algebra generated by the identity matrix I_2 and by a. Every character of A is determined uniquely by its value on a. Since $a^2 = 0$, the only possible value that a character can attain on a is 0. Therefore, $\Delta = \{\varphi_0\}$, where φ_0 is the unique character determined by $\varphi_0(I_2) = 1$ and $\varphi_0(a) = 0$. The Gelfand transform takes a to $\hat{a} \equiv 0$, and so it is not injective.

Example 9.3.2. Let X be a compact Hausdorff space and let $A = C(X)$. We shall show that $\Delta \cong X$, and that the Gelfand transform is an isometric isomorphism (or the identity, if the proper identifications are made).

Every $x \in X$ induces a nonzero multiplicative linear functional $\delta_x : f \mapsto f(x)$. The map $x \mapsto \delta_x$ is a continuous map, and it is injective because, by Tietze's extension theorem[2], for every pair of points $x \neq y$ in X there exists $f \in C(X)$ such that $f(x) \neq f(y)$. Since both spaces are compact and Hausdorff, if we show that this continuous and injective map is also onto Δ, then it will follow that it is a homeomorphism.

Let J be an ideal in $C(X)$ that is not contained in any ideal of the form $\ker \delta_x$. We will show that this forces J to be $C(X)$. For every $x \in X$, there exists $f \in J$ such that $|f(x)| > 0$, and hence $|f| > 0$ in a neighborhood of x. By compactness, we can find $f_1, \ldots, f_n \in J$ such that

$$0 < g = \sum |f_i|^2 = \sum \overline{f_i} f_i \in J,$$

because J is an ideal. But then g is an invertible element in the ideal J, which forces $J = C(X)$. It follows that the only maximal ideals are of the form $\ker \delta_x$, and thus the only characters are point evaluations, as claimed.

Let $h : X \to \Delta$ be given by $h(x) = \delta_x$. Then $\Gamma(f)(\delta_x) = \delta_x(f) = f(x)$ or

$$\Gamma(f) = f \circ h^{-1},$$

which is an isometric isomorphism. If we identify every point x with $\delta_x \in \Delta$, then the Gelfand transform is the identity.

---

[2]See Theorem 20.4 in [13].

Example 9.3.3. Let $A = \ell^1(\mathbb{Z})$. For $n \in \mathbb{Z}$, let e_n denote the "standard basis vector" with 1 in the nth slot and 0 elsewhere. An easy calculation verifies the identity $e_m * e_n = e_{m+n}$ for all $m, n \in \mathbb{Z}$, which implies, in particular, that $e_n^{-1} = e_{-n}$ for all n. It follows that A is the closure of the algebra generated by the unit $1 = e_0$, by e_1, and by $e_{-1} = e_1^{-1}$. Therefore, if $\varphi \in \Delta$, then φ is completely determined by $\lambda = \varphi(e_1)$. Note that $\lambda \neq 0$ because e_1 is invertible. Now, φ is multiplicative and contractive, so $|\lambda^n| = |\varphi(e_n)| \leq 1$ for all $n \in \mathbb{Z}$, hence $\lambda = e^{i\theta} \in \mathbb{T}$.

Conversely, let us now verify that every $e^{i\theta} \in \mathbb{T}$ gives rise to a character by way of

$$\varphi_\theta : a = (a_n)_{n \in \mathbb{Z}} \mapsto \sum_{n \in \mathbb{Z}} a_n e^{in\theta}.$$

Indeed, the series on the right hand side converges absolutely, and if two such series are multiplied we see that

$$\left(\sum_{n \in \mathbb{Z}} a_n e^{in\theta} \right) \left(\sum_{n \in \mathbb{Z}} b_n e^{in\theta} \right) = \sum_{n \in \mathbb{Z}} c_n e^{in\theta},$$

where $c_n = \sum_{k \in \mathbb{Z}} a_k b_{n-k} = (a * b)_n$. So we have $\varphi_\theta(a)\varphi_\theta(b) = \varphi_\theta(a * b)$, and this completes the proof that every $e^{i\theta}$ gives rise to a character.

We therefore find that Δ is in a bijective correspondence with $\mathbb{T}$, given by

$$\Delta \ni \varphi \longleftrightarrow \varphi(e_1) = e^{i\theta} \in \mathbb{T}.$$

The map $\varphi \mapsto \varphi(e_1)$ is continuous by definition of the weak-$*$ topology. As a bijective continuous map between compact Hausdorff spaces, it is a homeomorphism.

The Gelfand transform maps $\ell^1(\mathbb{Z})$ onto the algebra $W \subseteq C(\mathbb{T})$ which consists of all (continuous) functions on $\mathbb{T}$ that have absolutely convergent Fourier series. This algebra W is called *the Wiener algebra*. The Gelfand transform in this case maps a summable sequence $a \in \ell^1(\mathbb{Z})$ to the function $f(e^{i\theta}) = \sum a_n e^{in\theta}$; it is the inverse of the "Fourier transform" that maps a function in the Wiener algebra to its sequence of coefficients. This operator in itself is sometimes referred to as a Fourier transform, and for legitimate reasons which we shall not go into now.

We note that the Gelfand transform is, in this case, injective, but not surjective; if it was then its image would have been all of $C(\mathbb{T})$, but this is not true[3]. The Gelfand transform is also not isometric, because its image cannot be closed since otherwise it would be surjective (why?). It is perhaps in such situations that the Gelfand transform is most useful.

---

[3]We have seen in Theorem 4.4.5 that there exists $f \in C(\mathbb{T})$ with a Fourier series that doesn't even converge pointwise, let alone uniformly.

Example 9.3.4. We can turn the tables, and start a discussion from the Wiener algebra

$$W = \left\{ f : \mathbb{T} \to \mathbb{C} : f(e^{i\theta}) = \sum a_n e^{in\theta} \text{ for some } a \in \ell^1(\mathbb{Z}) \right\}.$$

This is a subalgebra of $C(\mathbb{T})$ closed under pointwise multiplication. On W we introduce a norm $\|f\|_W = \sum |\hat{f}(n)|$. In other words, we simply give W the norm it inherits as the image of $\ell^1(\mathbb{Z})$ under the Gelfand transform: $\|\Gamma(a)\|_W = \|a\|_{\ell^1}$. We see that W is a Banach algebra; in fact, W is essentially $\ell^1(\mathbb{Z})$ in disguise.

Evaluation at a point on $\mathbb{T}$ is then clearly a well defined multiplicative linear functional on W. Conversely, it is easy to show that every character is a point evaluation at some point in $\mathbb{T}$. Now, if we let $\hat{f} = (\hat{f}(n))_{n \in \mathbb{Z}}$ denote the sequence of Fourier coefficients of a function $f \in W$, then there is a basic identity in harmonic analysis which reads

$$\widehat{fg}(n) = \hat{f} * \hat{g}(n).$$

From this, we see that

$$\sum \hat{f} * \hat{g}(n) e^{in\theta} = \sum \widehat{fg}(n) e^{in\theta}$$
$$= fg(\theta) = f(\theta)g(\theta) = \left(\sum \hat{f}(n) e^{in\theta} \right) \left(\sum \hat{g}(n) e^{in\theta} \right).$$

This is another way to see that $(a_n)_{n \in \mathbb{Z}} \mapsto \sum_{n \in \mathbb{Z}} a_n e^{in\theta}$ is multiplicative on $\ell^1(\mathbb{Z})$.

Example 9.3.5. Consider the disc algebra $A(\mathbb{D})$. Since $A(\mathbb{D})$ is the closure of the algebra generated by 1 and z, every character φ is determined by its value $\varphi(z)$ at z. Since characters are contractive, $\varphi(z) \in \overline{\mathbb{D}}$. Conversely, as $A(\mathbb{D}) \subseteq C(\overline{\mathbb{D}})$, every point in $\overline{\mathbb{D}}$ gives rise to a multiplicative evaluation functional on $C(\overline{\mathbb{D}})$, and this restricts to a character of $A(\mathbb{D})$. Thus $sp(A(\mathbb{D})) \cong \overline{\mathbb{D}}$ (it is not hard to check that this is a homeomorphism). The Gelfand transform then has a couple of interpretations. If we think of $A(\mathbb{D})$ as a subalgebra of $C(\overline{\mathbb{D}})$, then Γ is simply the inclusion into $C(\overline{\mathbb{D}})$. If we think of $A(\mathbb{D})$ as a subalgebra of $C(\mathbb{T})$, then Γ is the mapping that sends every $f \in A(\mathbb{D}) \subseteq C(\mathbb{T})$ to its unique analytic extension $\hat{f} \in C(\overline{\mathbb{D}}) \cap \mathcal{O}(\mathbb{D})$.

9.4 Applications: Wiener's $1/f$ theorem and the baby corona theorem

We shall now see a quick and nontrivial application of the above machinery to a result in harmonic analysis called *Wiener's $1/f$ theorem* or *Wiener's lemma*.

The proof we give is due to Gelfand and was one of the early successes of his theory of commutative Banach algebras [20]. The reader is invited to compare it to Wiener's original proof (see Lemma II_e in [50]).

Theorem 9.4.1 (Wiener's theorem). *Let f be a continuous function on $\mathbb{T}$ which does not vanish on $\mathbb{T}$, and suppose that the Fourier series of f is absolutely convergent. Then the Fourier series of $\frac{1}{f}$ is also absolutely convergent.*

Proof. Consider the Wiener algebra W introduced in Example 9.3.4. We saw that $sp(W) = \mathbb{T}$, where every point in $\mathbb{T}$ corresponds to a unique character given by point evaluation. Now if f does not vanish on $\mathbb{T}$, then f is not annihilated by any character. By Theorem 9.2.7 (3), f is invertible in W, that is, the inverse of f belongs to W. □

Here is another elegant illustration of the power of the theory.

Theorem 9.4.2 (Baby version of the corona theorem). *Let $f_1, \ldots, f_n \in A(\mathbb{D})$ be such that there is some $\delta > 0$, such that*

$$\sum |f_n(z)| > \delta, \quad \text{for all } z \in \mathbb{D}. \tag{9.2}$$

Then there exist $g_1, \ldots, g_n \in A(\mathbb{D})$ such that

$$f_1 g_1 + \cdots + f_n g_n = 1. \tag{9.3}$$

Proof. Consider

$$J = \{f_1 g_1 + \cdots + f_n g_n : g_1, \ldots, g_n \in A(\mathbb{D})\}.$$

Then J is an ideal in $A(\mathbb{D})$ that contains $f_1, \ldots, f_n$. By Example 9.3.5, every maximal ideal of $A(\mathbb{D})$ corresponds to point evaluation at a point of $\overline{\mathbb{D}}$. By assumption (9.2), there is no single maximal ideal that contains all the functions $f_1, \ldots, f_n$, and therefore J is not contained in any maximal ideal. But, as we have noted in Proposition 9.1.3, every proper ideal is contained in a maximal ideal. It follows that $J = A(\mathbb{D})$, and in particular, $1 \in J$ which is precisely the required conclusion. □

Remark 9.4.3. If we were to ask the same question in the setting of continuous or C^k or C^∞ functions, then it would have been trivial, since one can define

$$g_k = \frac{\overline{f_k}}{\sum_i |f_i|^2} \quad \text{for all } k = 1, \ldots, n,$$

and these functions g_k are at least as regular as the functions f_k and they satisfy $f_1 g_1 + \cdots + f_n g_n = 1$. However, if the functions $f_1, \ldots, f_n$ are assumed analytic, then it does not follow that the functions g_k constructed above are analytic. Constructively finding $g_1, \ldots, g_n \in A(\mathbb{D})$ that satisfy (9.3) given $f_1, \ldots, f_n \in A(\mathbb{D})$ is a difficult analytic task.

Let $H^\infty(\mathbb{D})$ be the algebra of all bounded holomorphic functions on the open disc $\mathbb{D}$ equipped with the supremum norm. One can pose a similar problem with $H^\infty(\mathbb{D})$ instead of $A(\mathbb{D})$. In this case, the problem of finding $g_1, \ldots, g_n \in H^\infty(\mathbb{D})$ corresponding to $f_1, \ldots, f_n \in H^\infty(\mathbb{D})$ satisfying (9.2) becomes substantially more difficult. The positive solution to this problem is called *the corona theorem*, which was proved by Carleson around 1962 [9]. The reason for this delightful name is as follows. One can show that the corona theorem (that is, the statement of Theorem 9.4.2 with $H^\infty(\mathbb{D})$ instead of $A(\mathbb{D})$) is equivalent to the fact that the point evaluations $f \mapsto f(z)$, for $z \in \mathbb{D}$, are dense in the maximal ideal space of $H^\infty(\mathbb{D})$ (see Exercise 9.6.6). The corona theorem answers the question of whether the maximal ideal space of $H^\infty(\mathbb{D})$ has a *corona* (latin for *crown*) around the unit disc reminiscent of the magnificent corona that surrounds the sun and is visible during an eclipse[4]. The corona theorem says that there is no corona — the unit disc fills up the interior of the maximal ideal space. We labeled Theorem 9.4.2 as a *baby version* of the corona theorem, to make clear the distinction between the celebrated, difficult result and the easy and elegant application of Gelfand theory that we presented here.

9.5   Gelfand theory for nonunital Banach algebras

9.5.1   The notion of spectrum in the nonunital setting

Recall that for a nonunital algebra A we define its unitization $\widetilde{A} = A \oplus_1 \mathbb{C}$, with product
$$(a, \lambda)(b, \mu) = (ab + \mu a + \lambda b, \lambda\mu),$$

norm $\|(a, \lambda)\| = \|a\| + |\lambda|$ and unit $1 = 1_{\widetilde{A}} = (0, 1)$. In Exercise 8.1.7 you checked that $\widetilde{A}$ is a Banach algebra.

Definition 9.5.1. Let A be a nonunital Banach algebra and let $a \in A$. The *spectrum* of a (with respect to A) is defined to be

$$\sigma(a) = \sigma_A(a) := \sigma_{\widetilde{A}}(a).$$

The spectrum of every element in a nonunital Banach algebra contains 0 (one way to justify this is to note that A is an ideal in $\widetilde{A}$). One can define the *spectral radius* of $a \in A$ as $r(a) = \sup\{|\lambda| : \lambda \in \sigma(a)\}$. It then follows that the spectral radius formula $r(a) = \lim \|a^n\|^{1/n}$ holds, because it holds in the unital Banach algebra $\widetilde{A}$.

Remark 9.5.2. The notion of spectrum that we use for a nonunital Banach algebras does not depend on the norm that we put on $\widetilde{A}$ (see Remark 4.3.3).

---

[4]Google it.

Definition 9.5.3. Let A be a nonunital Banach algebra. The **spectrum** of A is defined to be

$$sp(A) = \{\varphi : A \to \mathbb{C} : \varphi \text{ is a nonzero homomorphism}\}.$$

The elements of the spectrum are called **characters**.

Example 9.5.4. Let A be the Banach algebra generated by $\left(\begin{smallmatrix} 0 & 1 \\ 0 & 0 \end{smallmatrix}\right)$ in $M_2(\mathbb{C})$. Then it is not hard to see that there are no nonzero characters, hence $sp(A) = \emptyset$ (compare with Example 9.3.1).

To find the relationship between the spectrum of a nonunital Banach algebra and its unitization, we need the following lemma.

Lemma 9.5.5. *Let A be a nonunital algebra and let $\widetilde{A} = A \oplus \mathbb{C}$ be its unitization. Every homomorphism $\varphi : A \to B$ from A into a unital algebra B extends to a unique unital homomorphism $\tilde{\varphi} : \widetilde{A} \to B$.*

Proof. Define $\tilde{\varphi}((a, \lambda)) = \varphi(a) + \lambda 1_B$. It is evidently linear and unital. It is also multiplicative because

$$\begin{aligned}
\tilde{\varphi}((a, \lambda)(b, \mu)) &= \tilde{\varphi}((ab + \mu a + \lambda b, \lambda \mu)) \\
&= \varphi(ab + \mu a + \lambda b) + \lambda \mu 1_B \\
&= \varphi(a)\varphi(b) + \mu\varphi(a) + \lambda\varphi(b) + \lambda \mu 1_B \\
&= (\varphi(a) + \lambda 1_B)(\varphi(b) + \mu 1_B) \\
&= \tilde{\varphi}((a, \lambda))\tilde{\varphi}((b, \mu)).
\end{aligned}$$

Uniqueness is straightforward. ∎

In particular, if A is a commutative nonunital Banach algebra, every element of $sp(A)$ extends to an element of $sp(\widetilde{A})$. But A has another homomorphism that extends to a character of $\widetilde{A}$: the zero homomorphism $0 : A \to \mathbb{C}$. It extends to a character $\varphi_0 : \widetilde{A} \to \mathbb{C}$ given by $\varphi_0((a, \lambda)) = \lambda$. Conversely, every character of $\widetilde{A}$ restricts to a homomorphism of A, and the only one that restricts to the zero homomorphism of A is φ_0. To sum up:

Proposition 9.5.6. *If A is a nonunital commutative Banach algebra, then*

$$sp(\widetilde{A}) = \{\tilde{\varphi} : \varphi \in sp(A)\} \cup \{\varphi_0\}. \tag{9.4}$$

Corollary 9.5.7. *Let A be a nonunital commutative Banach algebra. Every $\varphi \in sp(A)$ is contractive (i.e., has norm less than or equal to 1).*

Proof. φ is contractive because its extension $\tilde{\varphi}$ is, by the unital theory. ∎

For a nonunital Banach algebra A with nonempty spectrum, we set $\Delta = sp(A)$. By liberally interpreting (9.4) we write $\widetilde{\Delta} = \Delta \cup \{\varphi_0\}$. With this identification $\Delta \subseteq \widetilde{A}_1^*$, and when we give it the weak-$*$ topology it becomes a subspace of a compact Hausdorff space, hence Δ is Hausdorff. In fact, if φ_n is a net in Δ that converges weak-$*$ to f, then f is also multiplicative, hence the closure of Δ in A_1^* is contained in $\Delta \cup \{0\}$. Therefore Δ is a locally compact Hausdorff space.

Definition 9.5.8. Let A be a nonunital commutative Banach algebra, and suppose that $\Delta \neq \emptyset$. The **Gelfand transform** of A is the map $\Gamma : A \to C_b(\Delta)$ given by

$$\Gamma(a)(\varphi) = \varphi(a), \quad \varphi \in sp(A).$$

As in the unital case, we shall write $\hat{a} = \Gamma(a)$.

Theorem 9.5.9. *Let A be a nonunital commutative Banach algebra, and suppose that $\Delta \neq \emptyset$. The Gelfand transform Γ is a contractive algebra homomorphism into a subalgebra of $C_0(\Delta)$ that separates points. For every $a \in A$,*

$$\|\hat{a}\| = r(a)$$

and

$$\sigma(a) = \hat{a}(\Delta) \cup \{0\}.$$

Proof. Most of this theorem follows from the unital case, after viewing Γ as the composition

$$A \to \tilde{A} \xrightarrow{\tilde{\Gamma}} C(sp(\tilde{A})) \xrightarrow{f \mapsto f|_\Delta} C(\Delta).$$

It remains to prove that $\Gamma(A) \subseteq C_0(\Delta)$. Let $a \in A$ and $\epsilon > 0$. Then $\{\varphi \in \Delta : |\hat{a}(\varphi)| \geq \epsilon\} = \{\varphi \in \tilde{\Delta} : |\hat{a}(\varphi)| \geq \epsilon\}$ is closed in $\tilde{A}_1^*$, hence compact. $\square$

9.5.2 Gelfand theory for $L^1(\mathbb{R})$

In this section, we develop Gelfand theory for $L^1(\mathbb{R})$. The crucial step is to determine the characters, and the Fourier transform plays a key role in the analysis. Recall that for every $f \in L^1(\mathbb{R})$ we define the Fourier transform[5] of f to be

$$\hat{f}(w) = \int_{-\infty}^{\infty} f(t)e^{-iwt}dt, \quad w \in \mathbb{R}.$$

It is a basic fact in harmonic analysis that $\hat{f} \in C_0(\mathbb{R})$ for every $f \in L^1(\mathbb{R})$, in fact the map $f \mapsto \hat{f}$ (also referred to as the Fourier transform) is a bounded linear map from $L^1(\mathbb{R})$ into $C_0(\mathbb{R})$ of norm 1 (see [14, Chapter 12] for the required theory of the Fourier transform).

---

[5] The reader might be puzzled that we are using the notation $\hat{f}$ for the Fourier transform, which is the notation that we use for the Gelfand transform. At this point, however, there is no danger of confusion between the two, since we have not yet determined the spectrum of $L^1(\mathbb{R})$. In the end, we shall see that the spectrum of $L^1(\mathbb{R})$ is $\mathbb{R}$, and that the Gelfand transform can be identified with the Fourier transform.

Theorem 9.5.10. *For every $w \in \mathbb{R}$, the map $\varphi_w : L^1(\mathbb{R}) \to \mathbb{C}$ defined by*

$$\varphi_w(f) := \hat{f}(w) = \int_{-\infty}^{\infty} f(t)e^{-iwt}dt, \quad \text{for all } f \in L^1(\mathbb{R}),$$

is a character of $L^1(\mathbb{R})$. Conversely, every character φ of $L^1(\mathbb{R})$ is of the form φ_w for some $w \in \mathbb{R}$. The map $w \mapsto \varphi_w$ is a homeomorphism, hence $sp(L^1(\mathbb{R})) \cong \mathbb{R}$.

Proof. We begin by showing that φ_w is a character of $L^1(\mathbb{R})$. It is evident that φ_w is a bounded linear functional, being given as integration against a bounded function. By the convolution theorem[6]

$$\widehat{f * g} = \hat{f}\hat{g}, \quad \text{for all } f, g \in L^1(\mathbb{R}).$$

Evaluating the above identity at $w \in \mathbb{R}$, we find that $\varphi_w(f * g) = \varphi_w(f)\varphi_w(g)$, whence φ_w is multiplicative. To see that φ_w is not the zero functional, recall the reflection formula[7]

$$\hat{\hat{f}}(t) = 2\pi f(-t),$$

which holds at every $t \in \mathbb{R}$ for every $f \in C_c^2(\mathbb{R})$, i.e. every function f that is twice continuously differentiable with compact support. Given $w \in \mathbb{R}$, we take $f \in C_c^2(\mathbb{R})$ such that $f(-w) \neq 0$. One can show that $\hat{f} \in C_0(\mathbb{R}) \cap L^1(\mathbb{R})$, and the reflection formula gives

$$\varphi_w(\hat{f}) = 2\pi f(-w) \neq 0.$$

This shows that φ_w is nonzero.

Thus we have a map $w \mapsto \varphi_w$ from $\mathbb{R}$ into $sp(A)$. To see that this map is injective, let $v \neq w$, and let $f \in C_c^2(\mathbb{R})$ be such that $f(-w) \neq f(-v)$. Then as above $\hat{f} \in L^1(\mathbb{R})$ and

$$\varphi_w(\hat{f}) = 2\pi f(-w) \neq 2\pi f(-v) = \varphi_v(\hat{f}),$$

so $\varphi_w \neq \varphi_v$. We note in passing that this argument shows that Fourier transforms of L^1 functions separate the points of $\mathbb{R}$.

Next, we show that $w \mapsto \varphi_w$ is surjective. Given $\varphi \in sp(L^1(\mathbb{R}))$, we need to show that $\varphi = \varphi_w$ for some $w \in \mathbb{R}$. Since $\varphi \in (L^1(\mathbb{R}))_1^*$, by Theorem 3.2.4 there exists a function $\phi \in L^\infty(\mathbb{R})_1$ such that

$$\varphi(f) = \int f(t)\phi(t)dt, \quad \text{for all } f \in L^1(\mathbb{R}).$$

We need to show that $\phi(t) = e^{-iwt}$ for some w. Let $f \in L^1(\mathbb{R})$ be such that $\varphi(f) \neq 0$. Put $f_s(t) = f(t - s)$, and define a function $F : \mathbb{R} \to \mathbb{C}$ by

---

[6] See Theorem 12.2.12 in [14].
[7] See Theorem 12.3.5 [14].

$F(s) = \varphi(f_s) = \int f(t-s)\phi(t)dt$. Then F is continuous and bounded. For every $g \in L^1(\mathbb{R})$,

$$\int F(s)g(s)ds = \int\int f(t-s)\phi(t)g(s)dtds$$

$$= \int \left(\int f(t-s)g(s)ds\right)\phi(t)dt$$

$$= \varphi(f*g) = \varphi(f)\varphi(g)$$

$$= \varphi(f)\int g(s)\phi(s)ds.$$

Since this holds for all $g \in L^1(\mathbb{R})$, it must be that

$$\varphi(f_s) = F(s) = \varphi(f)\phi(s) \tag{9.5}$$

for almost all $s \in \mathbb{R}$. Since F is continuous and $\varphi(f) \neq 0$, we conclude that ϕ can be chosen to be continuous. Plugging f_t for f in (9.5), we obtain

$$\varphi(f)\phi(s+t) = \varphi(f_{s+t}) = \varphi((f_t)_s) = \varphi(f_t)\phi(s) = \varphi(f)\phi(t)\phi(s).$$

We conclude that ϕ is a continuous and bounded function such that

$$\phi(s+t) = \phi(s)\phi(t). \tag{9.6}$$

It is well known[8] that the only bounded and continuous solutions of the functional equation (9.6) are of the form $\phi(t) = e^{iwt}$ for some $w \in \mathbb{R}$, as required.

Up to here, we showed that the map $\mathbb{R} \ni w \mapsto \varphi_w \in sp(L^1(\mathbb{R}))$ is bijective. It remains to show that it is a homeomorphism. For this, we need to show that $w_n \to w$ in $\mathbb{R}$ if and only if $\hat{f}(w_n) \to \hat{f}(w)$ for every $f \in L^1(\mathbb{R})$. This is elementary: the forward implication follows from the fact that $\hat{f} \in C_0(\mathbb{R})$ for every $f \in L^1(\mathbb{R})$, and the converse follows from the fact that Fourier transforms of L^1 functions separate the points of $\mathbb{R}$. $\qquad \square$

Now we see that under the identification of $sp(L^1(\mathbb{R}))$ with $\mathbb{R}$, the Gelfand transform $f \mapsto \hat{f} \in C_0(sp(L^1(\mathbb{R})))$ corresponds to the Fourier transform:

$$\Gamma(f)(\varphi_w) = \varphi_w(f) = \hat{f}(w).$$

Note that this does not mean in any way that "the Fourier transform boils down to the Gelfand transform" or something like that — we needed some understanding of the Fourier transform in order to identify the Gelfand transform in this case. Moreover, there are many issues related to the Fourier transform that cannot be dealt with within $L^1(\mathbb{R})$, for example, the Fourier

---

[8]For the benefit of readers not familiar with the solution of this functional equation, we overcome the urge to provide a proof or reference and leave this as a tractable yet rewarding exercise. Enjoy!

transform on $L^2(\mathbb{R})$ is of paramount importance in analysis and its applications, and is beyond the scope of Gelfand theory.

The above discussion can be generalized to the Banach algebras $L^1(G, \mu)$ where G is a locally compact abelian group, μ is the Haar measure, and the multiplication is given by convolution. The result is that characters of the Banach algebra $L^1(G)$ correspond precisely to characters of the group, that is, homomorphisms $\gamma : G \to \mathbb{T}$, via

$$\varphi_\gamma(f) = \hat{f}(\gamma) := \int_G f(t)\gamma(t)dt.$$

The function $\gamma \mapsto \hat{f}(\gamma)$ is called the **Fourier transform** on $L^1(G)$; under the identification $\gamma \leftrightarrow \varphi_\gamma$, it coincides with the Gelfand transform of f.

9.6   Additional exercises

Exercise 9.6.1. Let X be a compact Hausdorff space. For every closed $E \subseteq X$, the set
$$J_E = \left\{ f \in C(X) : f\big|_E = 0 \right\}$$
is a closed ideal. Prove that every closed ideal in $C(X)$ has this form.

Exercise 9.6.2. Let A, B be abelian and unital and $\varphi : A \to B$ be a unital homomorphism. Show that φ induces a continuous map $\varphi^* : sp(B) \to sp(A)$ given by $\varphi^*(\rho) = \rho \circ \varphi$. Note that we did not assume that φ is bounded.

Exercise 9.6.3. Suppose that A is commutative and unital.

1. The *radical of A* is defined as follows
$$rad(A) = \{x \in A : \lim_{n \to \infty} \|x^n\|^{1/n} = 0\}.$$

   Prove that $rad(A)$ is a closed ideal.

2. If $rad(A) = \{0\}$ then A is said to be *semi-simple*. Prove that $A/rad(A)$ is semi-simple.

3. Prove that if A and B are abelian and unital, if B is semi-simple, and if $\varphi : A \to B$ is a unital homomorphism, then φ is bounded.

Exercise 9.6.4. Suppose that A is unital, and let $\varphi : A \to M_n(\mathbb{C})$ be a surjective homomorphism. Prove that φ is bounded.

Exercise 9.6.5 (Joint spectrum). Let A be a unital and commutative Banach algebra, and let Δ denote its maximal ideal space. If $a = (a_1, \ldots, a_n) \in A^n$, we define the **joint spectrum** of a (relative to A) to be the set
$$\sigma(a) = \sigma_A(a) = \left\{ \lambda \in \mathbb{C}^n : \nexists b \in A^n \text{ such that } \sum b_i(a_i - \lambda_i) = 1 \right\}.$$

1. Prove that for every $a \in A^n$, $\sigma(a) = \hat{a}(\Delta)$, where $\hat{a} : \Delta \to \mathbb{C}^n$ is the map given by $\hat{a}(\varphi) = (\hat{a}_1(\varphi), \ldots, \hat{a}_n(\varphi))$.

2. Prove that $\sigma(a)$ is a nonempty compact set.

3. Suppose that A is generated as a Banach algebra by $a_1, \ldots, a_n$ and 1. Prove that $sp(A) \cong \sigma(a)$. Note in particular the case $n = 1$.

4. Let K be a compact subset in $\mathbb{C}^n$. Let $z_1, \ldots, z_n$ be the coordinate functions on K, and let $P(K)$ be the norm closure of the polynomials $\mathbb{C}[z_1, \ldots, z_n]$ in $C(K)$. Prove that the spectrum of $P(K)$ is homeomorphic to the **polynomially convex hull** of K, which is the following set

$$\text{Phull}(K) = \{z \in \mathbb{C}^n : |p(z)| \leq \sup_{w \in K} |p(w)| \text{ for all } p \in \mathbb{C}[z_1, \ldots, z_n]\}.$$

Exercise 9.6.6. Let $H^\infty(\mathbb{D})$ be the space of all bounded analytic functions on the unit disc $\mathbb{D}$ equipped with the supremum norm.

1. Prove that $H^\infty(\mathbb{D})$ is a Banach algebra.

2. Prove that there is a continuous map π from $sp(H^\infty(\mathbb{D}))$ onto the closed unit disc $\overline{\mathbb{D}}$ determined by $\pi(\rho) = \rho(\zeta)$, where ζ denotes the identity function $\zeta(z) = z$ on $\mathbb{D}$. Note that there are two things to prove: that π maps *into* $\overline{\mathbb{D}}$, and that π maps *onto* $\overline{\mathbb{D}}$. Thus, we can consider $\mathbb{D}$ as a subset of $sp(A)$.

3. Prove that $\pi^{-1}(z)$ is a singleton for every $z \in \mathbb{D}$ and that the restriction of π to $\pi^{-1}(\mathbb{D})$ is a homeomorphism.

4. It is a fact that π is not a homeomorphism. Can you see why? (a full treatment is quite beyond our scope).

5. Prove that the statement of Theorem 9.4.2 with $H^\infty(\mathbb{D})$ instead of $A(\mathbb{D})$ is equivalent to the disc $\mathbb{D}$ being dense in $sp(H^\infty(\mathbb{D}))$.

Exercise 9.6.7. Let B denote the uniform closure in $C_b(\mathbb{R})$ of the space spanned by all functions of the form $t \mapsto e^{i\lambda t}$, where $\lambda \in \mathbb{R}$.

1. Prove that B is a unital Banach algebra.

2. Prove that for all $f \in B$ and all $\epsilon > 0$, there exists $M > 0$ such that every interval $[a, a + M] \subset \mathbb{R}$ contains T such that $\sup_{t \in \mathbb{R}} |f(t + T) - f(t)| < \epsilon$ (functions that satisfy this are said to be **almost periodic**. In fact, every almost periodic function is in B).

3. Every $t_0 \in \mathbb{R}$ gives rise to a character of B by way of point evaluation: $f \mapsto f(t_0)$. Prove that these are **not** the only characters of B.

Exercise 9.6.8. Let a be an element in a unital Banach algebra A. Prove that there exists a commutative Banach subalgebra $B \subseteq A$ that contains a and the unit 1 such that

1. B is inverse closed, in the sense that if $b \in B \cap A^{-1}$ then $b^{-1} \in B$.

2. B is the smallest norm closed commutative subalgebra of A that is inverse closed and contains the unit and a.

3. $\sigma_B(b) = \sigma_A(b)$ for all $b \in A$.

Exercise 9.6.9. The purpose of this exercise is to prove the spectral mapping theorem for the holomorphic functional calculus (see Theorem 8.6.13). Let a be an element in a unital Banach algebra A, and let B be a commutative Banach subalgebra of A that contains a and the unit 1. Let $\Gamma \colon B \to C(sp(B))$ denote the Gelfand transform.

1. Show that $\Gamma(f(a)) = f(\Gamma(a))$ for every $f \in \mathrm{Hol}(a)$.

2. Deduce that $\sigma_B(f(a)) = f(\sigma_B(a))$.

3. Use Exercise 9.6.8 to establish the spectral mapping theorem (Theorem 8.6.13(5)).

Exercise 9.6.10. Let A be a unital Banach algebra and let $a \in A$. Let $g \in \mathrm{Hol}(a)$ and $f \in \mathrm{Hol}(g(a))$. Prove that $f \circ g \in \mathrm{Hol}(a)$ and that

$$f \circ g(a) = f(g(a)).$$

(**Suggestion:** use Exercise 8.7.14.)

Exercise 9.6.11. Let A be a unital commutative Banach algebra, and let $\Delta = sp(A)$ be the spectrum of A. Show that the Gelfand topology is unique, in the sense that the only topology on Δ under which it is compact, and such that for all $a \in A$ the Gelfand transform $\hat{a}$ is continuous, is equal to the Gelfand topology (i.e. the topology induced by the weak-$*$ topology on A^*).

Exercise 9.6.12. In this exercise, you will prove an extension of Wiener's $1/f$ theorem. Let $f \in W$ and let ϕ be a function holomorphic in a neighborhood of the image $f(\mathbb{T})$ of f (so the composition $\phi \circ f$ is well defined and continuous on $\mathbb{T}$). Prove that the Fourier series of $\phi \circ f$ is absolutely convergent.

Exercise 9.6.13. Here's another extension of Wiener's $1/f$ theorem due to Bochner and Phillips [1]. Let A be a unital commutative Banach algebra, and let W_A be the space of A valued functions on $\mathbb{T}$ of the form $f(t) = \sum_{n \in \mathbb{Z}} e^{int} a_n$, where $\{a_n\}_{n \in \mathbb{Z}}$ is a sequence in A such that $\|f\|_{W_A} := \sum \|a_n\| < \infty$. Prove that $(W_A, \| \cdot \|_{W_A})$ is a Banach algebra (this requires showing also that the "Fourier coefficients" of f are uniquely determined). Show that if $f \in W_A$ and $f(t) \in G(A)$ for all $t \in \mathbb{T}$, then $f^{-1} \colon \mathbb{T} \to A$ defined by $f^{-1}(t) = (f(t))^{-1}$ is continuous and that f^{-1} is has an absolutely convergent Fourier series.

Chapter 10

C*-algebras

Among all Banach algebras, there is a class of special algebras, called *C*-algebras*, that enjoy additional structure and have remarkable properties. C*-algebras are important because they correspond to closed subalgebras $A \subseteq B(H)$ that are *selfadjoint* in the sense that $a \in A \implies a^* \in A$, where a^* is the Hilbert space adjoint of the operator a (here H is a Hilbert space, of course). The theory of C*-algebras constitutes a major subfield of functional analysis. This chapter together with Chapters 11 and 12 may serve as a brief introduction to the theory of C*-algebras, however, we will be only scratching the surface. The textbook [14] is recommended for a thorough acquaintance with the theory of C*-algebras (see also [2, 3]).

10.1 Basic notions and first examples

Definition 10.1.1. 1. An *involution* on a complex algebra A is an operation $A \ni a \mapsto a^* \in A$ which has the following properties:

(a) $a^{**} = a$,

(b) $(\lambda a + b)^* = \bar{\lambda} a^* + b^*$,

(c) $(ab)^* = b^* a^*$,

for all $a, b \in A$, and all $\lambda \in \mathbb{C}$. An algebra with an involution defined on it is said to be a **-algebra* or an *involutive algebra*.

2. A *Banach *-algebra* is a Banach algebra A that has an isometric involution defined on it (i.e., $\|a^*\| = \|a\|$).

3. A *C*-algebra* is a Banach *-algebra A in which

$$\|a^* a\| = \|a\|^2 \tag{10.1}$$

for all $a \in A$.

The identity (10.1) is called the *C*-identity*. We defined a C*-algebra to be a Banach *-algebra to begin with, so $\|a\| = \|a^*\|$ is given, but it is not hard to show if we have an involution defined on a Banach algebra then (10.1) implies that the involution is isometric.

DOI: 10.1201/9781003297864-10

Example 10.1.2. Let H be a Hilbert space, and let $B(H)$ denote the algebra of all bounded operators on H. The adjoint operation $T \mapsto T^*$, defined by

$$\langle T^* g, h \rangle = \langle g, Th \rangle , \quad \text{for all } g, h \in H,$$

is an isometric involution that satisfies the C*-identity, hence $B(H)$ is a C*-algebra. To see that the C*-identity holds, let $h \in H$, and compute

$$\|Th\|^2 = \langle T^*Th, h \rangle \le \|T^*Th\| \|h\| \le \|T^*T\| \|h\|^2,$$

so

$$\|T\|^2 \le \|T^*T\| \le \|T^*\| \|T\|.$$

This implies that $\|T\| \le \|T^*\|$, and by symmetry (since $(T^*)^* = T^{**} = T$) we have $\|T\| = \|T^*\|$. Returning to the above equation, we find that $\|T\|^2 = \|T^*T\|$. We conclude that the algebra $B(H)$ of bounded operators on a Hilbert space is a C*-algebra.

Example 10.1.3. Let X be a locally compact Hausdorff space. On the Banach algebra $C_b(X)$ of bounded continuous functions on X we define an involution $f^*(x) = \overline{f(x)}$, that is, the involution is just pointwise complex conjugation. At every point $x \in X$ we have $|f(x)|^2 = |\overline{f(x)}f(x)|$, so $\|f\|^2 = \|f^*f\|$, and $C_b(X)$ is a C*-algebra. Every closed subalgebra of $C_b(X)$ that is invariant under conjugation, for example $C_0(X)$, is therefore also a C*-algebra. In particular, if X is a compact Hausdorff space then $C(X)$ is a commutative C*-algebra.

Definition 10.1.4. A **unit** in a C*-algebra A is an element $e \in A$ such that $ea = ae = a$ for all $a \in A$. A C*-algebra A is said to be **unital** if it has a unit. The element e is usually denoted as 1 or 1_A.

Looking back at the examples, we see that $B(H)$ is unital with unit $1 = I_H$, and when $\dim H = \infty$ then $K(H)$ is a nonunital C*-algebra (why?). The algebra $C_b(X)$ is unital because the constant function 1 is clearly an identity, but if X is not compact then $C_0(X)$ is not unital. If X is compact then $C(X) = C_b(X) = C_0(X)$ is unital.

Exercise 10.1.5. If A is a C*-algebra and if $e \in A$ is an element such that $ea = a$ for all $a \in A$ then e is an identity for A, $e^* = e$ and $\|e\| = 1$. In particular, if A has a unit, then the unit is selfadjoint and has norm 1.

Proposition 10.1.6. *Let A be a nonunital C*-algebra. Then A can be embedded isometrically as a closed two sided ideal of codimension 1 inside a unital C*-algebra $\tilde{A}$.*

Proof. We define $\tilde{A} = A \oplus \mathbb{C}$, with the same multiplication as in the unitization of Banach algebras described in Exercise 8.1.7. We already know that $\tilde{A}$ is a unital algebra and that A is a two sided ideal of codimension 1 in $\tilde{A}$. We define an involution $(a, \lambda)^* = (a^*, \bar{\lambda})$. It remains to define a norm that makes $\tilde{A}$ into a C*-algebra.

For every $a \in A$, let $L_a \in B(A)$ be given by $L_a(b) = ab$, $b \in A$. $\|L_a(b)\| = \|ab\| \leq \|a\| \|b\|$, so $\|L_a\| \leq \|a\|$. On the other hand, $\|L_a(a^*)\| = \|aa^*\| = \|a\| \|a^*\|$, and dividing by $\|a^*\|$ we see that $\|L_a\| = \|a\|$. In particular, the map $a \mapsto L_a$ is injective. Since A has no left unit (see Exercise 10.1.5), the map $(a, \lambda) \mapsto L_a + \lambda I_A$ is injective. We may therefore define a norm on $\tilde{A}$ by

$$\|(a, \lambda)\| = \|L_a + \lambda I_A\|,$$

and we identify $\tilde{A}$ with a subalgebra of $B(A)$. $\tilde{A}$ is closed in $B(A)$ because it is equal to $q^{-1}(\mathbb{C} \cdot \dot{I}_A)$, where q is the quotient map $q : B(A) \to B(A)/\{L_a : a \in A\}$. Therefore, $\tilde{A}$ is a Banach algebra.

It remains to prove that the C*-identity holds. It suffices to prove the inequality $\|x\|^2 \leq \|x^*x\|$ for all $x = (a, \lambda) \in \tilde{A}$, because then $\|x\|^2 \leq \|x^*x\| \leq \|x^*\| \|x\|$, and we obtain equality as in Example 10.1.2. To this end, let $b \in A_1$ and $\epsilon > 0$ be such that $\|xb\| = \|ab + \lambda b\| \geq \|x\| - \epsilon$. Then, using the C*-identity in A, we have

$$
\begin{aligned}
(\|x\| - \epsilon)^2 &\leq \|ab + \lambda b\|^2 \\
&= \|(ab + \lambda b)^*(ab + \lambda b)\| \\
&= \|L_{b^*}(L_{a^*} + \bar{\lambda}I)(L_a + \lambda I)b\| \\
&\leq \|b^*\| \|(a, \bar{\lambda})^*(a, \lambda)\| \|b\| \\
&\leq \|x^*x\|.
\end{aligned}
$$

This completes the proof. □

Example 10.1.7. Let A be a C*-algebra, and let $B \subseteq A$ be a subalgebra that is closed in the norm and closed under the adjoint operation. We say that B is a **C*-subalgebra**. A C*-subalgebra is clearly a C*-algebra in its own right.

Example 10.1.8. If H is a Hilbert space, then the ideal of compact operators $K(H)$ is a C*-subalgebra of the C*-algebra $B(H)$.

Example 10.1.9. If $a_1, \ldots, a_n$ are elements in a C*-algebra A, let us write $C^*(a_1, \ldots, a_n)$ for the **C*-algebra generated by** $a_1, \ldots, a_n$, which is the smallest C*-subalgebra of A that contains $a_1, \ldots, a_n$. It can be defined as the intersection of all C*-algebras containing $a_1, \ldots, a_n$, or as the norm closure in A of the *-algebra consisting of all linear combinations of all products involving the elements a_i and a_i^*, $i = 1, \ldots, n$. In particular, when A is unital we can define $C^*(1, a)$ — **the unital C*-algebra generated by** a.

Definition 10.1.10. A linear map $\pi : A \to B$ between C*-algebras is called a *-**homomorphism** if $\pi(aa') = \pi(a)\pi(a')$ and $\pi(a^*) = \pi(a)^*$ for all $a, a' \in A$. A bijective *-homomorphism is called a *-**isomorphism**. If A and B have units and $\pi(1_A) = 1_B$ then π is said to be **unit preserving** or simply **unital**. A *-homomorphism from a C*-algebra A into $B(H)$ is called a **representation**.

Example 10.1.11. A C*-subalgebra $A \subseteq B(H)$ is sometimes referred to as a **concrete C*-algebra**, in contrast with **abstract C*-algebras**, which are simply Banach ∗-algebras which satisfy the C*-identity. By a celebrated theorem of Gelfand and Naimark, every abstract C*-algebra is isometrically ∗-isomorphic to a concrete one. We shall prove this in Chapter 12 – a proof in which the majority of the material in this book will partake.

Example 10.1.12. Let (X, μ) be a measure space. The Banach space $L^\infty = L^\infty(X, \mu)$ becomes a Banach algebra under pointwise multiplication, and a C*-algebra when it is given the involution $f^*(x) = \overline{f(x)}$. By the Gelfand-Naimark theorem mentioned above, this C*-algebra is ∗-isomorphic to a concrete one. It is not hard to find such a representation (at least, under some mild assumption on μ). Let $H = L^2(X, \mu)$. Given $f \in L^\infty$, we define $M_f : H \to H$ by $M_f h = fh$. It is plain that $\|M_f\| \leq \|f\|_\infty$. If μ is semi-finite (see Section 3.1), then one can show that $\|M_f\| = \|f\|_\infty$. Moreover, it is easy to check that

$$M_f^* = M_{\overline{f}}, \quad M_{af+g} = aM_f + M_g \quad \text{and} \quad M_{fg} = M_f M_g.$$

The map $\pi : L^\infty \to B(H)$ given by $\pi(f) = M_f$ is an isometric, ∗-preserving homomorphism. The image of L^∞ under π is a closed subalgebra of $B(H)$ that is invariant under the ∗-operation, hence L^∞ is isometrically ∗-isomorphic to a concrete C*-algebra.

Example 10.1.13. On the Banach algebra $\ell^1(\mathbb{Z})$ we define an involution as follows: for every $a = (a_n) \in \ell^1(\mathbb{Z})$, we let $a^* = (b_n)$, where $b_n = \overline{a}_{-n}$. It is easy to check that this is an isometric involution, and it follows that $\ell^1(\mathbb{Z})$ becomes a Banach ∗-algebra in this way. But it is not a C*-algebra, because

$$\|(e_{-1} + e_0 - e_1)^*(e_{-1} + e_0 - e_1)\| = \| - e_{-2} + 3e_0 - e_2\| = 5$$

while

$$\|e_{-1} + e_0 - e_1\|^2 = 9.$$

In fact, there is no involution that one can define on $\ell^1(\mathbb{Z})$ that will make it into C*-algebra (see Exercise 10.7.1).

The existence of the adjoint on a C*-algebra gives rise to special classes of elements.

Definition 10.1.14. An element a in a C*-algebra is said to be

1. **selfadjoint** if $a = a^*$,

2. **normal** if $aa^* = a^*a$,

3. a **projection**, if $a = a^* = a^2$,

4. **unitary** if A is unital and $a^*a = aa^* = 1$.

Remark 10.1.15. Note that an element $a \in B(H)$ is a projection according to the above definition if and only if it is an orthogonal projection onto some subspace. In the context of C*-algebras, an element a that satisfies $a^2 = a$ is usually referred to as an *idempotent*; whereas the term *projection* is reserved for selfadjoint idempotents, that is, for elements that satisfy $a = a^* = a^2$.

Every element a in a C*-algebra can be written in a unique way as the linear combination $a = a_1 + ia_2$, where a_1, a_2 are selfadjoint. Indeed, just define

$$a_1 = \frac{a + a^*}{2} \ , \ a_2 = \frac{a - a^*}{2i};$$

the uniqueness is easy to prove. The selfadjoints a_1 and a_2 are referred to as the **real part** and **imaginary part** of a, respectively. Sometimes, one writes $a_1 = \operatorname{Re} a$ and $a_2 = \operatorname{Im} a$.

We will require the notion of the **exponential** of an element in a Banach space. Recall[1] that if A is a unital Banach algebra, then for every $a \in A$ we can define

$$\exp(a) = \sum_{n=0}^{\infty} \frac{a^n}{n!}.$$

One can show (see Exercise 8.7.3) that if $a, b \in A$ are two commuting elements (meaning that $ab = ba$), then

$$\exp(a + b) = \exp(a)\exp(b) = \exp(b)\exp(a).$$

By the properties of the adjoint, $\exp(a)^* = \exp(a^*)$. It follows that if a is selfadjoint then $\exp(ia)$ is unitary.

Exercise 10.1.16. Prove the above properties of the exponential.

10.2   Spectrum and Gelfand theory for C*-algebras

Lemma 10.2.1. *Let A be a unital commutative C*-algebra. Every $\varphi \in sp(A)$ is selfadjoint, in the sense that $\varphi(a^*) = \overline{\varphi(a)}$ for all $a \in A$.*

Proof. It suffices to prove that $\varphi(a) \in \mathbb{R}$ for every selfadjoint $a \in A$. Indeed, if that were the case, then writing $a = a_1 + ia_2$ as the combination of its real and imaginary parts, we find

$$\varphi\left((a_1 + ia_2)^*\right) = \varphi\left(a_1 - ia_2\right) = \varphi(a_1) - i\varphi(a_2) = \overline{\varphi\left(a_1 + ia_2\right)},$$

as required. Therefore, let us consider $a = a^* \in A$. To show that $\varphi(a) \in \mathbb{R}$,

---

[1]See Section 8.2 or more generally Section 8.6.

it suffices to show that $|e^{i\varphi(a)}| = 1$. By Proposition 9.2.4, φ is bounded and $\|\varphi\| \leq 1$. Thus,

$$e^{i\varphi(a)} = \sum_{n=0}^{\infty} \frac{(i\varphi(a))^n}{n!} = \varphi\left(\sum_{n=0}^{\infty} \frac{(ia)^n}{n!}\right) = \varphi(\exp(ia)).$$

But, by Exercise 10.1.16, $\exp(ia)$ is a unitary for every $a = a^*$, and so $\|\exp(ia)\| = 1$ whence $|\varphi(\exp(ia))| \leq 1$. On the other hand

$$1 = |\varphi(1)| \leq |\varphi(\exp(ia))||\varphi(\exp(-ia))| \leq 1,$$

and therefore

$$\left| e^{i\varphi(a)} \right| = |\varphi(\exp(ia))| = 1,$$

as required. □

Corollary 10.2.2. *For every selfadjoint a in a unital C*-algebra A,*

$$\sigma_A(a) \subseteq \mathbb{R}.$$

Proof. Let $B = C^*(1, a) \subseteq A$ be the C*-subalgebra generated by a and 1. Then $\sigma_A(a) \subseteq \sigma_B(a)$, so it suffices to prove that $\sigma_B(a) \subseteq \mathbb{R}$. In other words, we may assume that A is commutative. But then, by Theorem 9.2.7 on the Gelfand transform,

$$\sigma_A(a) = \{\varphi(a) : \varphi \in sp(A)\} \subseteq \mathbb{R},$$

where the last containment follows from the lemma. □

Theorem 10.2.3 (Spectral permanence theorem). *Let A and B be unital C*-algebras such that $1_B = 1_A \in B \subseteq A$. For every $b \in B$,*

$$\sigma_B(b) = \sigma_A(b).$$

The proof of the theorem depends on the following weaker result, which holds in the general setting of Banach algebras. For a subset $E \subseteq \mathbb{C}$ we let ∂E denote the topological boundary of E in $\mathbb{C}$.

Lemma 10.2.4 (Spectral permanence for Banach algebras). *Let A and B be unital Banach algebras such that $1_B = 1_A \in B \subseteq A$. For every $b \in B$,*

$$\partial\sigma_B(b) \subseteq \sigma_A(b).$$

Proof. Let $\lambda \in \partial\sigma_B(b)$, and assume for contradiction that $\lambda \notin \sigma_A(b)$. Let $\lambda_n \in \rho_B(b) \subseteq \rho_A(b)$ such that $\lambda_n \to \lambda$. By continuity of the inverse, $(a - \lambda_n)^{-1} \to (a - \lambda)^{-1}$ in A. But $(a - \lambda_n)^{-1} \in B$ for all n, so $(a - \lambda)^{-1} \in B$, because B is closed. But we started by assuming $\lambda \in \partial\sigma_B(b) \subseteq \sigma_B(b)$, so we have reached a contradiction. Therefore we must have that $\lambda \in \sigma_A(b)$, too. □

Proof of the spectral permanence theorem. It suffices to prove that whenever $b \in B$ is invertible in A, then the inverse b^{-1} is in B. Therefore, fix $b \in B \cap G(A)$. It is easy to check[2] that $b^* \in G(A)$, with $(b^*)^{-1} = (b^{-1})^*$, so $b^*b \in G(A)$. This means that $0 \notin \sigma_A(b^*b)$. Now $(b^*b)^* = b^*b$ is selfadjoint, so $\sigma_B(b^*b) \subseteq \mathbb{R}$ by Corollary 10.2.2. Therefore

$$\sigma_B(b^*b) = \partial\sigma_B(b^*b) \subseteq \sigma_A(b^*b).$$

It follows that b^*b is invertible in B, in other words $(b^*b)^{-1} \in B$. But now

$$(b^*b)^{-1}b^*b = 1,$$

so $(b^*b)^{-1}b^* \in B$ is a left inverse for b. Similarly, $b^*(bb^*)^{-1}$ is a right inverse for b, and we conclude that b is invertible in B. □

Theorem 10.2.5. *Let A be a unital commutative C*-algebra and let $\Delta = \mathrm{sp}(A)$. Then the Gelfand transform is an isometric *-isomorphism of A onto $C(\Delta)$.*

Proof. To establish that the Gelfand transform Γ is a *-homomorphism, we need only show that it preserves adjoints, in the sense that $\Gamma(a^*) = \Gamma(a)^* = \overline{\Gamma(a)}$ for all $a \in A$. Lemma 10.2.1 implies that for every $a \in A$ and $\varphi \in \Delta$,

$$\widehat{a^*}(\varphi) = \varphi(a^*) = \overline{\varphi(a)} = \overline{\widehat{a}(\varphi)},$$

and this shows $\Gamma(a^*) = \Gamma(a)^*$. It follows that $\Gamma(A)$, which we already know to be a subalgebra of $C(\Delta)$ that contains the constants and separates points, is also selfadjoint. By the Stone–Weierstrass theorem [14, Theorem 1.4.5], $\Gamma(A)$ is dense in $C(\Delta)$. The proof will be complete once we show that Γ is isometric. This is where the C*-identity will come in.

Let $a \in A$. By the C*-identity, together with commutativity,

$$\|a^2\| = \|(a^2)^*a^2\|^{1/2} = \|(a^*a)(a^*a)\|^{1/2} = \|(a^*a)^*(a^*a)\|^{1/2} = \|a^*a\| = \|a\|^2.$$

By induction, we find that for all $n \in \mathbb{N}$,

$$\|a^{2^n}\| = \|(a^{2^{n-1}})^2\| = \|a^{2^{n-1}}\|^2 = \|a\|^{2^n}.$$

Theorem 9.2.7 together with the spectral radius formula (Theorem 8.4.3) show

$$\|\widehat{a}\| = r(a) = \lim_{n\to\infty} \|a^n\|^{1/n} = \lim_{n\to\infty} \|a^{2^n}\|^{1/2^n} = \|a\|.$$

Thus, the Gelfand transform is isometric. □

Corollary 10.2.6. *For every normal a in a unital C*-algebra A,*

$$\|a\| = r(a).$$

---

[2]For example: $b^*(b^{-1})^* = (b^{-1}b)^* = 1^* = 1$.

Proof. The norm, and hence also the spectral radius, do not depend on the C*-algebra in which we consider the element a. Hence, we may replace A with the commutative unital C*-algebra generated by a, and now the result follows from the previous theorem. □

Remark 10.2.7. We see that every unital commutative C*-algebra "is" of the form $C(X)$ for some compact Hausdorff space X. More precisely, one can define a *contravariant functor* F from the category of unital commutative C*-algebras with unital homomorphisms, to the category of compact Hausdorff spaces, with continuous maps, by

$$F(A) = sp(A) \, , F(\varphi) = \varphi^*\big|_{sp(B)}$$

for A a unital commutative C*-algebra and $\varphi : A \to B$ a unital homomorphism. One can show that F is a *contravariant equivalence* of these categories.

Exercise 10.2.8. Let X, Y be compact Hausdorff spaces. Prove that every unital homomorphism from $\varphi : C(X) \to C(Y)$ is given as $\varphi(f) = f \circ h$ for some continuous $h : Y \to X$. Prove that φ is an isomorphism if and only if h is a homeomorphism. Conclude that $C(X)$ and $C(Y)$ are isomorphic (as algebras) if and only if X and Y are homeomorphic.

10.3 The continuous functional calculus for normal operators

If a is a normal element in a unital C*-algebra A, then the **continuous functional calculus** for a is a way of making sense of $f(a)$ for any function f continuous on the spectrum of a. As an element in a unital algebra, a can be plugged into complex polynomials as in (8.4). However, since a is in a C*-algebra, then one may also define $p(a)$ whenever $p \in \mathbb{C}[z, \bar{z}]$ is a polynomial in both z and $\bar{z}$

$$p(z) = \sum_{m,n} c_{mn} z^m \bar{z}^n, \tag{10.2}$$

by putting

$$p(a) = \sum_{m,n} c_{mn} a^m a^{*n}. \tag{10.3}$$

Of course, Equation (10.3) makes sense for every element a in a C*-algebra, but the map $p \mapsto p(a)$ will not be a homomorphism if a is not normal. In fact, the functional calculus for a non-normal element can fail at a much more basic level: we want the functional calculus to be a homomorphism from *functions* on the spectrum – not from formal expressions – into A, but if a is not normal then this cannot always be done.

Example 10.3.1. Consider the element $a = \left(\begin{smallmatrix} 0 & 1 \\ 0 & 0 \end{smallmatrix}\right) \in M_2(\mathbb{C})$. Then $\sigma(a) = \{0\}$. The polynomials $p(z) = z$ and $q(z) = z^2$ agree, as functions on $\sigma(a)$, with the polynomial $r(z) = 0$. But $p(a) = a$ while $r(a) = q(a) = 0$. This kind of problem does not arise when a is normal.

Theorem 10.3.2 (The continuous functional calculus). *Let A be a unital C*-algebra and let $a \in A$ be normal. Let $B = C^*(1, a)$ be the unital C*-algebra generated by a, and put $X = \sigma(a)$. There exists an isometric $*$-isomorphism Φ from $C(X)$ onto B, that extends the polynomial functional calculus, and maps every polynomial of the form (10.2) to the corresponding element of the form (10.3).*

Proof. Since a is normal, B is a commutative C*-algebra. We now prove that $sp(B) \cong X$. Indeed, consider the Gelfand transform of a as a continuous function $\hat{a} : sp(B) \to \mathbb{C}$. We know from Theorem 9.2.7 on the Gelfand transform that $\hat{a}(sp(B)) = \sigma_B(a)$ and from the spectral permanence theorem that $\sigma_B(a) = \sigma(a) = X$. Thus, $\hat{a}$ is surjective. But it is also injective, because every character is uniquely determined by its value on a. It follows that $\hat{a}$ is a homeomorphism from $sp(B)$ onto X. Therefore, $\hat{a}$ gives rise to an isometric $*$-isomorphism $\alpha : C(X) \to C(sp(B))$, given by

$$\alpha : f \mapsto f \circ \hat{a}.$$

Note that for the function $\zeta(z) = z$, $\alpha(\zeta) = \hat{a}$. Since α is a $*$-isomorphism, $\alpha(p) = \sum_{m,n} c_{mn} \hat{a}^m \overline{\hat{a}}^n$ for $p(z) = \sum_{m,n} c_{mn} z^m \overline{z}^n$. By Theorem 10.2.5, the Gelfand transform $\Gamma : B \to C(sp(B))$ is an isometric $*$-isomorphism. Now the map

$$\Phi = \Gamma^{-1} \circ \alpha$$

is an isometric $*$-isomorphism between $C(X)$ and B, as required. $\qquad\square$

Remark 10.3.3. Usually, one simply identifies $sp(B)$ and X, and then thinks of the inverse of the Gelfand transform as the functional calculus. When $f \in C(\sigma(a))$, it is common to write $f(a)$ for $\Phi(f)$, and this agrees with any other reasonable way of defining $f(a)$; for example, if f is holomorphic, then this definition of $f(a)$ agrees with the holomorphic functional calculus studied in Section 8.6.

Corollary 10.3.4. *A normal element in a unital C*-algebra is selfadjoint if and only if its spectrum is contained in the real line.*

Proof. A normal element corresponds by the above isomorphism to the identity function $\zeta(z) = z$ on the spectrum. The identity function on a subset X of the complex plane is selfadjoint (under conjugation) if and only if $X \subseteq \mathbb{R}$. $\quad\square$

Theorem 10.3.5 (The spectral mapping theorem). *Let a be a normal operator in a unital C*-algebra and let $f \in C(\sigma(a))$. Then $\sigma(f(a)) = f(\sigma(a))$. Moreover, if g is a continuous function on $f(\sigma(a))$, then*

$$g(f(a)) = g \circ f(a), \tag{10.4}$$

where $g(f(a))$ is g evaluated at $f(a)$, defined by the functional calculus of $f(a)$, and $g \circ f(a)$ is defined by applying the functional calculus of a to the function $g \circ f$.

Proof. Since $f \mapsto f(a)$ is a *-isomorphism,

$$\sigma(f(a)) = \sigma(f) = f(\sigma(a)).$$

Now, if

$$g(z) = \sum_{m,n} c_{mn} z^m \bar{z}^n$$

is a polynomial in z and $\bar{z}$, then

$$g \circ f(z) = \sum_{m,n} c_{mn} f(z)^m \overline{f(z)}^n.$$

Therefore, both sides of (10.4) are equal to

$$\sum_{m,n} c_{mn} f(a)^m (f(a)^*)^n.$$

By the Stone–Weierstrass theorem, there is a sequence of polynomials g_n in z and $\bar{z}$ that converges uniformly to g on $\sigma(f(a))$. Then, $g_n(f(a)) \to g(f(a))$ by continuity of the functional calculus. On the other hand, $g_n \circ f$ converges uniformly to $g \circ f$ on $\sigma(a)$, so $g_n \circ f(a) \to g \circ f(a)$. Since $g_n(f(a)) = g_n \circ f(a)$ for all n, we obtain (10.4) in the limit. $\square$

10.4 Application: the square root of a positive operator and polar decomposition

Definition 10.4.1. A normal element a in a unital C*-algebra is said to be **positive**, denoted $a \geq 0$, if $\sigma(a) \subset [0, \infty)$.

By the spectral permanence theorem, a normal element is positive independently of which C*-algebra we choose to view it in. By Corollary 10.3.4 a positive element must be selfadjoint.

Theorem 10.4.2. *Let a be a normal element in a unital C*–algebra A. Then $a \geq 0$ if and only if there exists a positive element $b \in A$ such that $b^2 = a$. When this occurs, the element b is unique and is contained in the C*-algebra generated by a and 1.*

Remark 10.4.3. The element b in the above theorem is referred to as the **positive square root** (or sometimes simply as the **square root**) of a and is denoted $b = \sqrt{a}$ or $b = a^{1/2}$.

Proof. Suppose that $a \geq 0$. Let $g(t) = \sqrt{t}$ be the positive square root function defined on $\sigma(a)$. We remark that $g \in C(\sigma(a))$, because $\sigma(a) \subseteq [0, \infty)$. Let $b = g(a)$ be given by the continuous functional calculus. Then b is normal because the functional calculus is a *-homomorphism. By the spectral mapping theorem, $\sigma(b) \subseteq [0, \infty)$. Thus, $b \geq 0$. By the functional calculus

$$b^2 = (g(a))^2 = (g^2)(a) = a.$$

Conversely, if $a = b^2$ with $b \geq 0$, then a is normal (and in fact it is selfadjoint), and by the spectral mapping theorem, it satisfies $\sigma(a) = \{t^2 : t \in \sigma(b)\} \subseteq [0, \infty)$.

Suppose now that $a \geq 0$, and let $c \geq 0$ in A be such that $c^2 = a$. Let $b = g(a)$ be as in the first paragraph. We will prove that $b = c$. By Theorem 10.3.5 applied to the functions $f : t \mapsto t^2$ and $g : t \mapsto \sqrt{t}$ (with c in place of a),

$$b = g(a) = g(c^2) = g \circ f(c) = c.$$

□

Exercise 10.4.4. For $a \geq 0$, prove that $\sqrt{a}$ is contained in the C*-algebra generated by a (the point is that $\sqrt{a}$ is in the possibly nonunital C*-algebra generated *only* by a).

Theorem 10.4.5. *Let T be an operator on a Hilbert space H. The following conditions are equivalent:*

1. *$T \geq 0$ (in other words, T is normal and $\sigma(T) \subseteq [0, \infty)$).*

2. *There exists a positive $S \in B(H)$ such that $T = S^2$.*

3. *There exists some $R \in B(H)$ such that $T = R^*R$.*

4. *For all $h \in H$, $\langle Th, h \rangle \geq 0$.*

Proof. Conditions (1) and (2) are equivalent, by Theorem 10.4.2. (2) clearly implies (3), and $\langle R^*Rh, h \rangle = \langle Rh, Rh \rangle = \|Rh\|^2 \geq 0$ for every $h \in H$, so (3) implies (4).

Suppose that (4) holds. We first need to show that T is normal; we will in fact show that it is selfadjoint. If $\langle Th, h \rangle \geq 0$ for all h, then this inner product is, in particular, a real number. Hence

$$\langle h, T^*h \rangle = \langle Th, h \rangle = \overline{\langle Th, h \rangle} = \langle h, Th \rangle.$$

Since this holds for all $h \in H$ (and because we are working over the complex numbers) $T = T^*$. So T is selfadjoint, hence normal. From selfadjointness, it also follows that $\sigma(T) \subseteq \mathbb{R}$. Suppose that $\lambda < 0$. Then

$$\langle (T - \lambda I)h, h \rangle = \langle Th, h \rangle + |\lambda|\|h\|^2 \geq |\lambda|\|h\|^2.$$

On the other hand

$$|\langle (T - \lambda I)h, h \rangle| \le \|(T - \lambda I)h\|\|h\|.$$

Combining the two inequalities we find that $\|(T - \lambda I)h\| \ge |\lambda|\|h\|$, so $T - \lambda I$ is bounded below. Recall that this means that $T - \lambda I$ is injective and has a closed range. But then as $T - \lambda I$ is selfadjoint, it follows that

$$\mathrm{Im}(T - \lambda I) = \overline{\mathrm{Im}((T - \lambda I)^*)} = \ker(T - \lambda I)^{\perp} = H,$$

and we conclude that $T - \lambda I$ is invertible so $\lambda \notin \sigma(T)$. This shows that $\sigma(T) \subseteq [0, \infty)$. $\square$

Remark 10.4.6. Condition (4) only makes sense for operators on Hilbert spaces, and not for elements in an abstract C*-algebra. Conditions (1), (2) and (3) make sense in any C*-algebra, and are in fact equivalent in general. As mentioned in the proof, we have already seen that (1) and (2) are equivalent in general. However, if there is no Hilbert space to pass through, it is not straightforward to show that (3) implies $T \ge 0$. This will be proved in Theorem 12.1.5.

To state another important decomposition theorem, we need a new definition.

Definition 10.4.7. An operator $U \in B(H)$ is said to be a ***partial isometry*** if the restricted operator $U|_{\ker U^{\perp}}$ is an isometry from $\ker U^{\perp}$ onto $\mathrm{Im}\, U$. The space $\ker U^{\perp}$ is called the ***initial space*** of U and the space $\mathrm{Im}\, U$ is called the ***final*** space of U.

Exercise 10.4.8. If U is a partial isometry, then U^*U is the orthogonal projection onto the initial space of U, and UU^* is the orthogonal projection onto the final space of U.

Exercise 10.4.9. For an operator $U \in B(H)$, the following are equivalent.

1. U is a partial isometry.

2. $UU^*U = U$.

3. U^*U is a projection.

4. U^* is a partial isometry.

Theorem 10.4.10 (The polar decomposition). *Let $T \in B(H)$. Then there exists a unique partial isometry U with $\ker U = \ker T$ and a unique positive operator P with $\ker P = \ker T$ such that $T = UP$. The operator P is given by $P = (T^*T)^{1/2}$, and it is contained in $C^*(T)$.*

Proof. Put $P = (T^*T)^{1/2}$. This is well defined because $T^*T \geq 0$, by Theorem 10.4.5. Then

$$\|Ph\|^2 = \langle P^2 h, h \rangle = \langle T^*Th, h \rangle = \|Th\|^2.$$

In particular, $\ker P = \ker T$. Moreover, the equality of norms implies that the map $Ph \mapsto Th$ is a well defined isometric linear map from $\operatorname{Im} P$ to $\operatorname{Im} T$. It therefore extends continuously to an isometry U from $\overline{\operatorname{Im} P} = \ker P^\perp$ to $\overline{\operatorname{Im} T}$. Setting $U = 0$ on $\ker P$ completes the construction. This settles the question of existence of the polar decomposition.

As for uniqueness: the assumptions (and Exercise 10.4.8) imply that $PU^*U = P$, so that $T^*T = PU^*UP = P^2$, and thus P must be equal to $(T^*T)^{1/2}$, the unique positive square root of T^*T. The only partial isometry U with initial space $\ker T^\perp$ that maps $Ph \mapsto Th$ must be the one that we defined in the existence part of the proof, so there is only one such partial isometry U that will satisfy $UP = T$. □

Remark 10.4.11. The operator $P = (T^*T)^{1/2}$ is denoted $|T|$ is called the **absolute value** of T. The decomposition $T = U|T|$ is called the **polar decomposition** of T. We have noted that $|T| \in C^*(T)$. As for U, it is in general not contained in $C^*(T)$, but we may see later that it is always contained in the *von Neumann algebra* generated by T.

Remark 10.4.12. When H is finite dimensional, it is easy to see that U can be adjusted on $\ker T$ so as to be unitary. The terminology *polar decomposition* is also used for the (nonunique) decomposition $T = UP$ where $P \geq 0$ and U is unitary.

10.5 Nonunital commutative C*-algebras

If A is a nonunital commutative C*-algebra, we let $\widetilde{A} = A \oplus \mathbb{C}$ be its unitization equipped with the norm that makes it into a C*-algebra as in Proposition 10.1.6. The spectrum of elements in A and the spectrum $\Delta = sp(A)$ of A are defined as in Section 9.5; the particular norm we use does not affect these concepts (see Remark 9.5.2).

Proposition 10.5.1. *Let A be a commutative C*-algebra. If $A \neq 0$, then $\Delta := sp(A) \neq \emptyset$.*

Proof. If A is unital, then we already know that $A \cong C(\Delta)$. If A is nonunital, we consider the commutative unital $\widetilde{A} = C(\Delta \cup \{\varphi_0\})$ (by Proposition 9.5.6). Since $\widetilde{A}$ is not one dimensional, $\Delta \neq \emptyset$. □

The unitization of a commutative C*-algebra allows us to import many of the results from the unital part of the theory to the general case. For

example, every character φ of a commutative nonunital C*-algebra is a
*-homomorphism. Indeed, we consider $\tilde{\varphi}$ on $\tilde{A}$. We proved that $\tilde{\varphi}$ satis-
fies $\tilde{\varphi}(a^*) = \overline{\tilde{\varphi}(a)}$, and by plugging elements of A in this identity we get
$\varphi(a^*) = \overline{\varphi(a)}$.

We will not spell out the nonunital version of every result, but we record
the most important one.

Theorem 10.5.2. *Let A be a nonunital commutative C*-algebra . The
Gelfand transform is an isometric *-isomorphism onto $C_0(\Delta)$.*

Proof. This follows from the unital case in a straightforward way. Alternatively, starting from Theorem 9.5.9 the isometricity and selfadjointness follow
as in the unital case, and then surjectivity follows by invoking the appropriate version of the Stone–Weierstrass theorem. The details are left as an
exercise. □

Exercise 10.5.3. Let A be a commutative C*-algebra with spectrum Δ. We
know that $A \cong C_0(\Delta)$. Prove that Δ is compact if and only if A is unital.
Prove that if A is nonunital, the spectrum $\tilde{\Delta}$ of $\tilde{A}$ is equal to the one-point
compactification of Δ.

Exercise 10.5.4. Recall the notion of positive elements in a unital C*-
algebra. We say that an element a in a nonunital C*-algebra A is **positive** if
$\sigma(a) \subseteq [0, \infty)$, where the spectrum of a is defined as in Definition 9.5.1. Prove
that in a commutative C*-algebra A, every positive element has a unique
positive square root in A.

10.6 Application: the Stone–Čech compactification

Definition 10.6.1. A topological space X is said to be **completely regular**
if it is Hausdorff and if for every closed set $F \subset X$ and every $x \notin F$, there
exists a continuous function $f : X \to [0, 1]$ such that $f\big|_F \equiv 0$ and $f(x) = 1$.

Every locally compact Hausdorff space is completely regular, and in particular, every compact Hausdorff space is (see [35, Section 33]). For every
completely regular space, there exists a certain compactification with remarkable properties, called the Stone–Čech compactification. We will now see how
the theory of commutative C*-algebras can be used to obtain the existence
of this compactification (see [35, Theorem 38.2] for the standard topological
proof).

Theorem 10.6.2. *Let X be a completely regular space. There exists a compact
Hausdorff space βX and a continuous function $h : X \to \beta X$ such that*

1. h is a homeomorphism of X onto $h(X)$.

2. $h(X)$ is dense in βX.

3. Every continuous bounded function on X extends uniquely to a continuous function on βX; more precisely, for every $f \in C_b(X)$, there exists a unique $\tilde{f} \in C(\beta X)$ such that $\tilde{f} \circ h = f$.

Moreover, the pair $(\beta X, h)$ is unique in the following sense: if (Y, g) is a pair with the properties of $(\beta X, h)$, then there exists a homeomorphism $u : \beta X \to Y$ such that $u \circ h = g$.

Proof. It is easy to check that $C_b(X)$ is a commutative unital C*-algebra. Hence $C_b(X)$ is *-isomorphic to the algebra of continuous functions on its spectrum. Write $\beta X = sp(C_b(X))$, endowed with the Gelfand topology (i.e., the weak-* topology).

We know that βX is a compact Hausdorff space, as the spectrum of a commutative unital C*-algebra. Let $h : X \to \beta X$ be the map sending every point x to the evaluation functional $\delta_x \in \beta X$. It is easy to show, using complete regularity, that h is injective. We observed in the past that h is continuous, but this does not imply immediately that h is a homeomorphism, since X is not necessarily compact. To see that it is a homeomorphism, we need to show that $F \subseteq X$ is closed if and only if $h(F)$ is closed in $h(X)$. One direction follows from continuity; the other will follow at once from the following identity

$$h(\overline{F}) = h(X) \cap \overline{h(F)}.$$

Let $x \in X$. We need to show that $x \in \overline{F}$ if and only if δ_x (which is always in $h(X)$) is in $\overline{h(F)}$. But take note: $x \in \overline{F}$ if and only if $f(x) = 0$ for every $f \in C_b(X)$ that vanishes on F (this follows from the complete regularity of X). On the other hand, $\delta_x \in \overline{h(F)}$ if and only if $g(\delta_x) = 0$ for every $g \in C(\beta X)$ which vanishes on $h(F)$. But by the Gelfand isomorphism $C_b(X) = C(\beta X)$, every such g has the form $\hat{f}$ for some $f \in C_b(X)$. Thus, the conditions for $x \in \overline{F}$ and $h(x) = \delta_x \in \overline{h(X)}$ are the same.

Next, $h(X)$ is dense in βX, for otherwise (by using complete regularity of βX) we could find some $g = \hat{f} \in C(\beta X)$ that vanishes on $h(X)$ but not on some point in $\beta X \setminus X$; this is a contradiction because if $g = \hat{f}$ vanishes on $h(X)$ then $f \equiv 0$.

If we define $\tilde{f} = \hat{f} \in C(\beta X)$, then we see that $\tilde{f}$ is continuous and satisfies $\tilde{f} \circ h = f$. This extension is unique because $h(X)$ is dense.

Finally, if (Y, g) is another such pair, then $C(Y)$ must be isomorphic to $C_b(X)$, via $f \leftrightarrow f|_{g(X)} \circ g$. If we denote the composition of the consecutive isomorphisms $C(\beta X) \to C_b(X) \to C(Y)$ by φ, then φ gives a homeomorphism $h : Y \cong \beta X$ given by $h(y) = \varphi^*(\delta_y) = \delta_y \circ \varphi$ that preserves point evaluations at X. $\square$

Remark 10.6.3. Note that if X is a locally compact Hausdorff space, then $C_b(X)$ is a unital C*-algebra that contains $C_0(X)$ as an ideal. $C_b(X)$ is a unitization of $C_0(X)$, but it is different from the unitization $\widetilde{C_0(X)}$. In some sense, $\widetilde{C_0(X)}$ is the minimal unitization and $C_b(X)$ is the maximal one, similar to how the one-point compactification can be considered as the minimal compactification of X and the Stone–Čech compactification the maximal one.

10.7 Additional exercises

Exercise 10.7.1. Consider the Banach algebra $\ell^1(\mathbb{Z})$ with the usual norm and the convolution product. Show that there is no involution whatsoever that one can define on $\ell^1(\mathbb{Z})$ that makes it into a C*-algebra.

Exercise 10.7.2. Let A be a unital commutative Banach $*$-algebra. We say that the involution is **symmetric** if the Gelfand transform is a $*$-homomorphism, in the sense that $\Gamma(a^*) = \Gamma(a)^* := \overline{\Gamma(a)}$ for all $a \in A$. For example, by Theorem 10.2.5, the involution is symmetric in the case that A is a C*-algebra.

1. Prove that the involution in a commutative Banach $*$-algebra A is symmetric if and only if $1 + a^*a$ is invertible for every $a \in A$.

2. Define two involutions on $\ell^1(\mathbb{Z})$ that make it into a Banach $*$-algebra, one which makes it symmetric, and the other which does not.

Exercise 10.7.3. Let A be a unital C*-algebra.

1. Prove that a normal $a \in A$ is unitary if and only if $\sigma(a) \subseteq \mathbb{T}$.

2. Prove that every element in A can be written as the linear combination of four unitaries.

3. Prove that every selfadjoint element $a \in A$ has a unique decomposition $a = a_+ - a_-$ where $a_+, a_- \geq 0$, $\|a_\pm\| \leq \|a\|$, and $a_+a_- = a_-a_+ = 0$.

4. Prove that the above decomposition into positive and negative parts holds also if A is not unital.

Exercise 10.7.4 (Fuglede's theorem). Let $A, B \in B(H)$, and suppose that $AB = BA$ and that A is normal. In this exercise, you will prove *Fuglede's theorem*, which states that the above assumptions imply that $A^*B = BA^*$.

1. Prove that $\exp(A - A^*)$ is unitary.

2. Noting that $\exp(A)B = B\exp(A)$, show that

$$\exp(A^*)B\exp(-A^*) = \exp(A^* - A)B\exp(A - A^*),$$

implying (why?) that $\|\exp(A^*)B\exp(-A^*)\| \leq \|B\|$.

3. Prove that for every bounded linear functional f on $B(H)$, the function

$$h(z) = f(\exp(zA^*)B\exp(-zA^*))$$

is an entire holomorphic function.

4. Use the above results to prove that $A^*B = BA^*$ (and therefore $AB^* = B^*A$ as well).

5. Give an example that shows that the conclusion might fail if A is not assumed normal.

6. One can generalize: consider the case $A_1B_1 = B_2A_2$ where A_1, A_2 are normal. It is not true that $A_1^*B_1 = B_2A_2^*$ in general, but it is true under an additional assumption, which is either $A_1 = A_2$ or $B_1 = B_2$. Figure out which condition works and prove it.

Exercise 10.7.5. Let A be a commutative unital C*-algebra and $a = (a_1, \ldots, a_n) \in A^n$. Define

$$\sigma_A(a) = \{(\phi(a_1), \ldots, \phi(a_n)) : \phi \in sp(A)\}.$$

1. Prove that if B is a C*-subalgebra of A that contains 1 and $a_1, \ldots, a_n$, then $\sigma_B(a) = \sigma_A(a)$. This allows us, for a commuting normal tuple a in a unital C*-algebra, to define $\sigma(a)$ to be $\sigma_C(a)$, where $C = C^*(1, a)$.

2. Prove that two commuting tuples of normal operators on a Hilbert space $a \in B(H)^n$ and $b \in B(H)^n$ satisfy $\sigma(a) = \sigma(b)$ if and only if there is a *-isomorphism $\pi : C^*(1, a) \to C^*(1, b)$ such that $\pi(a_i) = b_i$ for $i = 1, \ldots, n$ (a and b are then said to be *algebraically equivalent*).

3. Give an example of normal tuples that are algebraically equivalent but not unitarily equivalent.

Exercise 10.7.6. Is the Banach limit an element of the maximal ideal space of $\ell^\infty(\mathbb{N})$?

Exercise 10.7.7 (The maximal ideal space of ℓ^∞ and ultrafilters). Exercise on $\ell^\infty \cong C(\beta\mathbb{N})$. A family of subsets $\mathcal{F} \subseteq 2^{\mathbb{N}}$ is said to be a **filter** if the following conditions hold:

1. The empty set is not contained in $\mathcal{F}$.

2. If $A \in \mathcal{F}$ and $A \subseteq B$ then $B \in \mathcal{F}$.

3. If $A, B \in \mathcal{F}$ then $A \cap B \in \mathcal{F}$.

A filter is said to be an **ultrafilter** if it satisfies, in addition to the above, that for every $A \subseteq \mathbb{N}$, either A is in $\mathcal{F}$ or its complement $\mathbb{N} \setminus A$ is in $\mathcal{F}$.

1. Let $A \subseteq \mathbb{N}$ be a nonempty set. Prove that $\mathcal{F} = \{B \subseteq \mathbb{N} : A \subseteq B\}$ is a filter. Such a filter is said to be a **principal** filter.

2. Give an example of a nonprincipal filter.

3. Prove that every filter is contained in an ultrafilter. Conclude that there exist principal as well as nonprincipal ultrafilters.

4. Let $\mathcal{F}$ be an ultrafilter. A bounded sequence $a \in \ell^\infty = \ell^\infty(\mathbb{N})$ is said to **converge along** $\mathcal{F}$ **to a limit** L, denoted

$$L = \lim_{\mathcal{F}} a_n,$$

if for every neighborhood U of L, the set $\{n \in \mathbb{N} : a_n \in U\}$, is in $\mathcal{F}$. Prove the following shocking fact: given an ultrafilter $\mathcal{F}$, every bounded sequence $a \in \ell^\infty$ has a unique limit along $\mathcal{F}$.

5. Prove that taking the limit $\lim_{\mathcal{F}} a_n$ of a bounded sequence a along an ultrafilter $\mathcal{F}$ defines a character of ℓ^∞. Conversely, prove that for every character $\varphi \in sp(\ell^\infty)$, there exists an ultrafilter $\mathcal{F}$ such that $\varphi(a) = \lim_{\mathcal{F}} a_n$ for all $a \in \ell^\infty$.

Exercise 10.7.8 (The reduced group C*-algebra). Let G be a group. Here is a natural and interesting construction of a C*-algebra associated with G. The Hilbert space $\ell^2(G)$ has a standard orthonormal basis $\{\delta_g\}_{g \in G}$ given by

$$\delta_g(h) = \begin{cases} 1 & h = g, \\ 0 & \text{else.} \end{cases}$$

For $f \in G$, we define $L_g : \ell^2(G) \to \ell^2(G)$ by

$$L_f \delta_g = \delta_{fg}.$$

Consider the C*-algebra

$$C_r^*(G) := C^*(\{L_g : g \in G\})$$

generated by $\{L_g : g \in G\}$. This C*-algebra is called **the reduced group C*-algebra of** G. Obviously, $C_r^*(G)$ is commutative if and only if G is an abelian group.

In the case that G is abelian, $C_r^*(G) \cong C(X)$ for a compact space X. Prove that X is naturally identified with the space of **characters** of G, that is, with the space of all homomorphisms $\varphi : G \to \mathbb{T}$ of G into the unit circle, endowed with the topology of pointwise convergence:

$$\varphi_\alpha \to \varphi \iff \varphi_\alpha(g) \to \varphi(g) \text{ for all } g \in G.$$

Chapter 11

The spectral theorem and von Neumann algebras

In this chapter, we shall prove the spectral theorem for normal operators on a Hilbert space. The spectral theorem is a cornerstone of operator theory and functional analysis. We shall also use it as a natural link to a brief study of von Neumann algebras, which constitute a special class of C*-algebras that provide a suitable framework for the analysis of normal operators.

11.1 The spectral theorem for normal operators

11.1.1 What does the spectral theorem say?[1]

The spectral theorem is the basic structure theorem for normal operators, telling us what a general normal operator "looks like". Recall the spectral theorem for normal operators on a finite dimensional Hilbert space: if T is a normal operator acting on a finite dimensional space $H = \mathbb{C}^n$, then T is unitarily equivalent to a diagonal operator, that is, there exists a unitary operator U on H such that

$$UTU^* = \begin{pmatrix} \lambda_1 & & \\ & \ddots & \\ & & \lambda_n \end{pmatrix}, \tag{11.1}$$

where $\lambda_1, \ldots, \lambda_n$ are the points in the spectrum of T (some of the points are possibly repeated). It is interesting to point out that this diagonalization theorem follows effortlessly from the continuous functional calculus. Indeed, the characteristic function χ_λ of a point $\lambda \in \sigma(T)$ is continuous on the spectrum of T, and since the functional calculus is a *-isomorphism $P_\lambda = \chi_\lambda(T)$ is a nontrivial orthogonal projection. Letting $\zeta(z) = z$ be the identity function, we have that

$$\zeta = \sum_{\lambda \in \sigma(T)} \lambda \chi_\lambda,$$

---

[1]The title of this section is borrowed from Halmos's paper [25], which readers are likely to enjoy.

DOI: 10.1201/9781003297864-11

and therefore $T = \zeta(T) = \sum_{\lambda \in \sigma(T)} \lambda \chi_\lambda(T)$ or, briefly,

$$T = \sum_{\lambda \in \sigma(T)} \lambda P_\lambda \tag{11.2}$$

from which we easily read off (11.1).

More generally, if T is a compact normal operator on a Hilbert space, then T is unitarily equivalent to a diagonal operator (an infinite diagonal matrix, acting by multiplication on ℓ^2), the diagonal of which corresponds to the eigenvalues of T, which form a sequence converging to 0, so that

$$UTU^* = \begin{pmatrix} \lambda_1 & & \\ & \lambda_2 & \\ & & \ddots \end{pmatrix} ; \tag{11.3}$$

see [11, Theorem 10.3.7]. Thus, the spectral theorem for compact normal operators gives us a simple model of what a general operator in the class of interest looks like. With just a little more effort and a careful use of the basic spectral theory of compact operators on a Banach space (see the beginning of Section 5.5) one can obtain this spectral theorem, too, from the continuous functional calculus. It is interesting to note that this approach is bound to fail for general normal operators. For example, if λ is not an isolated point in the spectrum then χ_λ is not a continuous function on $\sigma(T)$, and if T has connected spectrum containing more than one point then $C(\sigma(T))$ will contain no nontrivial projections at all.

One can define an operator T to be **diagonalizable** if it is unitarily equivalent to a diagonal operator as in (11.3), where the diagonal elements form a bounded sequence of complex numbers (not necessarily converging to 0). It is not hard to see that a diagonalizable operator is a bounded normal operator (which is not necessarily compact). However, a general bounded normal operator need not be unitarily equivalent to a diagonal operator.

Example 11.1.1. The operator $T: L^2[0,1] \to L^2[0,1]$ given by $Tf(x) = xf(x)$ is a selfadjoint bounded operator. In particular, T is normal. It is an easy exercise to see that this operator has no eigenvalues, so it cannot be unitarily equivalent to a diagonal operator.

The above example shows that there is no hope to prove that all normal operators are diagonalizable. On the other hand, the operator in the example is rather well understood, and it suggests a natural generalization of the concept of a diagonal operator. It turns out that the general case is not significantly more complicated than this.

To introduce the class of operators that will serve as a model for general normal operators, we recall some notions from measure theory (see Section 1.5). Let (X, μ) be a measure space and consider the Hilbert space $L^2 = L^2(X, \mu)$. As we described in Example 10.1.12, every $f \in L^\infty = L^\infty(X, \mu)$ defines a bounded operator $M_f : h \mapsto fh$ on L^2.

Exercise 11.1.2. Let (X, μ) be a semi-finite measure space and $f \in L^\infty = L^\infty(X, \mu)$. Prove the following facts (or look them up; see, e.g., [28, Example 2.4.11]).

1. $\|M_f\| = \|f\|_\infty$, where $\|f\|_\infty$ is the essential supremum of f (see Example 1.5.17).

2. $M_f^* = M_{\bar{f}}$.

3. If $g : X \to \mathbb{C}$ and $h \mapsto gh$ defines a bounded operator on L^2, then g is essentially bounded: $\|g\|_\infty < \infty$.

4. If $f, g \in L^\infty$, then $M_f M_g = M_{fg}$ and $M_f + M_g = M_{f+g}$.

5. M_f is selfadjoint if and only if f is real valued almost everywhere.

6. M_f is unitary if and only if $|f| = 1$ almost everywhere.

The operator M_f, where $f \in L^\infty$, is called a ***multiplication operator***. By the second and fourth items above, multiplication operators are normal. The spectral theorem says that this collection exhausts (up to unitary equivalence) all normal operators. A concise yet precise statement of the spectral theorem is this: *every normal operator is unitarily equivalent to a multiplication operator.* The proof of the spectral theorem is a marvelous application of the algebraic methods of Gelfand theory to operator theory.

11.1.2   Statement and proof of the spectral theorem

If $\mathcal{F} \subseteq H$, then it is common to let $[\mathcal{F}]$ denote the closure of the linear span of $\mathcal{F}$.

Definition 11.1.3. A vector $h \in H$ is said to be a ***cyclic vector*** for $T \in B(H)$ if
$$[C^*(I, T)h] := \overline{\operatorname{span}}\{Ah : A \in C^*(I, T)\} = H.$$
If T has a cyclic vector then T itself is said to be ***cyclic***.

Theorem 11.1.4 (The spectral theorem for normal cyclic operators). *Let T be a cyclic normal operator on a Hilbert space H. Then there exists a regular Borel probability measure μ on $X = \sigma(T)$ and a unitary $U \colon L^2(X, \mu) \to H$ such that*
$$U^*TU = M_\zeta,$$
where $\zeta \in L^\infty(X, \mu)$ is the function $\zeta(z) = z$.

Proof. Suppose that $h \in H$ is a cyclic unit vector for T and let $X = \sigma(T)$. By the continuous functional calculus (Theorem 10.3.2), there is an isometric *-isomorphism $\Phi \colon C(X) \to C^*(I, T)$ which satisfies $\Phi(p) = p(T, T^*)$ for every $p \in \mathbb{C}[z, \bar{z}]$. Recall that we write $f(T) = \Phi(f)$ for $f \in C(X)$.

Define a linear functional $\rho : C(X) \to \mathbb{C}$ by

$$\rho(f) = \langle f(T)h, h \rangle$$

for $f \in C(X)$. Then ρ is a bounded positive linear functional on $C(X)$, and $\rho(1) = 1$. By the Riesz representation theorem[2], there exists a unique regular Borel probability measure μ on X such that $\rho(f) = \int f d\mu$ for all $f \in C(X)$. This is the sought after measure μ appearing in the statement of the theorem. We now need to construct a unitary $U : L^2(X, \mu) \to H$ such that U intertwines T and M_ζ.

We define $U : L^2(X, \mu) \to H$ by first requiring that $Uf = f(T)h$ for all $f \in C(X)$. Using the properties of the functional calculus, we compute:

$$\|f\|_2^2 = \int |f|^2 d\mu = \rho(|f|^2)$$
$$= \langle |f|^2(T)h, h \rangle = \langle f(T)^* f(T)h, h \rangle = \|f(T)h\|^2,$$

for all $f \in C(X)$, and so the map $f \mapsto f(T)h$ is isometric, and in particular well defined, on $C(X) \subset L^2(X, \mu)$. Now, the space $C(X)$ is a dense subspace of $L^2(X, \mu)$, and since we assumed that h is cyclic, $\{f(T)h : f \in C(X)\}$ is a dense subspace of H. Thus, the map U extends to a unitary $L^2(X, \mu) \to H$.

Finally, consider the continuous function $\zeta \in C(X)$ given by $\zeta(z) = z$. Then for every $f \in C(X)$,

$$TUf = Tf(T)h = U(\zeta f) = UM_\zeta f.$$

So $TU|_{C(X)} = UM_\zeta|_{C(X)}$, and because $C(X)$ is dense in $L^2(X, \mu)$ we may conclude that $TU = UM_\zeta$, and the proof is complete. □

Remark 11.1.5. In the proof above, we defined a measure $\mu = \mu_h$ by

$$\int f d\mu_h = \langle f(T)h, h \rangle \text{ for all } f \in C(\sigma(T)),$$

where $h \in H$ was assumed to be a cyclic vector for T. In fact, the same construction makes very good sense also when h is not necessarily cyclic. The measure μ_h is then sometimes referred to as the **spectral measure associated** to h (or T). Warning: the term "spectral measure" will appear again below and will then mean something different. In any case, it is an instructive exercise to see what the measure μ_h looks like when T is a selfadjoint matrix and h is an arbitrary vector.

We now turn to the general case. The idea is simple: every operator breaks up into a direct sum of cyclic operators. To explain this in detail we need a few preparations.

---

[2]See Theorem 3.4.1 and the following remark.

Definition 11.1.6. A subspace $M \subseteq H$ is said to be ***invariant*** for a subset of operators $S \subseteq B(H)$ if $TM \subseteq M$ for all $T \in S$. A subspace M is said to be ***reducing*** for S if both M and $M^{\perp}$ are invariant.

Exercise 11.1.7. Prove that a subspace M is reducing for T if and only if M is invariant under both T and T^*. Prove also that M is reducing for T if and only if it is reducing for $C^*(T)$. Deduce that h is a cyclic vector for T if and only if the smallest reducing subspace for T that contains h is H itself.

The following lemma follows immediately from the exercise.

Lemma 11.1.8. *If $\mathcal{A}$ is a C*-algebra of $B(H)$ and $F \subseteq H$ then*

$$[\mathcal{A}F] := \overline{\text{span}}\{ah : a \in A, h \in F\}$$

is a reducing subspace for A.

We say that H is the ***direct sum*** of a family of Hilbert spaces $\{H_i\}_{i \in \Lambda}$, and we write $H = \bigoplus_{i \in \Lambda} H_i$, if for all $i \in \Lambda$ we have that H_i is a closed subspace of H, the spaces H_i and H_j are orthogonal for $i \neq j$, and if every $h \in H$ can be written in a unique way as a norm convergent sum $h = \sum_{i \in \Lambda} h_i$ where $h_i \in H_i$ for all i. If $T_i \in B(H_i)$ and $\sup_i \|T_i\| < \infty$, then we can define the ***direct sum*** of the T_is to be $T = \oplus_i T_i$ defined by $T(\sum_i h_i) = \sum_i T_i h_i$. In this case $\|T\| = \sup_i \|T_i\|$, and every H_i is reducing for T. Conversely, if H_i is reducing for T for all i, then $T_i := T|_{H_i}$ is a bounded operator on H_i and $T = \oplus_i T_i$. Finally, if $T = \oplus_i T_i$ then $\sigma(T)$ is the equal to the closure of $\cup_i \sigma(T_i)$, and in particular $\sigma(T_i) \subseteq \sigma(T)$ (see Exercise 11.5.1).

A standard application of Zorn's lemma shows that given any operator on a Hilbert space H, there exists a decomposition of the space into a direct sum of cyclic reducing subspaces.

Lemma 11.1.9. *Let $T \in B(H)$. Then there exists a family of vectors $\{h_i\}_{i \in \Lambda}$ such that*

$$H = \bigoplus_{i \in \Lambda} H_i,$$

where $H_i = [C^*(I, T)h_i]$ is reducing for T. In other words, every operator is the direct sum of cyclic operators.

It follows from the above lemma together with Theorem 11.1.4 that every normal operator is a direct sum of multiplication operators. This, in fact, means that it is a multiplication operator.

Theorem 11.1.10 (The spectral theorem). *Every normal operator T on a Hilbert space H is unitarily equivalent to a direct sum of multiplication operators M_{f_i} on $L^2(X_i, \mu_i)$ where μ_i is a regular Borel probability measure on $X_i \subseteq \sigma(T)$. Moreover, there exists a locally compact Hausdorff space X, a semi-finite regular Borel measure μ on X, a function $f \in L^\infty(X, \mu)$ and a unitary operator $U \colon L^2(X, \mu) \to H$, such that*

$$U^*TU = M_f.$$

When H is separable μ can be chosen to be a regular Borel probability measure.

Proof. By Lemma 11.1.9 we have that $T = \oplus_i T_i$ where every T_i is a cyclic operator on $H_i \subseteq H$. By Theorem 11.1.4 T_i is unitarily equivalent to a multiplication operator M_{f_i} on (X_i, μ_i) as stated. We have that $X_i = \sigma(T_i) \subseteq \sigma(T)$. If $U_i \colon L^2(X_i, \mu_i) \to H_i$ implements the unitary equivalence between T_i and M_{f_i}, then the direct sum

$$U := \oplus_{i \in \Lambda} U_i \colon \bigoplus_{i \in \Lambda} L^2(X_i, \mu_i) \longrightarrow \bigoplus_{i \in \Lambda} H_i = H$$

is a unitary that implements an equivalence of $\oplus_{i \in \Lambda} T_i$ and $\oplus_{i \in \Lambda} M_{f_i}$. We summarize what we obtained so far: T is unitarily equivalent to a direct sum of multiplication operators. We need to show that the direct sum $\oplus_{i \in \Lambda} M_{f_i}$ can in fact be described as a multiplication operator on an L^2 space.

Let $X = \bigsqcup_{i \in \Lambda} X_i$ be the disjoint union of the X_i and provide it with the σ-algebra Σ in which a set $E \subseteq X$ is measurable if and only if $E \cap X_i$ is μ_i measurable for all $i \in \Lambda$. This is readily seen to be a σ-algebra and on it we may define the measure $\mu = \bigsqcup_i \mu_i$ by the formula

$$\mu(E) = \sum_{i \in \Lambda} \mu_i(E \cap X_i)$$

for all $E \in \Sigma$. It follows from the definition of (X, μ) that $L^2(X, \mu)$ can be identified with the direct sum $\bigoplus_{i \in \Lambda} L^2(X_i, \mu_i)$. If we define $f \colon X \to \mathbb{C}$ by $f\big|_{X_i} = f_i$, then f is measurable and $\|f\|_\infty = \sup_i \|f_i\| = \sup_i \|T_i\| = \|T\|$, therefore $f \in L^\infty(X, \mu)$. It is plain that M_f acts on $L^2(X, \mu)$ by multiplying each summand $L^2(X_i, \mu_i)$ by f_i, that is, $M_f = \oplus_{i \in \Lambda} M_{f_i}$.

The space X becomes a locally compact Hausdorff space when it is given the direct product topology, namely, the topology in which $E \subseteq X$ is open if and only if $E \cap X_i$ is open in X_i for all $i \in \Lambda$. One may then check that the measure μ is regular.

Now suppose that H is separable. Then the family $\{H_i\}_{i \in \Lambda}$ is countable. Let $\{c_i\}_{i \in \Lambda}$ be a family of positive numbers such that $\sum_i c_i = 1$. It is easy to check that M_{f_i} on $L^2(X_i, \mu_i)$ is unitarily equivalent to M_{f_i} on $L^2(X_i, c_i \mu_i)$, so that T is unitarily equivalent to M_f on $\bigoplus_{i \in \Lambda} L^2(X_i, c_i \mu_i) = L^2(X, \mu)$, where we now define the probability measure μ by

$$\mu(E) = \sum_i c_i \mu_i(E \cap X_i).$$

$\square$

11.2 Von Neumann algebras and the Borel functional calculus

The spectral theorem for a normal operator T allows us very naturally to extend the functional calculus from the algebra of continuous functions $C(\sigma(T))$

to the significantly larger algebra $\mathcal{B}(\sigma(T))$ consisting of bounded Borel measurable functions on $\sigma(T)$. Since the continuous functional calculus is an isomorphism, the range of the Borel functional calculus will be larger than the C*-algebra $C^*(I, T)$ generated by T. What is this range? This kind of question is best treated within the framework of von Neumann algebras.

The theory of von Neumann algebras was introduced by von Neumann in [49] and later developed in a series of papers titled "On Rings of Operators"[3] by Murray and von Neumann. In [36], the motivation for Murray and von Neumann's investigations was presented thusly:

> *"First, the formal calculus with operator-rings leads to them. Second, our attempts to generalize the theory of unitary group-representations essentially beyond their classical frame have always been blocked by the unsolved questions connected with these problems. Third, various aspects of the quantum mechanical formalism suggest strongly the elucidation of this subject. Fourth, the knowledge obtained in these investigations gives an approach to a class of abstract algebras without a finite basis, which seems to differ essentially from all types hitherto investigated."*

The theory of von Neumann algebras is a major branch of functional analysis, with many important connections, as anticipated, with operator theory, group representations, ergodic theory and physics. Von Neumann algebras have turned out to be connected in unforeseen ways with knot theory, quantum cryptography and computer science. Applications aside, the theory itself is rich and beautiful and two Fields Medals[4] were awarded for work on von Neumann algebras. Our goal in this chapter is just to give a taste of the subject; a more thorough treatment can be found in the following references: [1, 28, 29].

11.2.1   Topologies on $B(H)$ and von Neumann algebras

Recall that the **weak operator topology** (see Example 6.3.17), also referred to as **WOT**, is the weak topology on $B(H)$ generated by all the functionals of the form
$$w_{g,h}(T) = \langle Tg, h \rangle,$$
where $g, h \in H$.

The **strong operator topology**, or **SOT**, is the topology on $B(H)$ generated by the semi-norms
$$p_h(T) = \|Th\|$$
where $h \in H$ (see Example 6.3.10).

---

[3]Von Neumann algebras were originally referred to simply as "rings of operators".
[4]Awarded to Alain Connes and Vaughan Jones.

The norm topology, the WOT and the SOT all make $B(H)$ into a locally convex space. When $\dim H = \infty$ these are three different topologies. If $T \in B(H)$, then the maps $A \mapsto AT$ and $A \mapsto TA$ are WOT and SOT continuous. Multiplication is not jointly continuous, WOT nor SOT, but it is SOT continuous when restricted to bounded sets (Exercise 11.5.3). The adjoint operation is WOT continuous. Indeed, if $T_n \xrightarrow{WOT} T$, then

$$\langle T_n^* x, y \rangle = \langle x, T_n y \rangle \to \langle x, Ty \rangle = \langle T^* x, y \rangle.$$

On the other hand, the adjoint operation is not SOT continuous. For example, if S denotes the unilateral shift on $\ell^2(\mathbb{N})$ then $T_k := (S^*)^k$ converges SOT to 0 as $k \to \infty$, but $T_k^* = S^k$ does not (see Example 6.3.17).

Theorem 11.2.1. *The closed unit ball $B(H)_1$ of $B(H)$ is WOT compact.*

Proof. The proof is similar to the proof of Alaoglu's theorem and is left to the reader. □

Exercise 11.2.2. Show that $B(H)_1$ with the strong operator topology is not compact.

Finally, we come to the definition of a von Neumann algebra.

Definition 11.2.3. A *von Neumann algebra* is a concrete C*-algebra $\mathcal{M} \subseteq B(H)$ that contains $1 = I_H$ and is closed in the strong operator topology.

Example 11.2.4. The algebra $B(H)$ of all bounded operators on a Hilbert space is a von Neumann algebra. If H is finite dimensional then every C*-subalgebra of $B(H)$ is a von Neumann algebra. When $\dim H = \infty$ the algebra $K(H)$ of compact operators is not a von Neumann algebra: this is clear because it does not contain the unit. One can also show that $K(H)$ is not SOT closed. Indeed, by Example 6.3.11, if $\{e_i\}_{i \in I}$ is an orthonormal basis for H and for every finite subset $F \subset I$ we let $P_F h = \sum_{i \in F} \langle h, e_i \rangle e_i$ be the projection onto $\text{span}\{e_i\}_{i \in F}$, then $\{P_F\}_{F \in \Lambda}$ is a net in $K(H)$ that converges in the strong operator topology to $I_H \notin K(H)$.

It might appear that the reasons that $K(H)$ is not a von Neumann algebra both have to do with I not being in $K(H)$. So consider instead the C*-algebra $K(H) + \mathbb{C}I$. Again it is not SOT closed: if $M \subset H$ is any infinite dimensional subspace with infinite dimensional complement, then $P_M \notin K(H) + \mathbb{C}I$, but one can easily find a net of compact projections P_F that converges to P_M in the SOT.

There is in fact an incredibly rich supply of von Neumann algebras, but it is not so easy to exhibit examples let alone show that they are distinct. Consider the following example.

Example 11.2.5. In Example 10.1.12 we saw that $L^\infty(X, \mu)$ can be represented faithfully as a C*-algebra of multiplication operators $M = \{M_f : f \in$

$L^\infty\}$ on $L^2(X, \mu)$. It is customary to identify L^∞ with M. We can ask whether L^∞ is a von Neumann algebra. The answer is positive, but it is not so easy to prove it directly. For concreteness, suppose that $X = [0, 1]$ and μ is the Lebesgue measure. If $\{M_{f_\alpha}\}_\alpha$ is a SOT convergent net, we want to show that $T = SOT - \lim M_{f_\alpha}$ is also a multiplication operator. One obstacle is that we don't know that the net is bounded — it might not be. We avoid wrestling with this analytic challenge because *the double commutant theorem* to be proved below will provide a very elegant algebraic solution to this problem.

Exercise 11.2.6. Prove directly that ℓ^∞ is a von Neumann algebra.

Example 11.2.7. If $\mathcal{A} \subseteq B(H)$ is a unital C*-algebra, then its SOT closure is a von Neumann algebra. It is not straightforward to prove this directly. Consider, for example, proving that $\overline{\mathcal{A}}^{SOT}$ is closed under the involution. The map $T \mapsto T^*$ is not SOT continuous, so a naive attempt will fail. Similarly to the case of L^∞, this analytic difficulty will be easily resolved by the double commutant theorem. In any case, it is clear that the intersection of von Neumann algebras is a von Neumann algebra, so one can always define the smallest von Neumann containing a certain set. If $T \in B(H)$, we let $W^*(T)$ denote the smallest von Neumann algebra that contains T. It is true – but not trivial – that $W^*(T)$ is equal to the SOT closure of $C^*(I, T)$.

11.2.2   The Borel functional calculus

By the continuous functional calculus, for every normal operator T and a continuous function $f \in C(\sigma(T))$, one can define an operator $f(T) \in C^*(I, T)$. Moreover, we have the precise description

$$C^*(I, T) = \{f(T) : f \in C(\sigma(T))\}.$$

In this section we will extend the functional calculus to all bounded Borel functions, that is, we will show how to define $f(T)$ whenever f is a bounded Borel measurable function defined on $\sigma(T)$. This assignment (called the **Borel functional calculus**) will have nice properties similar to those of the continuous functional calculus, with the main differences being (i) the map $f \mapsto f(T)$ is not necessarily isometric, and (ii) $f(T)$ will not necessarily lie in $C^*(I, T)$, but rather in $W^*(T)$ – the von Neumann algebra generated by T.

Let $\mathcal{B}(X)$ denote the algebra of all bounded Borel measurable functions on a compact space X, equipped with the supremum norm and the adjoint operation $f^* = \overline{f}$. It is not hard to verify that $\mathcal{B}(X)$ is a C*-algebra (the only thing that is not immediate is that the limit of a sequence of Borel measurable functions is Borel measurable – this is a standard result in measure theory).

Theorem 11.2.8 (The Borel functional calculus). *Let T be a normal operator on a Hilbert space H, and write $X = \sigma(T)$. There exists a contractive *-homomorphism $\mathcal{B}(X)$ into $W^*(T)$ that extends the continuous functional calculus. If ϕ_n is a bounded sequence in $\mathcal{B}(X)$ that converges pointwise to ϕ, then $\phi_n(T) \to \phi(T)$ in the strong operator topology.*

Proof. By the spectral theorem, T is unitarily equivalent to a direct sum of multiplication operators M_{f_i} on $L^2(X_i, \mu_i)$ where μ_i is a regular Borel measure on $X_i \subseteq \sigma(T)$. In fact, the proof shows that $f_i(z) = z$ for all $i \in \Lambda$ and all $z \in X_i = \sigma(T)$. Since a unitary equivalence $U: \bigoplus_{i \in \Lambda} L^2(X_i, \mu_i) \to H$ induces an isometric $*$-isomorphism which is also a SOT homeomorphism $A \mapsto U^*AU$ from $B(H)$ onto $B(\oplus_i L^2(X_i, \mu_i))$, we may as well assume that $T = \oplus_i M_{f_i} = M_f$ on $L^2(X, \mu) = \bigoplus_{i \in \Lambda} L^2(X_i, \mu_i)$ as in the spectral theorem.

Given $\phi \in \mathcal{B}(X)$, the operator $\phi(T)$ is defined to be

$$\phi(T) = M_{\phi \circ f} = \oplus_i M_{\phi \circ f_i} = \oplus_i M_\phi.$$

This makes sense because ϕ being bounded and Borel measurable implies that $\phi \circ f_i = \phi \in L^\infty(X_i, \mu_i)$. The properties of multiplication operators imply that $\phi \mapsto \phi(T)$ is a $*$-homomorphism. It is also clear that $\|\phi(T)\| \leq \|\phi\|_\infty$. The subtle parts of the theorem are those concerning the strong operator topology.

Suppose that we have a bounded sequence ϕ_n converging pointwise to ϕ. We wish to prove that $\phi_n(T) \xrightarrow{SOT} \phi(T)$ and this amounts to showing that $\phi_n(T)g \to \phi(T)g$ for all $g = (g_i)_{i \in \Lambda} \in \bigoplus_{i \in \Lambda} L^2(X_i, \mu_i)$. Fixing g, we compute

$$\|(\phi_n(T) - \phi(T))g\|^2 = \sum_{i \in \Lambda} \|(\phi_n - \phi)g_i\|^2_{L^2(\mu_i)}$$
$$= \sum_{i \in \Lambda} \int_{X_i} |\phi_n - \phi|^2 |g_i|^2 d\mu_i$$

By the dominated convergence theorem in the space $L^1(X_i, \mu_i)$, the ith term in the sum converges to 0 as $n \to \infty$. A second application of the dominated convergence theorem, this time in the space $\ell^1(\Lambda)$, shows that the entire sum converges to 0 as $n \to 0$. This shows that $\phi_n(T)g \to \phi(T)g$ as required.

It remains to prove that $\phi(T) \in W^*(T)$ for every $\phi \in \mathcal{B}(X)$. It suffices to show that $\phi(T)$ is in the SOT closure of $C^*(I, T)$. A typical basic neighborhood of $\phi(T)$ has the form

$$U(\phi(T); g^1, \ldots, g^m) := \{A \in B(H) : \|Ag^k - \phi(T)g^k\| < 1, \ k = 1, \ldots, m\}.$$

Write $g^k = (g_i^k)_{i \in \Lambda} \in \bigoplus_{i \in \Lambda} L^2(X_i, \mu_i)$. Since $\sum_i \|g_i^k\|^2 < \infty$, there are at most countably many indices i such that $\|g_i^k\| \neq 0$, so assume without generality that $\Lambda = \mathbb{N}$. Define a regular Borel measure $\nu = \sum_{i \in \mathbb{N}} 2^{-i} \mu_i$. By a consequence of Lusin's theorem[5], there is a bounded sequence of continuous functions $\phi_n : X \to \mathbb{C}$, with $|\phi_n| \leq |\phi|$ that converge ν-almost everywhere to ϕ. Then $\phi_n(T) \in C^*(I, T)$, and the same argument as above shows that $\phi_n(T)g^k \to \phi(T)g^k$ for $k = 1, \ldots, m$. Thus, there exists n such that $\phi_n(T) \in U(\phi(T); g^1, \ldots, g^m)$. Therefore, $\phi(T) \in \overline{C^*(I, T)}^{SOT} \subseteq W^*(T)$. $\quad\square$

---

[5] See Theorem 2.24 and its corollary in [13].

Remark 11.2.9. The Borel functional calculus is actually *onto* $W^*(T)$, but we shall not require this result, nor do we have all the tools required to prove it. On the other hand, the Borel functional calculus is typically not an iso-morphism.

Exercise 11.2.10. In the above proof, we assumed that T is the direct sum of certain multiplication operators. However, T might be unitarily equivalent to a different direct sum of multiplication operators. Would we get a different functional calculus if we used a different representation of T as a direct sum of multiplication operators? Of course not! Prove that the definition of the Borel functional calculus does not depend on the direct sum of multiplication operators that we chose to define it by.

11.3 Spectral measure and the integral formulation of the spectral theorem

Fix a normal operator $T \in B(H)$. For the characteristic function χ_S of a Borel set $S \subseteq X = \sigma(T)$ we define

$$E(S) = \chi_A(S) \in W^*(T).$$

Since $\bar{\chi}_S = \chi_S$ and $\chi_S^2 = \chi_S$, and since the Borel functional calculus is a *-homomorphism, the operator $E(S)$ satisfies $E(S)^* = E(S)$ and $E(S)^2 = E(S)$. Thus, $E(S)$ is a projection (recall that in the context of C*-algebras and operators on a Hilbert space, we use the term *projection* for orthogonal projections; see Remark 10.1.15). The properties of the Borel functional calculus also imply the following properties.

Exercise 11.3.1. Prove that $E(\cdot)$ satisfies the following properties:

1. $E(S_1 \cap S_2) = E(S_1)E(S_2)$ for every pair of Borel sets S_1, S_2.

2. $E(\emptyset) = 0$ and $E(X) = I$.

3. $E(\cup_n S_n) = \sum_n E(S_n)$ for every disjoint family of Borel sets $S_1, S_2, \ldots,$ where the sum converges in the strong operator topology.

Definition 11.3.2. A map from the Borel subsets of a topological space X into the projections on some Hilbert space, that satisfies the three properties listed above, is called a **spectral measure**. The spectral measure E con-structed from the functional calculus above is called **the spectral measure associated with** T.

There is an important formulation of the spectral theorem in terms of the spectral measure. To state this version of the spectral theorem requires developing a theory of integration against spectral measures. If $s = \sum_{k=1}^{N} c_k \chi_{S_k}$ is a simple function, then we define the integral $\int s(\lambda) dE_\lambda$ to be

$$\int s(\lambda) dE_\lambda = \sum_{k=1}^{N} c_k E(S_k).$$

We often write $\int s dE$ for the integral. Because $\int s dE$ is, by definition, nothing but the Borel functional calculus applied to a simple function $\int s dE = s(T)$, we find that the integral defines a $*$-homomorphism from the $*$-algebra of simple Borel functions on $X = \sigma(T)$ into $B(H)$. Now given $\phi \in \mathcal{B}(X)$, we define

$$\int \phi dE = \lim_{n \to \infty} \int s_n dE$$

where $\{s_n\}$ is a sequence of simple functions that converges uniformly to ϕ (it is routine to show that this limit does not depend on the choice of the sequence of simple functions). One can show that this defines a reasonable notion of an operator valued integral, which enjoys many of the standard properties of any integration theory (linearity, monotonicity) and also enjoys the unusual property that it is multiplicative with respect to the integrand.

Since the bounded Borel functional calculus is contractive, we conclude that the functional calculus is given by "integration against the spectral measure"

$$\phi(T) = \int \phi dE = \int \phi(\lambda) dE_\lambda. \tag{11.4}$$

In particular, one has the formula (11.5), which is sometimes referred to as the **spectral decomposition** of T, and is a very pleasing generalization of the spectral theorem for normal operators on a finite dimensional space in the form (11.2). We summarize the above discussion.

Theorem 11.3.3 (Spectral decomposition). *Let $T \in B(H)$ be normal. Then there exists a unique spectral measure E defined on the Borel subsets of $\sigma(T)$ such that (11.4) holds for every bounded Borel function ϕ on $\sigma(T)$. In particular, T is given by the integral*

$$T = \int_{\sigma(T)} \lambda dE_\lambda. \tag{11.5}$$

Since the integral in the spectral decomposition (11.5) is a norm limit of sums of the form $\sum_k c_k E(S_k)$, the equality (11.5) implies that every normal operator T can be approximated in the norm by linear combinations of projections in the von Neumann algebra that it generates.

Corollary 11.3.4. *Every von Neumann algebra M is equal to the norm closure of the linear span of projections in M.*

Proof. Every $T \in B(H)$ can be written as $T = A + iB$ where $A = \frac{T+T^*}{2}$ and $B = \frac{T-T^*}{2i}$ are selfadjoint and belong to the von Neumann algebra generated by T. Hence it suffices to show that a selfadjoint operator T can be approximated by projections in $W^*(T)$. But since a selfadjoint operator is normal, this follows from (11.5) as explained above. $\square$

The integrals appearing in (11.4) and (11.5) can also be interpreted in a weaker sense. The importance of this weak interpretation is that it enables us to define a spectral integral when we are given a spectral measure which is not known to come from the functional calculus of a normal operator (see Remark 11.3.6). Let E be the spectral measure associated with a normal operator T. For every $x, y \in H$, we can define a set function

$$\mu_{x,y}(S) = \langle E(S)x, y \rangle. \tag{11.6}$$

The properties of the spectral measure E imply that $\mu_{x,y}$ is a Borel measure on $X = \sigma(T)$. One can show that a finite positive measure on a compact Hausdorff space must be regular,[6] thus it follows that $\mu_{x,y}$ is regular. For every characteristic function $\phi = \chi_A$ we have by (11.6) that

$$\int \phi \, d\mu_{x,y} = \langle \phi(T)x, y \rangle. \tag{11.7}$$

It follows that (11.7) holds for simple functions, and by taking limits we obtain (11.7) for bounded Borel functions. Thus (11.4) holds in the sense that $\phi(T)$ is the unique operator that satisfies (11.7) for all $x, y \in H$. Writing formally $d\mu_{x,y}(\lambda) = \langle dE_\lambda x, y \rangle$, we interpret the spectral integral as a *weak integral*, that is $\int \phi(\lambda) dE_\lambda$ is the unique operator for which

$$\left\langle \left(\int \phi(\lambda) dE_\lambda \right) x, y \right\rangle = \int \phi(\lambda) \langle dE_\lambda x, y \rangle \tag{11.8}$$

for all $x, y \in H$.

Remark 11.3.5. The idea of the weak spectral integral can be used to show that the extension of the continuous functional calculus to the Borel functional calculus is a manifestation of a general extension phenomenon. Let X be a compact Hausdorff space. Given a unital representation $\pi \colon C(X) \to B(H)$, one can define for every $x, y \in H$ a functional

$$f \mapsto \langle \pi(f)x, y \rangle.$$

By the Riesz representation theorem, there exists a unique regular Borel measure $\mu_{x,y}$ on X such that

$$\langle \pi(f)x, y \rangle = \int f \, d\mu_{x,y} \tag{11.9}$$

---

[6]See [, Theorem 2.18].

for every $f \in C(X)$. Now that we have a measure, we can integrate also Borel functions. For every $\phi \in \mathcal{B}(X)$ we may define the form

$$\langle x, y \rangle_\phi = \int \phi d\mu_{x,y}$$

for every $x, y \in H$. Using (11.9) and an approximation argument one sees that $\langle x, y \rangle_\phi$ is a bounded sesquilinear form on H, so there exists a bounded operator $\tilde{\pi}(\phi)$ on H such that $\int \phi d\mu_{x,y} = \langle \tilde{\pi}(\phi)x, y \rangle$. With a little more work[7] one shows that $\tilde{\pi}$ is a representation of $\mathcal{B}(X)$ that extends π, and it has the additional nice property that if ϕ_n is a bounded sequence tending pointwise to ϕ, then $\tilde{\pi}(\phi_n) \to \tilde{\pi}(\phi)$ in the SOT. In particular, if T is a normal operator, $X = \sigma(T)$ and $\pi: C(X) \to B(H)$ is the continuous functional calculus $\pi(T) = f(T)$, then the Borel functional calculus is recovered as $\phi(T) = \tilde{\pi}(\phi)$.

Remark 11.3.6. The above process can be reversed. One can start with a spectral measure E defined on the Borel sets of a compact Hausdorff space X and taking values in $B(H)$. Then one can define the spectral integral $\int \phi(\lambda)dE_\lambda$ in the weak sense as follows. First, one defines a measure $\mu_{x,y}$ for all $x, y \in H$ by (11.6). Then, writing $d\mu_{x,y}(\lambda) = \langle dE_\lambda x, y \rangle$, one defines $\int \phi(\lambda)dE_\lambda$ to be the operator that satisfies (11.8) for all $x, y \in H$. It can be shown that $\phi \mapsto \int \phi(\lambda)dE_\lambda$ is a representation of $\mathcal{B}(X)$. If X is a compact subset of the complex plane, then $T = \int \lambda dE_\lambda$ defines a normal operator with spectrum in X, and $\int \phi(\lambda)dE_\lambda$ is equal to $\phi(T)$. See, e.g., [2], [28] or [42] for more details.

11.4 The double commutant theorem

Definition 11.4.1. For a set $\mathcal{S} \subseteq B(H)$, the **commutant** of $\mathcal{S}$ is the set $\mathcal{S}'$ defined by

$$\mathcal{S}' = \{T \in B(H) : TS = ST \text{ for all } S \in \mathcal{S}\}.$$

Note that $\mathcal{S}'$ is a WOT closed algebra that contains the identity I. Moreover, if $\mathcal{S} = \mathcal{S}^*$, then $\mathcal{S}'$ is also selfadjoint.

Exercise 11.4.2. Prove that $\mathcal{S} \subseteq \mathcal{S}''$ and that $\mathcal{S}' = \mathcal{S}'''$.

Lemma 11.4.3. *Let $\mathcal{A} \subseteq B(H)$ be a *-subalgebra, let $P \in B(H)$ be a projection, and set $M = PH$. Then M is invariant for $\mathcal{A}$ if and only if $P \in \mathcal{A}'$.*

Proof. We first note that or $A \in B(H)$, $AM \subseteq M$ if and only if $PAx = Ax$ for all $x \in M$, which happens if and only if $PAP = AP$. Therefore, $A^*M \subseteq M$ if and only if $PA^*P = A^*P$, or $PAP = PA$.

---

[7] See [2, Section 2.7] for some details.

Now we prove the equivalence. If M is invariant for $\mathcal{A}$, then for all $A \in \mathcal{A}$ the space M is invariant under both A and A^*, so $AP = PAP = PA$. Thus $\mathcal{A}M \subseteq M$ implies $P \in \mathcal{A}'$. Conversely, if $P \in \mathcal{A}'$, then $PA = AP$ for all $A \in \mathcal{A}$, so $PAP = AP^2 = AP$ and therefore M is invariant.   □

Definition 11.4.4. The **null space** of a subset $\mathcal{S} \subseteq B(H)$ is the set

$$\text{null } \mathcal{S} = \cap_{T \in \mathcal{S}} \ker T.$$

Definition 11.4.5. A subset $\mathcal{S} \subseteq B(H)$ is said to be **nondegenerate** if $[\mathcal{S}H] = H$.

Exercise 11.4.6. A $*$-algebra $\mathcal{A} \subseteq B(H)$ is nondegenerate if and only if $\text{null}(\mathcal{A}) = \{0\}$.

Theorem 11.4.7 (von Neumann's double commutant theorem). *Let $\mathcal{A} \subseteq B(H)$ be a nondegenerate $*$-subalgebra. Then*

$$\mathcal{A}'' := (\mathcal{A}')' = \overline{\mathcal{A}}^{WOT} = \overline{\mathcal{A}}^{SOT}.$$

Proof. As $\mathcal{A}''$ is WOT closed and contains $\mathcal{A}$, it holds that

$$\overline{\mathcal{A}}^{SOT} \subseteq \overline{\mathcal{A}}^{WOT} \subseteq \mathcal{A}''.$$

Now let $T \in \mathcal{A}''$. To show that $T \in \overline{\mathcal{A}}^{SOT}$, we have to find, for every $x_1, \ldots, x_n \in H$ and every $\epsilon > 0$, an operator $A \in \mathcal{A}$ such that $\|Tx_i - Ax_i\| < \epsilon$ for all i.

Let us first assume that $n = 1$. Let P be the orthogonal projection onto the subspace $[\mathcal{A}x_1]$. Then by Lemma 11.4.3, $P \in \mathcal{A}'$; therefore, $P^\perp = I - P \in \mathcal{A}'$, too. Now, $AP^\perp x_1 = P^\perp A x_1 = 0$, so the assumption that $\mathcal{A}$ is nondegenate together with the exercise above imply that $P^\perp x_1 = 0$, or $x_1 \in [\mathcal{A}x_1]$. Now since $T \in \mathcal{A}''$ and $P \in \mathcal{A}'$, we have

$$Tx_1 = TPx_1 = PTx_1 \in [\mathcal{A}x_1].$$

But this means that for every $\epsilon > 0$ there exists some $A \in \mathcal{A}$ such that $\|Ax_1 - Tx_1\| < \epsilon$, so we are done with the case $n = 1$.

Now assume that $n \geq 2$. Define

$$H^{(n)} = H \oplus \cdots \oplus H$$

(n times), and for every operator $A \in B(H)$, define

$$A^{(n)} = A \oplus \cdots \oplus A$$

in $B(H^{(n)}) = M_n(B(H))$. Put $\mathcal{A}^{(n)} = \{A^{(n)} : A \in \mathcal{A}\}$. Then $\mathcal{A}^{(n)}$ is a $*$-subalgebra of $B(H^{(n)})$. A calculation (see Exercise 11.5.2) shows that

$$(\mathcal{A}^{(n)})' = M_n(\mathcal{A}')$$

and that
$$(\mathcal{A}^{(n)})'' = (M_n(\mathcal{A}'))' = (\mathcal{A}'')^{(n)}.$$

Now put $x = (x_1, \ldots, x_n)$, where $x_1, \ldots, x_n \in H$. We will apply the case where $n = 1$ to the algebra $\mathcal{A}^{(n)}$, the operator $T^{(n)}$ and the vector x. By the case $n = 1$, for every $\epsilon > 0$ there exists $A^{(n)} \in \mathcal{A}^{(n)}$ such that $\|T^{(n)}x - A^{(n)}x\| < \epsilon$, and this implies $\|Tx_i - Ax_i\| < \epsilon$ for all $i = 1, \ldots, n$, as required. $\qquad\square$

Corollary 11.4.8. *A unital $*$-subalgebra $\mathcal{A} \subseteq B(H)$ is a von Neumann algebra if and only if $\mathcal{A} = \mathcal{A}''$.*

Corollary 11.4.9. *For every $T \in B(H)$ it holds that*

$$W^*(T) = \overline{C^*(I,T)}^{SOT} = C^*(I,T)'' = \{T, T^*\}''.$$

Corollary 11.4.10. *Let $T \in B(H)$, and suppose that $T = U|T|$ is its polar decomposition. Then both $|T|$ and U are in $W^*(T)$.*

Proof. We have already seen in Theorem 10.4.10 that $|T| \in C^*(I,T) \subseteq W^*(T)$. By Corollary 11.4.9, to show that $U \in W^*(T)$ it suffices to show that U commutes with every $S \in \{T, T^*\}'$. Given such S, we have

$$SU|T| = ST = TS = U|T|S = US|T|.$$

It follows that $SUx = USx$ for every $x \in \text{Im}\,|T|$. To finish the proof we need to show that $SUx = USx$ for every $x \in \ker T = \ker |T| = \text{Im}\,|T|^\perp$. By the definition of U, $Ux = 0$ for $x \in \ker T = \text{Im}\,|T|^\perp$, so $SUx = 0$. On the other hand, if $x \in \ker T$, then $TSx = STx = 0$, so $S(\ker T) \subseteq \ker T$. Thus $USx = 0$ for $x \in \ker T$, and we are done. $\qquad\square$

Example 11.4.11. If $A = M_n(\mathbb{C})$, then $A' = \mathbb{C}I_n$, and $A'' = (\mathbb{C}I_n)' = M_n(\mathbb{C})$. More generally, if $A = B(H)$, then $A' = \mathbb{C}I_H$, and $A'' = (\mathbb{C}I_H)' = B(H)$. The first equality follows from the fact that A has no invariant subspaces, thus the only projections in A' are 0 and I_H. Since a von Neumann algebra is generated by its projections, $A' = \mathbb{C}I_H$.

Similarly, if $A = K(H)$ (the algebra of compact operators), then $K(H)' = \mathbb{C}I_H$ and $\overline{K(H)}^{SOT} = K(H)'' = (\mathbb{C}I_H)' = B(H)$.

Example 11.4.12. Let (X, μ) be a probability space, and consider the algebra $\mathcal{A} = \{M_f : f \in L^\infty(X, \mu)\}$. In Example 11.2.5 we stated that $\mathcal{A}$ is a von Neumann algebra, but did not prove it. We shall now see that $\mathcal{A}' = \mathcal{A}$, and it will follow that $\mathcal{A} = \mathcal{A}''$ is a von Neumann algebra. It is common to abuse notation and write $(L^\infty)' = L^\infty$.

Of course, $\mathcal{A} \subseteq \mathcal{A}'$ because $\mathcal{A}$ is commutative. Let $T \in \mathcal{A}'$. If T was in $\mathcal{A}$, then $T = M_f$ and we would be able to recover f as $f = M_f 1 = T1$. Here, 1 denotes the constant function with value 1 everywhere on X, which is in $L^2(X, \mu)$ because μ is a probability measure. We therefore define $f := T1$, and hope to be able to show that f – originally only known to be in L^2 – is bounded, and that $T = M_f$. But this is easy: for every function $h \in L^\infty \cap L^2$,

$$fh = M_h f = M_h T1 = T M_h 1 = Th,$$

therefore $\|fh\|_2 \leq \|T\|\|h\|$. It follows that $f \in L^\infty$, and the equation above shows that T agrees with M_f on the space $h \in L^\infty \cap L^2$, which is dense in L^2. Since both operators T and M_f are bounded and agree on a dense subspace, they are equal, so $T = M_f \in \mathcal{A}$. This shows that $\mathcal{A}' \subseteq \mathcal{A}$, and the proof is complete.

Surely one can prove directly that $\mathcal{A}$ is a strongly closed algebra, but the above simple computation shows the beauty and power of the double commutant theorem: an analytical problem (involving, perhaps, the limits of convergent unbounded nets) is reduced to a rather simple minded algebraic problem: computing the commutant.

11.5   Additional exercises

Exercise 11.5.1 (Direct sums of Hilbert spaces). In this exercise, you will work out the details of direct sums of Hilbert spaces used in the proof of the spectral theorem. You will need to understand series with possibly uncountably many members (see Exercise 6.7.2).

Let $\{H_i\}_{i \in \Lambda}$ be a family of Hilbert spaces, where Λ is an arbitrary index set. Define a new space $\bigoplus_{i \in \Lambda} H_i$ as follows:

$$\bigoplus_{i \in \Lambda} H_i = \left\{ (h_i)_{i \in \Lambda} : h_i \in H_i \text{ for all } i \in \Lambda \text{ and } \sum_{i \in \Lambda} \|h_i\|_{H_i}^2 < \infty \right\}.$$

On $\bigoplus_{i \in \Lambda} H_i$ define an inner product

$$\left\langle (g_i)_{i \in \Lambda}, (h_i)_{i \in \Lambda} \right\rangle = \sum_{i \in \Lambda} \langle g_i, h_i \rangle_{H_i}.$$

1. Prove that this inner product is well defined, and that $\bigoplus_{i \in \Lambda} H_i$ is a Hilbert space. The space $\bigoplus_{i \in \Lambda} H_i$ is called the **direct sum** of the Hilbert spaces $\{H_i\}_{i \in \Lambda}$. Note that every space H_i can be considered to be a closed subspace of $\bigoplus_{i \in \Lambda} H_i$.

2. Sometimes the direct sum $\bigoplus_{i \in \Lambda} H_i$ is called the **external direct sum**, to make a distinction from the following almost indistinguishable situation. Suppose that G is a Hilbert space, and that $\{G_i\}_{i \in \Lambda}$ is a collection of mutually orthogonal closed subspaces such that every $g \in G$ is equal to

the convergent sum $g = \sum_{i \in \Lambda} g_i$ for some (necessarily unique) $g_i \in G_i$. Prove that G is unitarily equivalent to the external direct sum $\bigoplus_{i \in \Lambda} G_i$ by a unitary that "maps G_i to G_i" in an obvious sense. We then denote $G = \bigoplus_{i \in \Lambda} G_i$. This is sometimes called the **internal direct sum**. But after thinking through this exercise, the reader should be convinced that there is no essential difference.

3. If $T_i \in B(H_i)$ and $\sup_{i \in \Lambda} \|T_i\| < \infty$, then we can define an operator $T \in B(\oplus_i H_i)$ by $T((h_i)) = (T_i h_i)$. We then write $T = \oplus_i T_i$. Conversely, if $T \in B(\oplus_i H_i)$ reduces every H_i, then $T = \oplus_{i \in \Lambda} T_i$ where $T_i = T\big|_{H_i}$.

4. Verify that $(\oplus_i T_i)^* = \oplus_i T_i^*$ and that $(\oplus_i S_i)(\oplus_i T_i) = \oplus_i S_i T_i$.

5. Prove that if $T = \oplus_i T_i$ then $\sigma(T)$ is equal to the closure of $\cup_i \sigma(T_i)$.

Exercise 11.5.2. Let H be a Hilbert space, and define, with the notation of Exercise 11.5.1,

$$H^{(n)} = \underbrace{H \oplus H \oplus \cdots \oplus H}_{n \text{ times}} := \bigoplus_{i \in \{1,\ldots,n\}} H_i$$

where $H_i = H$ for all $i = 1, \ldots, n$. For $T \in B(H)$ write $T^{(n)}$ for $T \oplus \cdots \oplus T = \oplus_{i=1}^n T$ and for a C*-algebra $\mathcal{A} \subseteq B(H)$ we write

$$\mathcal{A}^{(n)} = \{T^{(n)} : T \in \mathcal{A}\}.$$

1. Explain how $B(H^{(n)})$ can be identified with $M_n(B(H))$.

2. Prove that $\mathcal{A}^{(n)}$ is a C*-algebra, and that $\overline{\mathcal{A}^{(n)}}^{SOT} = (\overline{\mathcal{A}}^{SOT})^{(n)}$.

3. Prove that $(\mathcal{A}^{(n)})' = M_n(\mathcal{A}')$ and that $(\mathcal{A}^{(n)})'' = (\mathcal{A}'')^{(n)}$.

Exercise 11.5.3. Prove that the restriction of multiplication to bounded subsets of $B(H)$ is jointly continuous in the SOT, but not in the WOT. Harder: prove that multiplication is not jointly continuous in the SOT.

Exercise 11.5.4. Show that for H separable, $B(H)_1$ is metrizable in both the weak and the strong operator topologies. The metric for $B(H)_1$ with the weak operator topology is

$$d_w(A, B) = \sum_{m,n=1}^{\infty} 2^{-m-n} |\langle (A - B)x_m, x_n \rangle|,$$

where $\{x_n\}$ is a dense sequence in the unit ball of H. Fill in the rest of the details.

Exercise 11.5.5. An operator $D \in B(H)$ is said to be **diagonalizable** if there exists an orthonormal basis $\{e_i\}_{i \in I}$ of H and a corresponding family of scalars $\{\lambda_i\}_{i \in I}$ such that $De_i = \lambda_i e_i$ for all $i \in I$.

1. Prove that every diagonalizable operator is normal.

2. Prove that an operator $T \in B(H)$ is normal if and only if for every $\epsilon > 0$ there exists a diagonalizable operator $D \in B(H)$ such that $\|D - T\| < \epsilon$.

In fact, one can also arrange that $D - T$ has small norm *and* is compact, but that might be too difficult for an exercise; see [, Corollary II.4.2].

Exercise 11.5.6. Let (X, μ) be a σ-finite measure space and $f \in L^\infty(X, \mu)$. Prove that the spectrum of a multiplication operator M_f is equal to the *essential range* of f, defined by

$$\text{ess-ran}(f) = \{\lambda \in \mathbb{C} : \forall \epsilon > 0. \mu(\{x \in X : |f(x) - \lambda| < \epsilon\}) > 0\}.$$

Exercise 11.5.7. True or false: two multiplication operators are unitarily equivalent if and only if they have the same spectrum.

Exercise 11.5.8. State and prove a spectral theorem for commuting normal tuples.

Exercise 11.5.9. Let T be a normal operator on a Hilbert space. Prove that for every accumulation point $\lambda \in \sigma(T)$ there exists an orthonormal sequence $\{e_n\}$ such that $\|Te_n - \lambda e_n\| \to 0$. Deduce from this and from the spectral theorem the fact that the spectrum of a compact normal operator is either finite or consists of a sequence converging to zero. Conclude that a compact normal operator is diagonalizable.

Exercise 11.5.10. Let T be a normal operator on a Hilbert space and let E denote its spectral measure. Prove that T is compact if and only if $E(\{z : |z| > \epsilon\})$ is a finite rank operator for every $\epsilon > 0$.

Exercise 11.5.11. Prove that if λ is an isolated point in the spectrum of a normal operator T, then λ is an eigenvalue. Deduce the spectral theorem for compact selfadjoint operators.

Exercise 11.5.12. Let T_1, T_2 be two cyclic normal operators. Then T_i is unitarily equivalent to M_ζ on $L^2(X_i, \mu_i)$, where μ_i is a probability measure on $X_i = \sigma(T_i) \subset \mathbb{C}$, and ζ is the identity function $\zeta(z) = z$. Prove that T_1 is unitarily equivalent to T_2 if and only if $X_1 = X_2$ and μ_1 and μ_2 are mutually absolutely continuous. (**Hint:** you may want to recall the Radon–Nikodym theorem.)

Exercise 11.5.13 (Maximal abelian algebras). An algebra $A \subset B(H)$ is said to be *maximal abelian* if it is abelian and if it is not properly contained in any abelian subalgebra of $B(H)$.

1. Prove that A is maximal abelian if and only if $A = A'$.

2. Prove that if μ is a probability measure then $L^\infty(\mu)$ is a maximal abelian, when considered as a subalgebra of $B(L^2(\mu))$.

3. True or false: a maximal abelian algebra must be a von Neumann algebra.

Exercise 11.5.14 (Fuglede's theorem revisited)**.** In this exercise you will give a different proof for Fuglede's theorem (see Exercise 10.7.4) that gives the following refined version: *if H is a Hilbert space and $B \in B(H)$ commutes with a normal operator $A \in B(H)$ then B commutes with $g(A)$ for every g which is a bounded Borel function on $\sigma(A)$.*

1. Explain why this implies the version of Fuglede's theorem stated in Exercise 10.7.4: *if A is normal then $AB = BA$ implies $A^*B = BA^*$.*

2. Explain why it suffices to prove the result for $g = \chi_E$ where $E \subseteq \sigma(A)$ is a Borel set.

3. Explain why it suffices to prove that the space $g(A)H$ is reducing for B.

4. Explain why, in fact, it suffices to prove that $g(A)H$ is invariant for B.

5. Now prove what remains to be shown (you may assume that A is a multiplication operator on some L^2 space (why?)).

Chapter 12

Representations of C*-algebras

In this chapter, we treat additional aspects of the theory of C*-algebras: positivity and order in C*-algebras, states, and representations. We then prove the fundamental Gelfand–Naimark theorem, which says that every C*-algebra can be faithfully represented on a Hilbert space. We close the chapter with a discussion of irreducible representations and present an application to group theory. This chapter serves as a culmination of many of the themes studied throughout the book.

12.1 Order on C*-algebras and states

Throughout this section, let A be a unital C*-algebra. Recall from Definition 10.4.1 that $a \in A$ is said to be **positive** if $a = a^*$ and $\sigma(a) \subseteq [0, \infty)$. We write $a \geq 0$ to indicate that a is positive. The set of positive elements in A is sometimes denoted by A_+. The notion of positivity allows us to define a partial order on the space A_{sa} of selfadjoint elements in A by

$$a \leq b \iff b - a \geq 0.$$

In particular, we write $a \leq 0$ if $-a \geq 0$, which is the same as $a = a^*$ and $\sigma(a) \subseteq (-\infty, 0]$.

Lemma 12.1.1. Let $a \in A_{sa}$, $\|a\| \leq 1$. Then $a \geq 0$ if and only if $\|1 - a\| \leq 1$.

Proof. By Gelfand theory for commutative C*-algebras (see Section 10.2), we may assume that a is a real valued function on a compact space X, in which case the conclusion is straightforward. We leave the details to the reader. □

Lemma 12.1.2. If $a, b \in A_+$ then $a + b \in A_+$.

Proof. If a and b were known to belong to the same commutative C*-algebra, this would be immediate, as in the previous lemma. Since a and b are not assumed to commute, a subtler argument is required.

Assuming, without loss of generality, that $\|a\|, \|b\| \leq 1$, we have that $\frac{1}{2}(a + b) \in A_{sa}$ and $\left\|\frac{1}{2}(a + b)\right\| \leq 1$. By Lemma 12.1.1, $\|1 - a\|, \|1 - b\| \leq 1$, so

$$\left\|1 - \frac{1}{2}(a + b)\right\| \leq \frac{1}{2}\|1 - a\| + \frac{1}{2}\|1 - b\| \leq 1.$$

DOI: 10.1201/9781003297864-12

Applying Lemma 12.1.1 once again, we find that $\frac{1}{2}(a+b) \geq 0$, and therefore $a + b \geq 0$. □

Corollary 12.1.3. *The relation $a \geq b$ is a partial order.*

Crucial in the theory of C*-algebras is the fact that for every $a \in A$, the element a^*a is positive. This is easy to show if A is a C*-subalgebra of some $B(H)$, as we saw in Theorem 10.4.5. In the next section, we will prove the Gelfand Naimark theorem, which says that every C*-algebra can be represented faithfully on some Hilbert space. However, the proof of the Gelfand–Naimark theorem depends on the positivity of a^*a, so we will not be able to deduce this fact from the easy result for operators. A key step is taken in the following lemma.

Lemma 12.1.4. *For all $a \in A$, if $a^*a \leq 0$, then $a = 0$.*

Proof. Suppose that $\sigma(a^*a) \subseteq (-\infty, 0]$. Recall that by Exercise 8.7.8, $\sigma(ab) \cup \{0\} = \sigma(ba) \cup \{0\}$ for every two elements a, b in a unital Banach algebra. It follows that $\sigma(aa^*) \subseteq (-\infty, 0]$, so $a^*a + aa^* \leq 0$ by Lemma 12.1.2. But if write $a = x + iy$ with $x, y \in A_{sa}$, we find that $a^*a + aa^* = 2x^2 + 2y^2$, and so $x^2 + y^2 \leq 0$, which can be written as $-x^2 \geq y^2$. But by Theorem 10.4.2 both x^2 and y^2 are positive, so x^2 is both positive and negative whence $\sigma(x^2) \subseteq (-\infty, 0] \cap [0, \infty) = \{0\}$. We conclude that $x = 0$ (because $\|x\|^2 = \|x^2\| = r(x^2) = 0$), and similarly one can show that $y = 0$, so that $a = 0$, as required. □

We now come to the crucial result that $a^*a \geq 0$ for all a. Clearly a^*a is selfadjoint, so the result amounts to $\sigma(a^*a) \subseteq [0, \infty)$. If a is assumed self-adjoint, this would follow immediately from the spectral mapping theorem. For general a the proof uses a clever application of the continuous functional calculus to the element a^*a.

Theorem 12.1.5. *For all $a \in A$, $a^*a \geq 0$.*

Proof. Define
$$f(t) = \begin{cases} \sqrt{t}, & t \geq 0 \\ 0, & t < 0 \end{cases}$$
and
$$g(t) = \begin{cases} 0, & t \geq 0 \\ \sqrt{-t}, & t < 0 \end{cases}.$$

We note that $f(t)^2 - g(t)^2 \equiv t$ and $gf \equiv 0$. By the functional calculus $a^*a = x^2 - y^2$ for $x = f(a^*a)$ and $y = g(a^*a)$. Our goal is to show $y^2 = 0$, which will give $a^*a = x^2 \geq 0$. Now, using the identities
$$a^*a = x^2 - y^2 \quad \text{and} \quad xy = yx = 0,$$

we get

$$(ay)^*ay = ya^*ay = y(x^2 - y^2)y = -y^4 \leq 0.$$

By Lemma 12.1.4, $ay = 0$, which in turn implies that $y^4 = 0$. Since y^2 is selfadjoint, we find that $y^2 = 0$, giving us $a^*a = x^2 \geq 0$ as required. □

Lemma 12.1.6. *For all $a, b \in A$, we have that $a^*a \leq \|a\|^2 \cdot 1$ and $b^*a^*ab \leq \|a\|^2 b^*b$.*

Proof. Indeed, $a^*a \geq 0$ and $r(a^*a) = \|a^*a\| = \|a\|^2$, so $\sigma(a^*a) \subseteq [0, \|a\|^2]$. It follows that $\sigma(\|a\|^2 \cdot 1 - a^*a) \subseteq [0, \|a\|^2]$, and therefore $\|a\|^2 \cdot 1 - a^*a \geq 0$, as required. Now, $\|a\|^2 \cdot 1 - a^*a = p^2$ for some $p \geq 0$ (by Theorem 10.4.2) and so

$$b^*(\|a\|^2 \cdot 1 - a^*a)b = b^*p^2b = (pb)^*pb \geq 0$$

which means that $b^*a^*ab \leq \|a\|^2 b^*b$ as claimed. □

Definition 12.1.7. A linear functional $f \in A^*$ is said to be **positive** if $f(a) \geq 0$ for all $a \geq 0$. A **state** is a positive linear functional f such that $f(1) = 1$. The set of all states on A is denoted $S(A)$.

It will follow from Theorem 12.3.2 that if $f \in S(A)$ then $\|f\| \leq 1$.

Example 12.1.8. Suppose that A is a unital C*-subalgebra of $B(H)$ and let $h \in H$ be a unit vector. Then

$$f(a) = \langle ah, h \rangle$$

is a state on A.

Example 12.1.9. Suppose that X is a compact Hausdorff space and let $A = C(X)$. If μ is a regular probability measure on X, then

$$\rho(g) = \int g d\mu$$

is a state on A. By the Riesz representation theorem, these are all the states on A (see Theorem 3.4.1 and the following remark).

We close this section with two lemmas that will be needed later.

Lemma 12.1.10. *If f is a state then f is selfadjoint, in the sense that $f(a^*) = \overline{f(a)}$.*

Proof. If $a \in A$ is selfadjoint then we can use the functional calculus to write $a = a_+ - a_-$ where $a_+, a_- \in A_+$, so $f(a) = f(a_+) - f(a_-) \in \mathbb{R}$. If a is a general element write $a = x + iy$ where x and y are selfadjoint. Now we have $f(a^*) = f(x - iy) = f(x) - if(y) = \overline{f(a)}$. □

Lemma 12.1.11. *Let $\rho : C(X) \to \mathbb{C}$ be a linear functional such that $\rho(1) = \|\rho\| = 1$. Then ρ is a state.*

Proof. We may assume that ρ is given by $\rho(g) = \int g d\mu$ for some $\mu \in M(X)$ with $\int d\mu = \mu(X) = 1$. Let $g \geq 0$. We need to show that $\int g d\mu \geq 0$. We may as well assume that $g \leq 1$. In this case, we have $\|1 - g\| \leq 1$. Since $\|\rho\| = \rho(1) = 1$, we have

$$1 \geq \left| \int (1 - g) d\mu \right| = \left| \int d\mu - \int g d\mu \right| = \left| 1 - \int g d\mu \right|.$$

It follows that $\int g d\mu \geq 0$. □

12.2 Representations and ∗-homomorphisms of C*-algebras

Recall that a ∗-homomorphism between two C*-algebras A and B is a homomorphism of algebras that preserves the involution and that a ∗-homomorphism from a C*-algebra A into $B(H)$ for some Hilbert space H is said to be a representation (Definition 10.1.10). Recall also that by the word *ideal* we mean a two sided ideal.

Lemma 12.2.1. *Let J be a proper closed ideal in a unital C*-algebra A. Assume that J is selfadjoint, in the sense that $a \in J \iff a^* \in J$. Then the operation $(a + J)^* = a^* + J$ makes the quotient A/J into a unital C*-algebra.*

Proof. By Proposition 9.1.1, the quotient A/J is a Banach algebra, and it is clear that $1 + J$ acts as a unit in A/J. Also, since J is selfadjoint, the operation $(a + J)^* = a^* + J$ is a well defined involution and

$$\|a^* + J\| = \inf\{\|a^* - b\| : b \in J\} = \inf\{\|a - b\| : b \in J\} = \|a + J\|.$$

It remains to prove that the C*-identity holds. We claim that for all $a \in A$,

$$\|a + J\| = \inf\{\|a(1 - b)\| : b \in J, 0 \leq b \leq 1\}.$$

Indeed, let $\epsilon > 0$, and let $k \in J$ such that

$$\|a - k\| < \|a + J\| + \epsilon.$$

Let $f \geq 0$ be a nondecreasing continuous function on $[0, \infty)$ such that $f(0) = 0$ and $f(t) = 1$ for all $t \geq \epsilon$. Now, $|t(1 - f(t))| \leq \epsilon$ for all $t \in [0, \infty)$, therefore by the continuous functional calculus $\|k^*k(1 - f(k^*k))\| \leq \epsilon$. Since f vanishes at the origin, it can be uniformly approximated by polynomials with no constant term. Setting $b = f(k^*k)$, it follows that $b \in J$. By the C*-identity,

$$\|k(1 - b)\|^2 = \|(1 - b)k^*k(1 - b)\| \leq \|k^*k(1 - b)\| \leq \epsilon,$$

thus

$$\|a(1-b)\| \le \|(a-k)(1-b)\| + \|k(1-b)\| < \|a+J\| + \epsilon + \sqrt{\epsilon},$$

for all a and all ϵ, proving the claim. We apply this to a^*a and then to a, obtaining

$$
\begin{aligned}
\|(a^* + J)(a + J)\| &= \|a^*a + J\| \\
&= \inf\{\|a^*a(1-b)\| : b \in J, 0 \le b \le 1\} \\
&\ge \inf\{\|(1-b)a^*a(1-b)\| : b \in J, 0 \le b \le 1\} \\
&= \inf\{\|a(1-b)\|^2 : b \in J, 0 \le b \le 1\} \\
&= \|a + J\|^2.
\end{aligned}
$$

On the other hand, as usual in a Banach $*$-algebra,

$$\|(a^* + J)(a + J)\| \le \|a^* + J\|\|a + J\| = \|a + J\|^2,$$

and the proof is complete. $\qquad\square$

Remark 12.2.2. The assumption of selfadjointness in Lemma 12.2.1 can be dropped: it is a significant and nontrivial fact that every closed ideal in a C*-algebra is selfadjoint (see Exercise 12.6.3). Thus, the quotient of a unital C*-algebra by a proper closed ideal is a unital C*-algebra. The result also holds if the word "unital" is removed.

Theorem 12.2.3. *Let* $\pi\colon A \to B$ *be a $*$-homomorphism between two unital C*-algebras. Then* $\|\pi\| \le 1$, *and* $\pi(A)$ *is a C*-subalgebra of* B. *Moreover, if* π *is injective, then* π *is isometric.*

Proof. We first prove the result under the assumption that π is unital, i.e. that $\pi(1_A) = 1_B$; in the final paragraph of the proof we shall explain what to do if this is not the case. So suppose that π is unital. If $a \in A$ is invertible in A then $\pi(a)$ is invertible in $\pi(A)$. This implies that $\sigma(\pi(a)) \subseteq \sigma(a)$. For every $a \in A$, a^*a and $\pi(a^*a)$ are selfadjoint, thus

$$\|\pi(a)\|^2 = \|\pi(a^*a)\| = \sup\{|\lambda| : \lambda \in \sigma(\pi(a^*a))\}.$$

But $\sigma(\pi(a^*a)) \subseteq \sigma(a^*a)$, so the right hand side is less than

$$\sup\{|\lambda| : \lambda \in \sigma(a^*a)\} = \|a^*a\| = \|a\|^2.$$

We find that $\|\pi(a)\| \le \|a\|$ for every $a \in A$. In other words, we established that $\|\pi\| \le 1$.

Now assume that π is not isometric. Find $b \in A$ such that $\|b\| = 1$ and $\|\pi(b)\| < 1$, and put $a = b^*b$. Then $a \in A_+$ satisfies that $\|a\| = \|b\|^2 = 1$ and $\|\pi(a)\| = \|\pi(b)\|^2 < 1$. Let $f \in C_{\mathbb{R}}([0,1])$ be such that $f \equiv 0$ on the interval $[0, \|\pi(a)\|]$, and such that $f(1) = 1$. Now, f is identically zero on $\sigma(\pi(a))$, so by the continuous functional calculus, $f(\pi(a)) = 0$. Since π is a homomorphism, we have $p(\pi(a)) = \pi(p(a))$ for every polynomial. It follows by continuity of π that $f(\pi(a)) = \pi(f(a))$, whence $\pi(f(a)) = 0$. But $f(a) \neq 0$ because f is not identically zero on $\sigma(a)$, therefore π is not injective.

It remains to show that $\pi(A)$ is always a C*-algebra. The only nontrivial part of this statement is that $\pi(A)$ is closed. In case π is injective then we showed that it is isometric and so its image $\pi(A)$ is closed. In general, we consider $K = \ker \pi$. Being the kernel of a $*$-homomorphism, K is a selfadjoint ideal. Since π is bounded, K is closed. By Lemma 12.2.1, the quotient A/K is a unital C*-algebra. Now π induces an injective and unit preserving $*$-isomorphism $\dot\pi \colon A/K \to B$. By what we showed $\dot\pi$ is isometric and so $\dot\pi(A/K) = \pi(A)$ is closed in B. This concludes the proof of the theorem under the assumption that π is unital.

Finally, we drop the assumption that $\pi(1_A) = 1_B$. Put $p = \pi(1_A)$. Then $p = p^2 = p^*$ is a projection. Defining $C = pBp$, we see that C is a C*-subalgebra of B, and it is a unital algebra with the unit $1_C = p$. Now, $\pi(A) = \pi(1)\pi(A)\pi(1) \subseteq C$, so we are free to think of π as a unital $*$-homomorphism from A into C. Now the first part of the proof shows that $\pi(A) \subseteq C \subseteq B$ is closed, that $\|\pi\| \leq 1$, and that π is isometric in case it is injective. $\qquad\square$

Exercise 12.2.4. Prove that Theorem 12.2.3 holds for not necessarily unital C*-algebras. (**Hint:** you might want to use Proposition 10.1.6.)

Theorem 12.2.3 contains three remarkable conclusions. First, the purely algebraic condition of being a $*$-homomorphism implies boundedness. Second, on top of being bounded, the map π also has a closed range. Third, the purely set theoretical assumption of injectivity implies that a $*$-homomorphism is isometric.

Corollary 12.2.5. *Let A be a unital C*-algebra with norm $\| \cdot \|$. If $\| \cdot \|'$ is another norm on the underlying $*$-algebra A that makes it into a C*-algebra, then $\| \cdot \|' = \| \cdot \|$.*

Example 12.2.6. Every norm on $\mathbb{C}^n$ induces an operator norm on the algebra M_n of $n \times n$ matrices that makes it into a Banach algebra, and all these norms are equivalent. But if we define involution as usual by the conjugate transposition, then the only norm that makes M_n into a C*-algebra is the one induced by the Euclidean norm.

If A is a nonunital Banach algebra, and if $\tilde{A} = A \oplus \mathbb{C}$ is its unitization, then there might be more than one norm on $\tilde{A}$ that makes it into a Banach algebra (see Example 8.1.9). For C*-algebras, the unitization $\tilde{A}$ defined in Proposition 10.1.6 is unique. To be precise:

Corollary 12.2.7. *Let A be a nonunital C*-algebra. There is a unique (up to isometric $*$-isomorphism) unital C*-algebra that contains A as a two sided ideal of codimension one.*

Exercise 12.2.8. Prove Corollary 12.2.7.

For the sequel, we shall require a few more definitions. Let A be a unital C*-algebra, and suppose that for all i in some index set Λ we have a representation $\pi_i \colon A \to B(H_i)$. Then, by Theorem 12.2.3, $\|\pi_i\| \leq 1$ for all $i \in \Lambda$, and we can form the representation $\pi = \oplus_i \pi_i$ on the Hilbert space $H = \oplus_i H_i$, given by

$$\pi(a) = \bigoplus_{i \in \Lambda} \pi_i(a) \colon (h_i)_{i \in \Lambda} \mapsto (\pi_i(a)h_i)_{i \in \Lambda}.$$

It is easy to see that π is a representation, called the **direct sum** of the family $\{\pi_i\}_{i \in \Lambda}$.

A representation $\pi \colon A \to B(H)$ is said to be **irreducible** if $\pi(A)$ has no nontrivial reducing subspaces (recall Definition 11.1.6). Since $\pi(A)$ is selfadjoint, we see that a representation is irreducible if and only it has no nontrivial invariant subspaces.

Suppose that $M \subseteq H$ is a nontrivial reducing subspace for π. Then $\pi_M \colon a \mapsto \pi(a)|_M$ is a representation, which is said to be a **subrepresentation** of π. Moreover, $M^\perp$ is also reducing, and $\pi = \pi_M \oplus \pi_{M^\perp}$. We see that a representation is irreducible if and only if it cannot be written as the direct sum of two nontrivial sub-representations.

If the closed subspace $[\pi(A)H]$ spanned by $\pi(a)h$ for all $a \in A$ and $h \in H$ is equal to H, then we say that π is **nondegenerate** (cf. Definition 11.4.5). If A is a unital C*-algebra, then π is nondegenerate if and only it is unital (that is, $\pi(1) = I_H$). If π is not nondegenerate then $[\pi(A)H]$ is an invariant subspace, and $\pi(\cdot)|_{[\pi(A)H]}$ is a nondegenerate representation.

Finally, two representations $\pi_i \colon A \to B(H_i)$ ($i = 1, 2$) are said to be **equivalent** if there is a unitary $U \colon H_1 \to H_2$ such that

$$U\pi_1(a)U^* = \pi_2(a)$$

for all $a \in A$.

Proposition 12.2.9. *Let A be a nonunital C*-algebra. A representation $\pi \colon A \to B(H)$ is irreducible if and only if $\pi(A)' = \mathbb{C}I_H$.*

Proof. By Lemma 11.4.3, P is a projection onto an invariant subspace for $\pi(A)$ if and only if P is in the commutant $\pi(A)'$ of $\pi(A)$. Now, if $\pi(A)' = \mathbb{C}I_H$, the only projections in the commutant are 0 and I_H, so π is irreducible. Conversely, if π is irreducible, then the only projections in $\pi(A)'$ are 0 and I_H. By Corollary 11.3.4, the von Neumann algebra $\pi(A)'$ is generated by its projections, and therefore $\pi(A)' = \mathbb{C}I_H$. □

12.3   The GNS construction and the Gelfand–Naimark theorem

Throughout this section A will denote a unital C*-algebra.

Definition 12.3.1. Let $f \in S(A)$. A **GNS representation** for f is a triplet (π, H, ξ) where H is a Hilbert space, $\xi \in H$ is a unit vector, and $\pi \colon A \to B(H)$ is a representation such that

1. ξ is **cyclic**, that is $[\pi(A)\xi] = H$.

2. For all $a \in A$,
$$f(a) = \langle \pi(a)\xi, \xi \rangle.$$

GNS is short for Gelfand, Naimark, and Segal. Note that a GNS representation must be unital, in the sense that $\pi(1) = I$.

Theorem 12.3.2 (The GNS construction). *Every state $f \in S(A)$ has a GNS representation. Moreover, the GNS representation of a state f is unique in the following sense: if (π, H, ξ) and (σ, K, η) are two GNS representations for f, then there exists a unitary $U \colon H \to K$ such that $U\xi = \eta$ and such that*

$$U\pi(a)U^* = \sigma(a), \quad \text{for all } a \in A. \tag{12.1}$$

Proof. **Existence.** We define a sesqui-linear form on A as follows

$$[a, b] = f(b^*a), \quad a, b \in A.$$

By Lemma 12.1.10, $[a, b] = \overline{[b, a]}$ for all $a, b \in A$. By the positivity of f and Theorem 12.1.5, this is a positive semidefinite form in the sense that

$$[a, a] = f(a^*a) \geq 0, \quad \text{for all } a \in A.$$

Since the form $[\cdot, \cdot] \colon A \times A \to \mathbb{C}$ satisfies all the properties of an inner product except (perhaps) definiteness, it follows that the Cauchy-Schwarz inequality holds:
$$|[a, b]| \leq [a, a]^{1/2}[b, b]^{1/2},$$

or

$$|f(b^*a)| \leq f(a^*a)^{1/2}f(b^*b)^{1/2}. \tag{12.2}$$

Now let
$$N = \{a \in A : f(a^*a) = 0\}.$$

By the Cauchy-Schwarz inequality (12.2),

$$N = \{a \in A : f(b^*a) = 0 \text{ for all } b \in A\}. \tag{12.3}$$

This shows that N is a left ideal in A, and in particular, it is a linear subspace. On the quotient space A/N we define an inner product

$$\langle a + N, b + N \rangle = f(b^*a).$$

This is indeed a well defined inner product, thanks to the alternative definition (12.3) of N. Now let H be the completion of the inner product space

$(A/N, \langle \cdot, \cdot \rangle)$. This is the Hilbert space on which we shall define the GNS representation of f.

First, for every $a \in A$, we define an operator $\pi(a)$ on the dense subspace A/N by
$$\pi(a)(b + N) = ab + N.$$

This operator is well defined since N is a left ideal. It is plain to see that $\pi(a)$ is a linear operator of A/N into A/N. Moreover, $\pi(a)$ is bounded, since

$$\begin{aligned}
\|\pi(a)(b + N)\|^2 &= \langle \pi(a)(b + N), \pi(a)(b + N) \rangle \\
&= f(b^* a^* a b^*) \\
&\leq \|a\|^2 f(b^* b) = \|a\|^2 \|b + N\|^2,
\end{aligned}$$

where we used Lemma 12.1.6 and the positivity of f. It follows that $\pi(a)$ can be extended to a bounded operator on A/N, hence it extends to a unique bounded operator on H. For the sake of brevity, we let $\pi(a)$ denote this extension, too. Now it is not difficult to verify, and we shall leave it to the reader, that $a \mapsto \pi(a)$ is a representation of A on H.

Define $\xi = 1 + N$. Since $\langle \xi, \xi \rangle = f(1^*1) = f(1) = 1$, ξ is a unit vector. Since $[\pi(A)\xi] = [A/N] = H$, the vector ξ is indeed cyclic. Finally, for every $a \in A$,
$$\langle \pi(a)\xi, \xi \rangle = \langle a + N, 1 + N \rangle = f(a),$$

as required.

Uniqueness. Suppose that (σ, K, k) is another GNS representation for f. We define $U : H \to K$ by specifying

$$U\pi(a)\xi = \sigma(a)\eta$$

for all $a \in A$. Clearly, $U\xi = U\pi(1)\xi = \sigma(1)\eta = \eta$. We compute

$$\begin{aligned}
\langle U\pi(a)\xi, U\pi(b)\xi \rangle &= \langle \sigma(a)\eta, \sigma(b)\eta \rangle \\
&= f(b^* a) = \langle \pi(a)\xi, \pi(b)\xi \rangle.
\end{aligned}$$

It follows that the map U extends to a unitary $U \colon H \to K$. Now, for all $a, b \in A$,

$$\begin{aligned}
U\pi(a)\pi(b)\xi = U\pi(ab)\xi &= \sigma(ab)\eta \\
&= \sigma(a)\sigma(b)\eta = \sigma(a)U\pi(b)\xi,
\end{aligned}$$

we see that the restrictions of $U\pi(a)$ and $\sigma(a)U$ to the dense subspace $\pi(A)\xi = A/N$ coincide. It follows that $U\pi(a)U^* = \sigma(a)$ for all $a \in A$. $\qquad\square$

The GNS construction has the following important corollary.

Corollary 12.3.3. *If $f \in S(A)$ then $\|f\| \leq 1$. Therefore $S(A)$ is a weak-* compact and convex subset of A_1^*.*

Proof. Using either Theorem 12.2.3 or the estimates obtained in the proof of Theorem 12.3.2, we obtain $|f(a)| \leq |\langle \pi(a)\xi, \xi \rangle| \leq \|\pi(a)\| \leq \|a\|$, so $\|f\| \leq 1$. Convexity and closedness are obvious, and compactness then follows Alaoglu's theorem. $\square$

Example 12.3.4. Let X be a compact Hausdorff space, and let ρ be a state of $C(X)$. Then the GNS representation of ρ can be described in concrete terms. Indeed, let ρ be given by integration against a regular Borel probability measure μ, that is,

$$\rho(f) = \int f d\mu$$

for all $f \in C(X)$. Let $H = L^2(\mu)$, and let $\pi \colon C(X) \to B(H)$ be given by

$$\pi(f) = M_f,$$

where M_f is the multiplication operator given by $M_f(h) = fh$ for all $h \in L^2(\mu)$. If we define ξ to be the constant function 1 on X, we have $[\pi(C(X))\xi] = [C(X)] = L^2(\mu)$, because $C(X)$ is dense in $L^2(\mu)$, so ξ is cyclic for $\pi(C(X))$. Finally,

$$\langle \pi(f)\xi, \xi \rangle = \langle f1, 1 \rangle = \int f d\mu = \rho(f).$$

Therefore $(\pi, L^2(\mu), 1)$ is the GNS representation of ρ.

Theorem 12.3.5. *For all $a \in A$ there exists a state $f \in S(A)$ such that*

$$f(a^*a) = \|a\|^2.$$

Proof. Let B denote the unital C*-algebra generated by a^*a and 1. By Theorem 10.2.5, $B \cong C(X)$ for a compact Hausdorff space X. In this identification, a^*a corresponds to a nonnegative function $g \in C(X)$. There exists a point $x \in X$ such that $g(x) = \|g\|_\infty$, and so the state δ_x satisfies $\delta_x(g) = \|g\|_\infty$. This gives a state $f_0 \in S(B)$ such that $f_0(a^*a) = \|a^*a\| = \|a\|^2$. By Hahn–Banach, f extends to a functional $f \in A^*$ such that $\|f\| = \|f_0\| = 1$ and $f(1) = f_0(1) = 1$. We claim that f is a state on A.

Indeed, let $b \in A_+$, and consider $\rho = f|_C$, where C is the commutative C*-subalgebra of A generated by 1 and b. Being the restriction of f, the functional ρ satisfies $\rho(1) = \|\rho\| = 1$, so Lemma 12.1.11 implies that ρ is a state. In particular, $f(b) = \rho(b) \geq 0$, as required. $\square$

Recall that an injective representation is isometric. It is common to refer to such a representation as **faithful**.

Theorem 12.3.6 (Gelfand–Naimark theorem). *Every unital C*-algebra A can be represented faithfully on a Hilbert space. In other words, for every unital C*-algebra A there exists a Hilbert space H and a unit preserving, isometric *-homomorphism $\pi \colon A \to B(H)$.*

Proof. For every nonzero $a \in A$, use Theorem 12.3.5 to find $f_a \in S(A)$ such that $f_a(a^*a) = \|a\|^2$. Let (π_a, H_a, ξ_a) be the GNS representation for f_a.

We now construct the Hilbert space

$$H = \bigoplus_{0 \neq a \in A} H_a$$

and the representation $\pi = \oplus_a \pi_a$ given by

$$\pi(b)(h_a)_a = (\pi_a(b)h_a)_a$$

for all $b \in A$ and $(h_a)_a \in H$. Every π_a is contractive so π is a well defined, contractive $*$-homomorphism. Finally, since

$$\|a\|^2 = f_a(a^*a) = \langle \pi(a^*a)\xi_a, \xi_a \rangle = \|\pi(a)\xi_a\|^2 \leq \|\pi_a(a)\|^2,$$

we conclude that π is isometric. $\qquad\qquad\qquad\qquad\qquad\qquad\qquad\square$

12.4   Pure states and irreducibility of representations

Definition 12.4.1. A state f on a C*-algebra A is said to be **pure** if f is an extreme point of the state space $S(A)$. The set of pure states on A is denoted $P(A)$

Theorem 12.4.2. *Let A be a unital C*-algebra, and let f be a state. Then f is a pure state if and only if its GNS representation (π, H, ξ) is irreducible.*

Proof. Suppose that π is reducible, say $H = H_1 \oplus H_2$ and $\pi = \pi_1 \oplus \pi_2$. Since ξ is cyclic, it follows that $\xi_i = P_i\xi \neq 0$ for $i = 1, 2$, where $P_i = P_{H_i}$ is the orthogonal projection onto H_i. Putting $\eta_i = \frac{1}{\|\xi_i\|}\xi_i$ and defining

$$g_i(a) = \langle \pi_i(a)\eta_i, \eta_i \rangle,$$

we have that $g_1, g_2 \in S(A)$ and that $f = \|\xi_1\|^2 g_1 + \|\xi_2\|^2 g_2$ is a convex combination of states. If we could rule out the possibility that $g_1 = g_2 = f$, then we would see that f is not pure. But if $g_1 = g_2 = f$, then for all $a \in A$,

$$0 = \langle \pi(a)\eta_1, \eta_1 \rangle - \langle \pi(a)\eta_2, \eta_2 \rangle$$

$$= \left\langle \pi(a)\xi_1, \frac{1}{\|\xi_1\|}\eta_1 \right\rangle - \left\langle \pi(a)\xi_2, \frac{1}{\|\xi_2\|}\eta_2 \right\rangle$$

$$= \left\langle \pi(a)\xi_1, \frac{1}{\|\xi_1\|}\eta_1 - \frac{1}{\|\xi_2\|}\eta_2 \right\rangle + \left\langle \pi(a)\xi_2, \frac{1}{\|\xi_1\|}\eta_1 - \frac{1}{\|\xi_2\|}\eta_2 \right\rangle$$

$$= \left\langle \pi(a)\xi, \frac{1}{\|\xi_1\|}\eta_1 - \frac{1}{\|\xi_2\|}\eta_2 \right\rangle.$$

It follows that $\frac{1}{\|\xi_1\|}\eta_1 - \frac{1}{\|\xi_2\|}\eta_2$ is orthogonal to $[\pi(A)\xi]$, hence $\frac{1}{\|\xi_1\|}\eta_1 = \frac{1}{\|\xi_2\|}\eta_2$. Since η_1 and η_2 live in orthogonal subspaces, this is impossible. We have thus ruled out the possibility that $g_1 = g_2 = f$, whence f is a proper convex combination of states so it is not pure.

For the converse, suppose that π is irreducible, and suppose that there are $g_1, g_2 \in S(A)$ and $t \in (0,1)$ such that $f = tg_1 + (1-t)g_2$. We need to show that $g_1 = g_2 = f$. It suffices to show that $g_1 = f$.

We define a sesqui-linear form $[\cdot, \cdot]$ on the dense linear subspace $\pi(A)\xi$ by

$$[\pi(a)\xi, \pi(b)\xi] = tg_1(b^*a).$$

It is straightforward to show that $[\cdot, \cdot]$ is a bounded semi-inner product, and hence that it extends continuously to a semi-inner product on H. By a standard application of the Riesz representation theorem in Hilbert space[1] there exists $T \in B(H)$ on H such that

$$[h_1, h_2] = \langle Th_1, h_2 \rangle$$

for all $h_1, h_2 \in H$. Unwinding the definitions, we see that

$$tg_1(b^*a) = [\pi(a)\xi, \pi(b)\xi] = \langle T\pi(a)\xi, \pi(b)\xi \rangle$$

for all $a, b \in A$. Plugging in $b = b_1^*b_2$ into the above and computing $g_1((b_2^*b_1)a) = g_1(b_2^*(b_1a))$ in two ways, we find that

$$\langle \pi(b_1)T\pi(a)\xi, \pi(b_2)\xi \rangle = \langle T\pi(b_1)\pi(a)\xi, \pi(b_2)\xi \rangle$$

for all $a, b_1, b_2 \in A$. Since ξ is cyclic for π, we conclude that $T \in \pi(A)'$. By Proposition 12.2.9, since $\pi(A)$ is irreducible, it follows that $T = rI_H$. But this means that g_1 is a scalar multiple of f; since both are states, we conclude that $g_1 = f$. $\qquad\square$

The connection between representations and states allows us to improve Theorem 12.3.6 considerably.

Theorem 12.4.3. *Let A be a unital C^*-algebra. For all $a \in A$ there exists a pure state $f \in P(A)$ such that*

$$f(a^*a) = \|a\|^2.$$

If (π_f, H_f, ξ_f) denotes the GNS representation of f, then π_f is an irreducible representation such that

$$\|\pi_f(a)\| = \|a\|.$$

Consequently, the direct sum

$$\pi = \bigoplus_{f \in P(A)} \pi_f$$

is a faithful representation of A on $\bigoplus_{f \in P(A)} H_f$.

---

[1]See, e.g., Exercise 5.7.11 in [14].

Proof. For a fixed $a \in A$, define

$$F = \{g \in S(A) : g(a^*a) = \|a\|^2\}.$$

By Theorem 12.3.5, $F \neq \emptyset$, and so it is a weak-$*$ compact face in $S(A)$ (Lemma 7.1.12). By the Krein–Milman theorem, F has an extreme point f, which must also be an extreme point of $S(A)$ (Exercise 7.1.10). In other words $f \in P(A)$. Now if (π_f, H_f, ξ_f) is the GNS representation of f, then

$$\|a\|^2 = f(a^*a) = \langle \pi_f(a^*a)\xi, \xi \rangle = \|\pi_f(a)\xi\|^2 \leq \|\pi_f(a)\|^2 \leq \|a\|^2,$$

thus $\|\pi_f(a)\| = \|a\|$. The rest follows as in the proof of Theorem 12.3.6. $\square$

12.5   Application: existence of irreducible unitary group representations

Let G be a group. A ***unitary representation*** of G is a group homomorphism U from G into the group $\mathcal{U}(H)$ of unitaries on some Hilbert space H. It is common practice to write U_g for $U(g)$ for $g \in G$. Thus, $G \ni g \mapsto U_g \in B(H)$ is a unitary representation if

$$U_g U_g^* = U_g^* U_g = I_H \quad \text{and} \quad U_{gh} = U_g U_h$$

for all $g, h \in G$. It is plain to see that for a unitary representation $U_{g^{-1}} = U_g^*$ for all g and that $U_e = I_H$, where e is the unit element in G.

Unitary representations are an incredibly useful gadget in the study of groups. In this section, we will illustrate the power of the theory of C*-algebras with a very quick application to group representation theory: we shall show that every group G has an *irreducible* unitary representation. In fact, we will show that there are enough irreducible representations to faithfully represent every element of G.

Unitary representations of groups are reminiscent of representations of C*-algebras. We shall use concepts like "direct sum" and "sub-representation" without defining them formally, confident that the reader who has made it this far in the book will be able to follow. We do, however, pause to define the key notion of this section.

Definition 12.5.1. A unitary representation $U \colon G \to \mathcal{U}(H)$ is said to be ***irreducible*** if there is no nontrivial proper closed subspace $M \subseteq H$ that is invariant under the family of operators $\{U_g : g \in G\}$.

Irreducible representations are of special importance and interest. It is not too hard to show that every unitary representation of a finite group on a finite dimensional space breaks up as the direct sum of irreducible representations, hence the study of unitary representations of finite groups reduces to the study of irreducible representations.

What about infinite groups? The active readers might, even as they read these lines, think they have a proof that every unitary representation decomposes into a direct sum of unitary representations. These readers are invited to pause and check whether their proof really goes through.

For infinite groups it is no longer true that every representation is the direct sum of irreducible representations — in fact, this fails already for the simplest example of an infinite group (see Exercise 12.6.14). Once one realizes that there are representations that do not contain any irreducible sub-representation, it is no longer clear that any nontrivial irreducible representations exist at all. Of course, the trivial representation $U\colon G \to \mathbb{T}$ given by $U_g = 1$ for all g is irreducible, but can we find enough irreducible representations to separate points?

Theorem 12.5.2. *Let G be a group. For very $g_1 \neq g_2 \in G$ there exists an irreducible representation $U\colon G \to \mathcal{U}(H)$ such that $U_{g_1} \neq U_{g_2}$.*

Proof. Consider the Hilbert space $\ell^2(G)$ with the orthonormal basis $\{\delta_h\}_{h \in G}$ given by

$$\delta_h(g) = \begin{cases} 1 & g = h, \\ 0 & \text{else.} \end{cases}$$

We define a unitary representation of G on $\ell^2(G)$. For $g \in G$, define $L_g\colon \ell^2(G) \to \ell^2(G)$ by

$$L_g \delta_h = \delta_{gh}.$$

It is easy to verify that L is a unitary representation with $L_{g^{-1}} = L_g^*$. This representation is faithful, in the sense that $L_g = I$ if and only if $g = e$. The representation $g \mapsto L_g$ is a common construction and is referred to as **the left regular representation** of G. The existence of the left regular representation shows that G has a faithful unitary representation. However, it is almost never irreducible (see Exercise 12.6.15).

To find an irreducible representation, we consider the C*-algebra A generated by $\{L_g : g \in G\}$. Let $g_1 \neq g_2 \in G$. By Theorem 12.4.3, there exists a (nondegenerate) irreducible representation $\pi\colon A \to B(H)$ such that $\pi(L_{g_1} - L_{g_2}) \neq 0$. As a unital *-homomorphism, π sends unitaries to unitaries. If we define $U\colon G \to \mathcal{U}(H)$ by $U_g = \pi(L_g)$, then U is a unitary representation of G and $U_{g_1} \neq U_{g_2}$.

Finally, since π is irreducible, then U is also irreducible. Indeed, if M is invariant under $U_h = \pi(L_h)$ for all $h \in G$, then it is invariant under $\pi(T)$ for every T in the C*-algebra A generated by $\{L_h : h \in G\}$, and so by the irreducibility of π it follows that M is trivial. $\square$

The construction and study of various C*-algebras associated with groups is an important area within the theory of C*-algebras. The C*-algebra A generated by $\{L_g : g \in G\}$ that we defined above is called **the reduced** *group C*-algebra of G*, and sometimes denoted $C_r^*(G)$. There are other C*-algebras that one may attach to a group, either to obtain information about the group or to construct a C*-algebra with certain desirable properties. Good references in which the reader can find further information are [8] or [14].

12.6 Additional exercises

Exercise 12.6.1. Let A be unital C*-algebra.

1. Prove that if $0 \le a \le b$ are invertible, then $b^{-1} \le a^{-1}$.

2. Find an example such that $0 \le a \le b$ but $a^2 \not\le b^2$.

Exercise 12.6.2. An **approximate identity** for a C*-algebra A is a net $\{e_\alpha\}_{\alpha \in \Lambda}$ such that $e_\alpha = e_\alpha^* \ge 0$ and $\|e_\alpha\| \le 1$, and such that $\lim_\alpha \|ae_\alpha - a\| = 0$ for all $a \in A$. (Sometimes one requires that $e_\alpha \le e_\beta$ when $\alpha \le \beta$; the net constructed in this exercise satisfies this, too). In this exercise you will prove that every C*-algebra has an approximate identity.

Suppose that A is nonunital, and let us work freely in the unitization $\tilde{A}$. Let Λ be the set of all finite sets of selfadjoint elements of A, ordered by inclusion (clearly Λ is a directed set). We will need the function

$$f_n(t) = nt(1 + nt)^{-1}.$$

1. For every $\alpha = \{x_1, \ldots, x_n\}$ define $e_\alpha = f_n(x_1^2 + \cdots + x_n^2)$ by the functional calculus. Prove that e_α is selfadjoint and that $\sigma(e_\alpha) \subseteq [0, 1]$ for all α.

2. Show that if $\alpha \le \beta$ then $e_\alpha \le e_\beta$. (**Hint:** it suffices to prove that $1 - e_\beta \le 1 - e_\alpha$ in $\tilde{A}$. For this, use the fact $a \le b \Rightarrow b^{-1} \le a^{-1}$ for invertible elements in a C*-algebra, see Exercise 12.6.1.)

3. Explain why in order to prove that $\lim_\alpha \|a - ae_\alpha\| = 0$ it suffices to prove this for selfadjoint a.

4. For a selfadjoint a, show that whenever $\alpha = \{a_1, \ldots, a_n\} \in \Lambda$ contains a and has at least m elements, then $\|a - ae_\alpha\| \le 1/4m$. For this, note that
$$(1 - e_\alpha)a^2(1 - e_\alpha) \le (1 - e_\alpha)(a_1^2 + \cdots + a_n^2)(1 - e_\alpha),$$
and identify the right hand side as $g_n(a_1^2 + \cdots + a_n^2)$ for an appropriate function $g_n(t)$.

5. Conclude that $\{e_\alpha\}_{\alpha \in \Lambda}$ is an approximate identity.

6. Prove that if A is separable, then one can find a sequence $\{a_n\}_{n=1}^\infty$ that is an approximate identity for A.

Exercise 12.6.3. Prove that every closed ideal J in a C*-algebra is self-adjoint. Conclude that the quotient of a (not necessarily unital) C*-algebra by a closed ideal is a C*-algebra. (**Hint:** the subspace $J \cap J^* \subseteq J$ is a C*-algebra, which is nonzero, because it contains JJ^*. Therefore, $J \cap J^*$ has an approximate unit. Use it.)

Exercise 12.6.4. Let A be a C*-subalgebra of $B(H)$, and let K be the ideal of compact operators in $B(H)$. Prove that if there exists a compact operator $k \in K$ and an element $a \in A$ such that $\|a - k\| < \|a\|$, then $A \cap K \neq \{0\}$.

Exercise 12.6.5. Let A be a C*-algebra. Suppose that J is a closed ideal and that B is a C*-subalgebra of A. Prove that $B + J$ is a C*-subalgebra in A and that $(B + J)/J \cong B/B \cap J$.

Exercise 12.6.6. Prove the "Cantor-Schroeder-Bernstein theorem for representations": if π and ρ are representations of a C*-algebra A such that π is unitarily equivalent to a sub-representation of ρ, and ρ is unitarily equivalent to a sub-representation of π, then π is unitarily equivalent to ρ.

Exercise 12.6.7 (The Toeplitz algebra). The purpose of this exercise is to study the structure of *the Toeplitz algebra*, which may be the most basic example of a noncommutative C*-algebra, after the compacts $K(H)$ and the full algebra of bounded operators $B(H)$.

Let S be the unilateral shift on $\ell^2(\mathbb{N})$, i.e. the operator given by $Se_n = e_{n+1}$ for $n = 0, 1, 2, \ldots$. The **Toeplitz algebra** is the C*-algebra $C^*(S)$ generated by S. Identify $\ell^2(\mathbb{N})$ with the Hilbert space

$$H^2 = \left\{ \sum_{n=0}^{\infty} a_n z^n : \sum |a_n|^2 < \infty \right\} \subset L^2(\mathbb{T}),$$

that is, H^2 is considered as the closed subspace of $L^2(\mathbb{T})$ that consists of those functions in L^2 whose negative Fourier coefficients vanish. Identify S with the multiplication operator $M_z : H^2 \to H^2$. Feel free to switch between S and M_z as convenient.

1. For every $f \in C(\mathbb{T})$ define an operator $T_f \in B(H^2)$ by $T_f h = P_{H^2} f h$ for $h \in H^2$, where P_{H^2} denotes the orthogonal projection of $L^2(\mathbb{T})$ onto H^2 (such an operator T_f is called a *Toeplitz operator*). Prove that $f \mapsto T_f$ is a linear $*$-preserving positive map ($f \geq 0 \Rightarrow T_f \geq 0$). Show, however, that it is not a representation.

2. Prove that for all $f, g \in C(\mathbb{T})$, the operator $T_{fg} - T_f T_g \in K := K(H^2)$, i.e. it is a compact operator. Conclude that $C^*(M_z)$ is not commutative, but it is *commutative modulo the compacts*.

3. Show that $C^*(S)$ contains all compact operators.

4. Show that every T_f satisfies $T_f = M_z^* T_f M_z$. Deduce that the matrix representation of T_f with respect to the orthonormal basis $1, z, z^2, \ldots$ has constant diagonals (such a matrix is called a *Toeplitz matrix*).

5. Use the fact that M_z is an isometry and the previous item to show that $\|T_f + A\| = \|T_f\|$ for all $A \in K$.

6. Prove that if $f, g \in C(\mathbb{T})$ and $A, B \in K$, $T_f + A = T_g + B$ implies $f = g$ and $A = B$.

7. Show that $C^*(M_z)$ is precisely the set of all operators of the form $T_f + A$ where $f \in C(\mathbb{T})$ and $A \in K$.

It is worth pausing to highlight that we have shown that the ideal K of compact operators is contained in $C^*(M_z)$ and that $C^*(M_z)/K \cong C(\mathbb{T})$. This is summarized by the *exact sequence*

$$0 \to K \to C^*(M_z) \to C(\mathbb{T}) \to 0$$

which is referred to as an *extension of $C(\mathbb{T})$ by the compacts*. Extensions of commutative C*-algebras by the compacts can be used to form a *K-homology group* for compact topological spaces; this goes under the name *Brown-Douglas-Fillmore theory*. See [14].

Exercise 12.6.8. Let A be a unital commutative C*-algebra.

1. Find all irreducible representations of A up to unitary equivalence.

2. Find all pure states of A (one can give two distinct proofs, either *-algebraic or direct).

3. What changes if A is not assumed unital?

Exercise 12.6.9 (Representations of the Toeplitz algebra). Find all the irreducible representations of the Toeplitz algebra defined in Exercise 12.6.7.

Exercise 12.6.10. Prove that every separable C*-algebra can be represented on a separable Hilbert space and that every finite dimensional C*-algebra can be represented on a finite dimensional Hilbert space.

Exercise 12.6.11. Let ξ be a unit vector in an infinite dimensional separable Hilbert space H. Define $\omega_\xi(T) = \langle T\xi, \xi \rangle$ for $T \in B(H)$.

1. Prove that ω_ξ is a pure state on $B(H)$.

2. Prove that any GNS representation of ω_ξ is equivalent to the identity representation.

3. Are all pure states on $B(H)$ of the form ω_ξ?

4. Now let ρ_ξ be the restriction of ω_ξ to the compacts $K(H)$. Repeat the above for ρ_ξ.

Exercise 12.6.12. Prove that if A is a unital C*-algebra and $\rho \in A^*$ satisfies $\rho(1) = 1 = \|\rho\|$, then ρ is a state. Use this to obtain the following special case of the *Krein extension theorem*: if B is a unital C*-subalgebra of a unital C*-algebra A, every positive functional on B can be extended to a positive functional on A. (**Hint:** see the proof of Theorem 12.3.5.)

Exercise 12.6.13. Let H be separable Hilbert space. *Calkin algebra* is the quotient algebra $C(H) := B(H)/K(H)$ of the algebra $B(H)$ by the compacts.

1. Prove that $C(H)$ is simple, in that it contains no ideals (no closed ideals and in fact no ideals whatsoever).

2. True or false: $C(H)$ is separable.

3. True or false: $C(H)$ can be represented on a separable Hilbert space.

Exercise 12.6.14. Let $G = \mathbb{Z}$ be the group of integers, and consider the left regular representation L of G on $\ell^2(G)$ as defined in the proof of Theorem 12.5.2. Show that L has no irreducible sub-representation. In particular, L cannot be written as the direct sum of irreducible representations.

Exercise 12.6.15. Prove that the left regular representation of a group G is *never* irreducible, unless G is the trivial group $G = \{e\}$.

Exercise 12.6.16. True or false: every Banach algebra be represented isometrically as a subalgebra of $B(H)$ where H is a Hilbert space.

Chapter 13

Unbounded operators

Among the most ubiquitous and important linear operators arising in analysis are differential operators. Consider the operator $Af = f'$ of differentiating a function of a single real variable. If we want to analyze this operator within the usual framework of functional analysis, we need to first specify the normed space of functions on which A is supposed to operate. We face a difficulty that A is not defined on the standard and convenient normed spaces such as the space of continuous functions or L^2.

One way to approach this difficulty is to carefully define a space on which A is a well defined bounded operator. For example, the space $X = C^1([0,1])$ of continuously differentiable functions becomes a Banach space if we equip it with the norm

$$\|f\| = \sup_{t\in[0,1]} \|f(t)\| + \sup_{t\in[0,1]} \|f'(t)\|,$$

and then the operator $Af = f'$ is a bounded operator from X into $C([0,1])$. It turns out that using continuously differentiable or everywhere differentiable functions is too naive an approach, which has limited applicability. It has been fruitful to extend the notion of differentiability, by considering *weak derivatives*, as we did in Section 3.3. Care is still required in defining the spaces on which the differentiation operators act. A systematic framework for specifying the spaces on which a given differential operator is an everywhere defined bounded linear operator is provided by the theory of *Sobolev spaces*. Working with Sobolev spaces requires a heavy dose of hard analysis and measure theory, and a significant portion of a graduate course in PDEs is typically dedicated to these spaces; see e.g. [7] or [18].

The *theory of distributions* provides another abstract setting within which one can study differential operators. Briefly, the space $\mathcal{D}'$ of distributions on $\mathbb{R}^d$ is the continuous dual of the topological vector space $\mathcal{D} = C_c^\infty(\mathbb{R}^d)$ consisting of compactly supported, infinitely differentiable functions. Here $\mathcal{D}$ is given the topology determined by declaring that $f_n \to f$ if and only if the supports of all of the functions f_n are contained in a single bounded set and for every $\alpha \in \mathbb{N}^d$ the mixed partial derivatives $\partial^\alpha f_n$ of f_n converge uniformly to $\partial^\alpha f$. It is immediate that differentiation of any order is continuous on $\mathcal{D}$; differential operators are then defined on $\mathcal{D}'$ by duality. For the theory of distributions and its applications to PDEs see [12] or [17].

There is yet another way to incorporate the study of differential operators in a functional analytic framework, which is to treat them as *unbounded operators*, that is, to continue to work within our favorite space but to keep track of the fact that an operator A is not defined on the entire space but only on a subspace D. In our simple example, the operator $Af = f'$ is easily seen to be well defined on a dense subspace $D = C^1([0,1]) \subset L^2[0,1]$, which we shall call the *domain* of A. On the domain D the operator A is unbounded (for example, $\|A(e^{int})\| = \|ine^{int}\| = n\|e^{int}\|$), and cannot be extended in a meaningful way to L^2. This situation requires dramatically changing the way we think of and work with operators.

In this chapter, we study the rudiments of the theory of unbounded operators. Our goal is to get acquainted with the key concepts and the most basic results, such as the spectral theorem for selfadjoint operators. For a more complete treatment of this theory the reader is referred to, e.g., [15, 31, 42, 45].

13.1 Unbounded operators in Banach and Hilbert spaces

13.1.1 Basic definitions and examples

Definition 13.1.1. Let X be a Banach space. An *operator in* X is a linear operator A from a linear subspace $D \subseteq X$ into X. The subspace D is called the *domain* of A and it is usually denoted $D(A)$.

The domain $D(A)$ of A is part of the definition of A. Sometimes, to emphasize the domain, one refers to the pair (A, D) or $(A, D(A))$ instead of simply A. In most cases of interest the domain $D(A)$ is dense in X, in which case we say that A is *densely defined*.

Remark 13.1.2. In some texts[1], a pair $(A, D(A))$ is called an "unbounded operator", or "(unbounded) operator". We use this terminology only informally, because there is nothing in the definition that excludes the possibility that A is a bounded operator. However, the formalism and results presented in this chapter are interesting mostly in the case where one handles densely defined operators that are not bounded. Indeed, if $D(A)$ is dense in X and A is bounded, in the sense that there exists a constant C such that $\|Ax\| \leq C\|x\|$ for all $x \in D(A)$, then A extends uniquely to all of X, thus essentially $A \in B(X)$.

Exercise 13.1.3. Provide an example of an unbounded operator A in a Banach space X that is truly unbounded in the sense that A is *everywhere defined* (meaning that $D(A) = X$) yet $A \notin B(X)$.

Definition 13.1.4. The *kernel* of an operator A is defined to be $\ker A = \{x \in D(A) : Ax = 0\}$. The *image* (or *range*) of A is defined to be the space $\operatorname{Im} A = \{Ax : x \in D(A)\}$.

---

[1]See, for example, [15].

Definition 13.1.5. Let A and B be two operators in a Banach space X. We say that B is an **extension** of A if $D(A) \subseteq D(B)$ and $B\big|_{D(A)} = A$. In this case we write $A \subseteq B$.

Definition 13.1.6. The **graph** of an operator is the linear space

$$\Gamma(A) = \{(x, Ax) : x \in D(A)\} \subseteq X \times X.$$

Exercise 13.1.7. Show that $A \subseteq B$ if and only if $\Gamma(A) \subseteq \Gamma(B)$.

Definition 13.1.8. An operator A is said to be **closed** if its graph $\Gamma(A)$ is a closed subspace of $X \times X$. It is said to be **closable** if it has a closed extension. The minimal closed extension of an operator A is called its **closure** and denoted $\overline{A}$.

Definition 13.1.9. If $(A, D(A))$ is a closed operator, then a linear subspace $D_0 \subseteq D(A)$ is said to be a **core** for A if the closure of the restriction $A\big|_{D_0}$ of A to D_0 is A.

The following exercise provides an alternative definition of closability.

Exercise 13.1.10. Show that an operator A is closable if and only if $\overline{\Gamma(A)}$ is the graph of a linear operator. In this case $\overline{\Gamma(A)} = \Gamma(\overline{A})$.

Example 13.1.11. Let $X = C([-1, 1])$ and define an operator A in X by $Af = f'$ for all $f \in D(A) = C^1([-1, 1])$. A is a well defined linear operator from its domain onto X. It is easy to see that the operator A is genuinely unbounded on its dense domain. We claim that A is closed. Indeed, suppose that $f_n \to f$ and $f'_n \to g$. Then $f_n(x) = \int_{-1}^{x} f'_n(t)dt + c_n$ and the fact that the sequences f_n and f'_n converge uniformly implies that $f(x) = \int_{-1}^{x} g(t)dt + \lim_n c_n$ is in $D(A)$ and $Af = g$. Thus, the graph $\Gamma(A)$ of A is closed.

Example 13.1.12. Let $H = L^2[-1, 1]$ and let B be "the same" operator as in the previous example, but now considered as an operator in H, that is $D(B) = C^1([-1, 1])$ and $Bf = f'$. Again, B is a densely defined unbounded operator in H, but B is *not* closed. For example, if $f_n = \chi_{[0,1]}x^{1+1/n}$, then letting $f(x) = \chi_{[0,1]}x$ and $g = \chi_{[0,1]}$ we see that $f_n \to f$ and $f'_n \to g$ but $f \notin D(B)$ and so $\Gamma(B)$ is not closed.

However, B is closable. To see, this, consider a sequence $(f_n, f'_n) \in \Gamma(B)$ that converges in $H \times H$. Let $g = \lim_n f'_n$. Then the sequence $f_n(x) - f_n(-1) = \int_{-1}^{x} f'_n(t)dt$ converges uniformly to a function $h(x) = \int_{-1}^{x} g(t)dt$. Since f_n converges in L^2, it follows that the sequence f_n converges uniformly to $f = h + c$ where $c = \lim_n f_n(-1)$.

This suggests what the closure of B should be. Recall that a function $f : [-1, 1] \to \mathbb{C}$ is *absolutely continuous* (AC) if and only if $f(x) = f(-1) + \int_{-1}^{x} g(t)dt$ for some $g \in L^1$, and in this case f is differentiable almost everywhere and $f' = g$ (see [13, Theorem 7.18] and the surrounding discussion). We define

$$D_1 = \{f \in L^2([-1, 1]) : f \in AC \text{ and } f' \in L^2\}$$

and then we define $B_1 f = f'$ for $f \in D_1$. The considerations from the previous paragraph show that $B_1 = \overline{B}$. It then follows that $C^1([-1,1])$ is a core for B_1.

Exercise 13.1.13. Provide an example of a densely defined operator that is not closable.

13.1.2   The adjoint of an operator

Definition 13.1.14. Let A be a densely defined operator in a Banach space X. The **adjoint** of A is an operator A^* in X^* defined as follows. First, the domain $D(A^*)$ of A^* is defined to be the space of all $\phi \in X^*$ such that the linear functional

$$D(A) \ni x \mapsto \langle Ax, \phi \rangle$$

is bounded on $D(A)$, i.e. there exists a constant C_ϕ such that

$$|\langle Ax, \phi \rangle| \le C_\phi \|x\|, \quad \text{for all } x \in D(A).$$

For every such ϕ the functional $x \mapsto \langle Ax, \phi \rangle$ extends to a unique bounded functional defined on X, and we define $A^*\phi$ to be this functional.

In other words, $(A^*, D(A^*))$ is the operator determined by the identity

$$\langle Ax, \phi \rangle = \langle x, A^*\phi \rangle, \quad \text{for all } x \in D(A), \phi \in D(A^*).$$

Note that the adjoint A^* is not defined for an operator A that is not densely defined.

Remark 13.1.15. When working in Hilbert spaces it is customary to use the inner product on H instead of the bilinear pairing between H and its dual H^*. Thus if A is an operator in a Hilbert space H, we put $D(A^*) = \{h \in H : g \mapsto \langle Ag, h \rangle_H \text{ is bounded}\}$ and we let A^* be the operator in H determined by the identity $\langle Ag, h \rangle_H = \langle g, A^*h \rangle_H$ for all $g \in D(A)$ and $h \in D(A^*)$. This leads to a slightly different notion of adjoint, but to no confusion.

Definition 13.1.16. Let A be a densely defined operator in a Hilbert space H. Then A is said to be **Hermitian** (or **symmetric**) if $A \subseteq A^*$. If $A = A^*$ then A is said to be **selfadjoint**.

Note that for a densely defined A, being symmetric is equivalent to

$$\langle Ax, y \rangle = \langle x, Ay \rangle, \quad \text{for all } x, y \in D(A). \tag{13.1}$$

Selfadjointness entails, in addition, the condition $D(A) = D(A^*)$.

Remark 13.1.17. Some authors define an operator to be Hermitian if it merely satisfies (13.1); in practice, all operators of interest are densely defined and this difference causes no harm. Selfadjoint operators are always understood to be densely defined, by definition.

The notions of a Hermitian or a selfadjoint operator only make sense in Hilbert spaces. We will continue to develop the theory in general Banach spaces when the going is easy, but invest most of our efforts below in the Hilbert space setting.

Proposition 13.1.18. *If A is a densely defined operator then A^* is closed.*

Proof. Note that $(\phi, \psi) \in \Gamma(A^*)$ if and only if

$$\langle Ax, \phi \rangle = \langle x, \psi \rangle \qquad \text{for all } x \in D(A).$$

It follows that $\Gamma(A^*)$ is closed.   □

Corollary 13.1.19. *A selfadjoint operator in a Hilbert space is closed.*

Example 13.1.20. Consider three operators A_1, A_2, A_3 in $L^2[0, 1]$ defined by $A_k f = i f'$ for $k = 1, 2, 3$ with domains

$$D(A_1) = \{f \in L^2[0, 1] : f \in AC \text{ and } f' \in L^2\},$$

$$D(A_2) = \{f \in L^2[0, 1] : f \in AC \text{ and } f' \in L^2 \text{ and } f(0) = f(1)\},$$

and

$$D(A_3) = \{f \in L^2[0, 1] : f \in AC \text{ and } f' \in L^2 \text{ and } f(0) = f(1) = 0\}.$$

Clearly, $A_3 \subseteq A_2 \subseteq A_1$. One can show that $A_3 = A_1^*$ (thus, A_1 is not Hermitian), $A_1 = A_3^*$ (thus A_3 is Hermitian but not selfadjoint) and that A_2 is selfadjoint. It follows then from Proposition 13.1.18 that all three are closed operators.

Let us consider this in a little more detail. If $f, g \in AC$ then integration by parts gives

$$\int_0^1 i f' \overline{g} = i f(1) \overline{g}(1) - i f(0) \overline{g}(0) - i \int_0^1 f \overline{g'} = i f(1) \overline{g}(1) - i f(0) \overline{g}(0) + \int_0^1 f \overline{i g'}.$$

If $f, g \in D(A_2)$, then $f(1)\overline{g}(1) = f(0)\overline{g}(0)$, so we find that $\langle A_2 f, g \rangle = \langle f, A_2 g \rangle$ for all $f, g \in D(A_2)$. This implies that $A_2 \subseteq A_2^*$. Likewise, we obtain that $\langle A_1 f, g \rangle = \langle f, A_3 g \rangle$ for all $f \in D(A_1)$ and $g \in D(A_3)$, implying that $A_1 \subseteq A_3^*$ and that $A_3 \subseteq A_1^*$. We will show that $A_2^* = A_2$ and leave the rest of the details to the reader.

Let $g \in D(A_2^*)$. Put $h = A_2^* g$. We do not yet know that g is absolutely continuous or that $h = i g'$. To see this, we consider

$$\int_0^1 i f' \overline{g} = \langle A_2 f, g \rangle = \langle f, A_2^* g \rangle = \int_0^1 f \overline{h}$$

for all $f \in D(A_2)$. Putting $H(x) = \int_0^x h(t) dt$, the right hand side integrates by parts to give

$$\int_0^1 i f' \overline{g} = f(1) \overline{H}(1) - \int_0^1 f' \overline{H}.$$

Taking $f = 1 \in D(A_2)$, we see that $H(1) = 0$, and we find that $\int_0^1 f'(ig - H) = 0$ for all $f \in D(A_2)$. In other words, $ig - H \in \operatorname{Im} A_2^\perp$. But $\operatorname{Im} A_2 = \{k \in L^2 : \int_0^1 k = 0\} = \mathbb{C}^\perp$, so $ig - H$ is constant. This means that $g(x) = -i \int_0^x h(t)dt + C$ is absolutely continuous. We saw above that $\int_0^1 h(t)dt = H(1) = 0$, so $g(0) = g(1) = C$. We conclude that $D(A_2^*) \subseteq D(A_2)$ and $ig' = h = A_2^*g$. To sum up: $A_2^* = A_2$, as claimed.

Exercise 13.1.21. Fill in the details of the above example (in particular $A_1^* = A_3$, $A_3^* = A_1$).

Exercise 13.1.22. For any one of the above operators, show that replacing AC by C^∞ in the definition of the domain leads to a core of the given operator.

The above example illustrates the role that domains play in the definition of unbounded operators. In practice, when the operator in question is a differential operator, describing the domain typically involves specifying certain boundary conditions. The boundary conditions of a differential equation are consequential to the solvability of the equation, the development of solutions in terms of eigenfunctions, and so forth.

Definition 13.1.23. An operator A in a Hilbert space H is said to be *essentially selfadjoint* if it is closable and its closure is selfadjoint.

Example 13.1.24. In $H = L^2[0,1]$ we define an operator $Bf = if'$ for all f in the domain

$$D(B) = \{f \in L^2[0,1] : f \in C^\infty([0,1]) \text{ and } f(0) = f(1)\}.$$

Then B is essentially selfadjoint, and $\overline{B}$ is equal to the operator A_2 from Example 13.1.20 (see Exercise 13.5.6). Equivalently, $D(B)$ is a core for the selfadjoint operator A_2.

Exercise 13.1.25. Show that for any operator A that is densely defined, $A^* = (\overline{A})^*$. Use that to deduce that if A is an essentially selfadjoint operator then its closure $\overline{A}$ is the unique selfadjoint extension of A.

Example 13.1.26. If μ is a semi-finite measure on a space X and $f : X \to \mathbb{C}$ a measurable function, then we let $H = L^2(X, \mu)$ and define $M_f h = fh$ on

$$D = \{h \in L^2(X, \mu) : fh \in L^2(X, \mu)\}$$
$$= \left\{h : X \to \mathbb{C} : h \text{ is measurable and } \int (1 + |f|^2)|h|^2 d\mu < \infty\right\}.$$

Then (M_f, D) is a closed operator, which is bounded if and only if f is essentially bounded, and selfadjoint if and only if f is (almost everywhere) real valued. Multiplication by bounded functions was treated in Exercise 11.1.2. Since this is a crucial example, we give some details in the unbounded case.

First, to see that D is dense in L^2, let g_N be the characteristic function of the set $\{x \in X : |f(x)| \leq N\}$, and note that $hg_N \in D$ for all $h \in L^2$ and that, by the dominated convergence theorem, $\int |hg_N - h|^2 d\mu \to 0$.

Next, to see that M_f is closed, suppose that $h_n \to h$ and that $fh_n \to g$, where convergence is in the norm of L^2. After passing to a subsequence, we may assume that we have convergence $h_n(x) \to h(x)$ and that $f(x)h_n(x) \to g(x)$ at almost every x. But fh_n then converges to fh almost everywhere as well, thus $fh = g \in L^2$. This means that $h \in D$ and $M_f h = g$, so M_f is closed.

Using measure theoretic arguments (similar to Proposition 3.2.5), one can show that a function $g \in L^2$ gives rise to a bounded functional on D given by

$$h \mapsto \langle fh, g \rangle = \int hf\bar{g}$$

if and only if $fg \in L^2$, that is, if and only if $g \in D$. Therefore $D(A^*) = D$ and it follows that $A^* = M_{\bar{f}}$. If f is real valued then A is selfadjoint.

13.1.3   Sums and products of operators

If A and B are operators in H, we can define the linear operators

$$A + B \quad \text{with domain} \quad D(A + B) = D(A) \cap D(B),$$

and

$$AB \quad \text{with domain} \quad D(AB) = \{x \in D(B) : Bx \in D(A)\}.$$

The domain of $A+B$ or AB might be $\{0\}$ even if A and B are densely defined. In some cases, $A + B$ and AB can and should be extended from the above domains to larger domains. In general, the analysis of the seemingly simple operations of sum and product is subtle and depends on the particulars of the case at hand. We shall require the following basic result.

Proposition 13.1.27. *Let A be a densely defined operator in a Banach space X, let $B \in B(X)$ and consider the operator $A + B$ with domain $D(A)$.*

1. *If A is closed, then $A + B$ is closed.*

2. *The domain of $(A + B)^*$ equals $D(A^*)$, and $(A + B)^* = A^* + B^*$.*

3. *If X is a Hilbert space and if A and B are selfadjoint, then $A + B$ is selfadjoint.*

Proof. Assume that A is closed and let $x_n \in D(A)$ be a sequence such that $x_n \to x$ and $Ax_n + Bx_n \to z$. Then $Bx_n \to Bx$ because B is bounded, hence $Ax_n \to y := z - Bx$. But A is closed, so $x \in D(A)$ and $Ax = y$, whence $(A + B)x = z$. This shows that $A + B$ is closed.

The functional $D(A) \ni x \mapsto \langle Bx, \phi \rangle$ is bounded for all $\phi \in X^*$. Therefore the functional

$$D(A) \ni x \mapsto \langle (A + B)x, \phi \rangle$$

is bounded precisely for $\phi \in D(A^*)$. In other words $D((A+B)^*) = D(A^*)$. The identity $(A+B)^* = A^* + B^*$ follows readily. The final assertion is then an immediate consequence. $\square$

Example 13.1.28. Let $H = L^2([0,1])$ and let $b\colon [0,1] \to \mathbb{R}$ be a bounded, real valued measurable function. On the domain $D(A_2)$ of Example 13.1.20, we define the operator

$$Lf = if' + bf.$$

Then $L = A_2 + M_b$ is a selfadjoint operator. This operator is a typical operator that arises in the theory of differential equation: a linear combination of derivatives and multiplication operators, defined on a subspace of functions that satisfy certain boundary conditions.

Exercise 13.1.29. Suppose that A, B and AB are densely defined in a Hilbert space H. Prove that

$$B^*A^* \subseteq (AB)^*,$$

and that if $B \in B(H)$ then

$$B^*A^* = (AB)^*.$$

13.2 Spectrum of unbounded operators

13.2.1 The spectrum of an operator in a Banach space

Definition 13.2.1. Let A be an operator in a Banach space X. The **resolvent set** $\rho(A)$ of A is the set of all $\lambda \in \mathbb{C}$ such that $A - \lambda I\colon D(A) \to X$ is bijective, and such that the linear inverse $(A - \lambda I)^{-1}\colon X \to D(A)$ is bounded. The **spectrum** of A is the complement of the resolvent set: $\sigma(A) = \mathbb{C} \setminus \rho(A)$.

This definition might seem a bit unfair: we are requiring the inverse of $A - \lambda I$ to be bounded, even though the operator A, and therefore $A - \lambda I$, need not be. Note, however, that if A is closed, then this boundedness would be automatic.

Proposition 13.2.2. *If A is closed and $A - \lambda I\colon D(A) \to X$ is bijective, then* $(A - \lambda I)^{-1}\colon X \to D(A)$ *is bounded.*

Proof. If A is closed, so is $A - \lambda I$ (by Proposition 13.1.27), meaning that $\Gamma(A - \lambda I)$ is closed in $X \times X$. But

$$\Gamma((A - \lambda I)^{-1}) = \{(y, x) : (x, y) \in \Gamma(A - \lambda I)\}$$

is then also closed. Therefore $(A - \lambda I)^{-1}$ is an everywhere defined operator on X with a closed graph. By the closed graph theorem, it is bounded. $\square$

From this point onwards, in order to lighten notation we shall write λ for λI. For every $\lambda \in \rho(A)$, the operator $R_\lambda = (A - \lambda)^{-1}$ is called the **resolvent** operator. The function $\lambda \mapsto R_\lambda$ is a holomorphic function on $\rho(A)$ with values in $B(X)$ that plays an important role in the analysis of unbounded operators on Banach spaces (see Exercise 13.5.1). Resolvent operators can only be defined for closed operators, for the following reason.

Proposition 13.2.3. *If A is not closed then $\rho(A) = \emptyset$, and so $\sigma(A) = \mathbb{C}$.*

Proof. If $\lambda \in \rho(A)$, then $(A - \lambda)^{-1}$ is a bounded map from X onto $D(A - \lambda) = D(A)$. It follows that the graph of $(A - \lambda)^{-1}$ is closed and therefore $A - \lambda$ is a closed operator. By Proposition 13.1.27, A is closed. $\square$

Example 13.2.4. Consider multiplication operator M_f from Example 13.1.26 with the domain D. Suppose that f is real valued. We can use Proposition 13.2.3 to give a different proof that (M_f, D) is closed. Indeed, we shall show that the resolvent set of M_f is not empty, and by the proposition, this means that M_f is closed. Let $\lambda \notin \mathbb{R}$, and consider the operator $(M_f - \lambda)^{-1} = M_{(f-\lambda)^{-1}}$ of multiplication by the function $(f - \lambda)^{-1}$. Now, since f is real valued, $|f - \lambda|$ is bounded below by Im λ, and so $(M_f - \lambda)^{-1}$ is bounded on L^2. A moment of thought shows that $(M_f - \lambda)^{-1}$ maps L^2 into D and that $(M_f - \lambda)(M_f - \lambda)^{-1} = I_{L^2}$. Even less thought shows that $(M_f - \lambda)^{-1}(M_f - \lambda) = I_D$. This means that $(M_f - \lambda) \colon D \to H$ is bijective and has a bounded inverse, so $\lambda \in \rho(A)$, as required.

Example 13.2.5. Let $H = L^2(X, \mu)$ and let $M_f h = fh$ be a multiplication operator as in Example 13.1.26, where $f \colon X \to \mathbb{C}$ a measurable function and μ is a semi-finite measure. Then the spectrum of M_f is equal to the **essential range** of f, defined by

$$\text{ess-ran}(f) = \Big\{ \lambda \in \mathbb{C} : \mu\left(\{x \in X : |f(x) - \lambda| < \epsilon\}\right) > 0 \text{ for all } \epsilon > 0 \Big\}.$$

Indeed, if $\lambda \notin \text{ess-ran}(f)$, then there exists $\epsilon > 0$ such that $\mu(\{x \in X : |f(x) - \lambda| < \epsilon\}) = 0$. This means that $g(x) = (f(x) - \lambda)^{-1}$ belongs to $L^\infty(X, \mu)$. It follows that $M_g = (M_f - \lambda)^{-1}$ is bounded and therefore $\lambda \notin \sigma(M_f)$. Conversely, if $\lambda \in \text{ess-ran}(f)$, then given $\epsilon > 0$ let

$$E = \{x \in X : |f(x) - \lambda| < \epsilon\}.$$

Then $\mu(E) > 0$. By semi-finiteness, there exists $F \subseteq E$ with $0 < \mu(F) < \infty$. Let $h = \mu(F)^{-1/2}\chi_F \in L^2(X, \mu)$. Then $\|h\| = 1$, but $\|(M_f - \lambda)h\| \le \epsilon$. This shows that $M_f - \lambda$ is not bounded below, therefore it cannot have a bounded inverse, whence $\lambda \in \sigma(M_f)$.

Example 13.2.6. Let us calculate the spectrum of the operators A_1, A_2, A_3 from Example 13.1.20. To decide if $\lambda \in \sigma(A_k)$, we need to check whether the operator $A_k - \lambda \colon D(A_k) \to H$ is bijective (since A_k is closed then by

Proposition 13.2.2 the inverse will be bounded). Our task is closely related to the solvability of the differential equation

$$if' - \lambda f = g \qquad (13.2)$$

under various boundary conditions. For $k = 1$, we are to consider the differential equation with no boundary condition. In this case, we expect that the homogeneous equation (i.e., when $g = 0$) will have many solutions, for any λ. Indeed, for all $\lambda \in \mathbb{C}$, if we put $f_\lambda(x) = e^{-i\lambda x}$ then $f_\lambda \in D(A_1)$ and $if' - \lambda f = 0$; in other words $A_1 f_\lambda = \lambda f_\lambda$. So f_λ is an eigenvector for A_1 with eigenvalue λ; in particular $\lambda \in \sigma(A_1)$. We conclude that $\sigma(A_1) = \mathbb{C}$.

When $k = 2$, we are studying a first order differential equation with a single linear constraint, so we expect this to be a well behaved problem. Most of the eigenvectors $f_\lambda(x) = e^{-i\lambda x}$ that we found for A_1 are no longer eigenvectors for A_2, because they are not in the domain $D(A_2)$. In fact, $f_\lambda \in D(A_2)$ is equivalent to $f(0) = f(1)$, and this happens if and only $\lambda = 2\pi n$ for some $n \in \mathbb{Z}$. So we have found an orthonormal basis of eigenvectors for A_2 with real eigenvalues in the domain $D(A_2)$. By Exercise 13.5.6, since A_2 is Hermitian and has a basis of eigenvectors, it is essentially selfadjoint and the spectrum of its closure is the closure of $\{2\pi n : n\mathbb{Z}\}$. But we already know that A_2 is selfadjoint, so $\sigma(A_2) = \{2\pi n : n\mathbb{Z}\}$.

We now turn to the operator A_3 (the methods below could also be used to calculate the spectrum of A_2 without recourse to Exercise 13.5.6). Now, the solution of the first order equation (13.2) needs to satisfy two constraints, so we expect that there will not be a solution, at least not for every $g \in L^2$. Suppose that $f \in D(A_3)$ solves (13.2). Multiplying both sides of (13.2) with the integrating factor $e^{i\lambda x}$, it becomes $(ife^{i\lambda x})' = ge^{i\lambda x}$, and we find that

$$f(x) = -i \int_0^x e^{-i\lambda(x-t)} g(t)dt + Ce^{-i\lambda x}.$$

Since f satisfies the boundary condition $f(0) = 0$, we get that $C = 0$. But the boundary condition $f(1) = 0$ now implies that $\int_0^1 e^{i\lambda t} g(t)dt = 0$. This is not satisfied for every $g \in L^2$, and that means that the operator $A_3 - \lambda : D(A_3) \to H$ is not surjective, thus $\lambda \in \sigma(A_3)$. The above argument works for every $\lambda \in \mathbb{C}$, therefore $\sigma(A_3) = \mathbb{C}$. It is worth recalling that A_3 is Hermitian, yet its spectrum is the entire complex plane.

Example 13.2.7. The spectrum of an unbounded densely defined operator might be empty. To see this, we consider an additional operator, $Af = if'$ on

$$D(A) = \{f \in L^2[0,1] : f \in AC \text{ and } f' \in L^2 \text{ and } f(0) = 0\}.$$

This is again a closed operator. The operator $A - \lambda : D(A) \to H$ is invertible if and only if the first order equation (13.2) subject to the initial condition $f(0) = 0$ has a unique solution $f \in D(A)$ for every $g \in L^2$. The reader's experience with ordinary differential equations suggests that for every λ there should be a unique solution f given any g. Indeed, using an integrating factor and invoking the initial condition, we see that for every $g \in L^2$, the function

$$f(x) = -i \int_0^x e^{-i\lambda(x-t)} g(t) dt$$

is the unique solution in $f \in D(A)$ to (13.2). This leads us to define $R_\lambda \colon H \to H$ by

$$R_\lambda(g) = -i \int_0^x e^{-i\lambda(x-t)} g(t) dt.$$

Then $\operatorname{Im} R_\lambda = D(A)$, and our discussion above shows that $(A - \lambda)R_\lambda = I_H$. One can check directly that $R_\lambda(A - \lambda) = I_{D(A)}$. This shows that a bounded inverse $(A - \lambda)^{-1}$ exists, therefore $\lambda \in \rho(A)$. It follows that $\sigma(A) = \emptyset$.

13.2.2   Selfadjoint operators and spectrum

Lemma 13.2.8. *If A is a densely defined operator in a Hilbert space H, then*

$$(\operatorname{Im} A)^\perp = \ker A^*.$$

If A^* is also densely defined then $\ker A = (\operatorname{Im} A^*)^\perp$.

Proof. This goes almost exactly like Proposition 5.3.8, with the difference that one needs to account for the domains. The details are left as an exercise. □

Theorem 13.2.9. *Let A be a selfadjoint operator in a Hilbert space H. Then $\sigma(A)$ is a nonempty closed subset of the real line.*

Proof. First, let us show that the spectrum is nonempty. Assume, for the sake of obtaining a contradiction, that $\sigma(A) = \emptyset$. Then $A - \lambda$ has a bounded inverse for all λ. In particular, taking $\lambda = 0$, we see that $A \colon D(A) \to H$ is bijective and its inverse $A^{-1} \colon H \to D(A)$ is bounded. But now

$$\langle Ax, y \rangle = \langle x, Ay \rangle \text{ for all } x, y \in D(A),$$

which follows from selfadjointness of A, implies that

$$\langle h, A^{-1}g \rangle = \langle A^{-1}h, g \rangle \text{ for all } h, g \in H,$$

so A^{-1} is a bounded selfadjoint operator. We now show that $A^{-1} - \lambda$ is invertible for all $\lambda \neq 0$. Indeed, recall that we are assuming that $\sigma(A) = \emptyset$, so the operator $(I - \lambda A) = -\lambda(A - \lambda^{-1}I)$ has a bounded inverse. Thus, the bounded operator $A^{-1} - \lambda = (I - \lambda A)A^{-1}$ is invertible. Since this is true for all $\lambda \neq 0$ we conclude that $\sigma(A^{-1}) = \{0\}$. But the norm of a normal operator is equal to its spectral radius (Corollary 10.2.6), so we must have that $\|A^{-1}\| = r(A) = 0$, a contradiction. To avoid the contradiction we must accept that $\sigma(A) \neq \emptyset$.

Next, we show that $\sigma(A) \subseteq \mathbb{R}$. Let $\lambda = a + ib$ for $a, b \in \mathbb{R}$ with $b \neq 0$. Our goal is to show that $\lambda \in \rho(A)$. We compute

$$\|(A - \lambda)h\|^2 = \|(A - a)h\|^2 - 2\operatorname{Re}(ib\langle(A - a)h, h\rangle) + |b|^2\|h\|^2.$$

Since $\langle(A - a)h, h\rangle$ is real, it follows that

$$\|(A - \lambda)h\|^2 = \|(A - a)h\|^2 + |b|^2\|h\|^2 \geq |b|^2\|h\|^2,$$

which shows that $A - \lambda$ is bounded below. Being bounded below implies three things. First, that $A - \lambda$ is injective. Second, using also the closedness of $A - \lambda$, it follows that $\operatorname{Im}(A - \lambda)$ is closed. Third, $A - \lambda$ being bounded below is the same thing as the map $(A - \lambda)^{-1} \colon \operatorname{Im}(A - \lambda) \to D(A)$ being bounded. Therefore, to show that $\lambda \in \rho(A)$, it remains to establish that $\operatorname{Im}(A - \lambda) = H$.

Replacing λ by $\overline{\lambda}$ in the above argument, we find that

$$\|(A - \overline{\lambda})h\| \geq |b|\|h\|,$$

so $(A - \lambda)^* = A - \overline{\lambda}I$ is injective. Using Lemma 13.2.8 we see that

$$\overline{\operatorname{Im}(A - \lambda)} = \operatorname{Im}(A - \lambda)^{\perp\perp} = (\ker(A - \lambda)^*)^\perp = H.$$

Since we already know that $\operatorname{Im}(A - \lambda)$ is closed, we have that $\operatorname{Im}(A - \lambda) = H$ and therefore $\lambda \in \rho(A)$. We have shown that $\sigma(A) \subseteq \mathbb{R}$.

It remains to show that $\sigma(A)$ is closed. It is enough to show that it is closed in $\mathbb{R}$. Now if $\lambda \in \mathbb{R} \setminus \sigma(A)$, then $(A - \lambda)^{-1}$ exists and is bounded. In particular, $A - \lambda$ is bounded from below: there exists $c > 0$ such that $\|(A - \lambda)h\| \geq c\|h\|$ for all $h \in D(A)$. Now, if $\mu \in \mathbb{R}$ and $|\mu - \lambda| < \frac{c}{2}$ then

$$\|(A - \mu)h\| \geq \|(A - \lambda)h\| - \|(\mu - \lambda)h\| \geq \frac{c}{2}\|h\|$$

and so $A - \mu$ is bounded below on $D(A)$, too. This means that $A - \mu$ is injective and has a closed range. Since $A - \mu$ is selfadjoint, it follows from Lemma 13.2.8 that $\operatorname{Im}(A - \mu) = H$. Thus $A - \mu$ has a bounded inverse, so $\mu \notin \sigma(A)$. To sum up, we have shown that if $\lambda \in \mathbb{R}$ is not in the spectrum, then there is a neighborhood of λ that is disjoint from the spectrum. In other words, the spectrum is closed. $\qquad\square$

Checking whether a given operator is Hermitian or not is typically a straightforward matter. Showing that an operator is selfadjoint is much more difficult. A large part of the theory of unbounded operators deals with the problem of deciding whether or not a Hermitian operator is selfadjoint, essentially selfadjoint, or at least has a selfadjoint extension. We shall not go into these delicate matters, and dedicate the remainder of the chapter to selfadjoint operators. But before doing so we exhibit an elegant criterion for selfadjointness in line with the ideas above.

Theorem 13.2.10. *Let A be a densely defined Hermitian operator in a Hilbert space H. Then A is selfadjoint if and only if*

$$\text{Im}(A - i) = \text{Im}(A + i) = H. \tag{13.3}$$

Proof. If A is selfadjoint then $\sigma(A) \subseteq \mathbb{R}$ by Theorem 13.2.9, and in particular (13.3) holds.

Conversely, assume that (13.3) holds. The assumption that A is Hermitian means that $A \subseteq A^*$. We need to show that $D(A) = D(A^*)$. For this, we take $h \in D(A^*)$, and we shall prove that $h \in D(A)$. Consider $(A^* + i)h$. By assumption, $A + i$ is surjective, so there exists $g \in D(A)$ such that $(A + i)g = (A^* + i)h$. But a second assumption is that $A - i$ is surjective, so by Lemma 13.2.8, $\ker(A^* + i) = \{0\}$; in particular $A^* + i$ is injective. Now $A^* + i$ is an extension of $A + i$, so we conclude that $g = h$. This completes the proof. □

Exercise 13.2.11. Use the above theorem to verify that the operator of multiplication by a real valued function in $L^2(\mu)$ (see Example 13.1.26) is selfadjoint. Now work a bit harder to show that the (appropriately defined) direct sum of selfadjoint operators is selfadjoint.

Exercise 13.2.12. Use the above theorem to verify that the operator A_2 from Example 13.2.6 is selfadjoint, and that A_3 is not.

13.3   The spectral theorem for unbounded selfadjoint operators

Among the most important operators in mathematical physics and PDEs are unbounded selfadjoint operators. In this section, we will prove the spectral theorem for unbounded selfadjoint operators – a key tool for working with them and one of the early crowning achievements of functional analysis.

13.3.1   The Cayley transform

The proof of the spectral theorem is based on bootstrapping the spectral theorem for unitary operators. The intuitive idea is as follows. Suppose that a selfadjoint operator A is given as a multiplication operator M_f on $L^2(X, \mu)$ as in Example 13.1.26. Let us see how to transform A into a unitary from which we can recover f. Define $\phi \colon \mathbb{R} \to \mathbb{T} \setminus \{1\}$ by

$$\phi(t) = \frac{t - i}{t + i}. \tag{13.4}$$

Then $v = \phi \circ f$ is a measurable function on X that takes values in $\mathbb{T} \setminus \{1\}$, so M_v is a unitary operator. Moreover, M_v acts as multiplication by

$$v = \phi \circ f = \frac{f - i}{f + i}.$$

It is not hard to see that the action of the multiplication operator M_v is given by

$$M_v = (M_f - i)(M_f + i)^{-1}. \tag{13.5}$$

Now, if

$$\psi(z) = i\frac{1 + z}{1 - z}, \tag{13.6}$$

then $\psi = \phi^{-1} \colon \mathbb{T} \setminus \{1\} \to \mathbb{R}$ and we recover f from v as $f = \psi \circ v$. We can recover $A = M_f$ as multiplication by

$$f = \psi \circ v = i\frac{1 + v}{1 - v},$$

or

$$M_f = i(1 + M_v)(1 - M_v)^{-1},$$

(though this is slightly trickier than (13.5) because there both sides of the equation are bounded operators, and here we are dealing with unbounded operators; see Proposition 13.3.3). The following definition and propositions will help us formalize the above ideas and then apply them to a general selfadjoint operator A.

Definition 13.3.1. Let A be a selfadjoint operator in a Hilbert space H. The ***Cayley transform*** of A is the operator $V = (A - i)(A + i)^{-1}$.

Proposition 13.3.2. *The Cayley transform of a selfadjoint operator is unitary.*

Proof. By Theorem 13.2.9, $\pm i \notin \sigma(A)$ and so $A \pm i$ is an invertible closed operator from $D(A)$ onto H. This means that V is defined on H and maps H onto H, via $D(A) = D(A \pm i)$. To see that V is isometric, note that every element in H has the form $(A + i)h$ for some $h \in D(A)$, and V maps this element to $(A - i)h$. Now

$$\|(A \pm i)h\|^2 = \|Ah\|^2 \pm 2\operatorname{Re} i\langle h, Ah\rangle + \|h\|^2 = \|Ah\|^2 + \|h\|^2,$$

so V is isometric (we used the fact that $\langle h, Ah\rangle$ is real). □

The next proposition says that we can recover a selfadjoint operator from its Cayley transform.

Proposition 13.3.3. *If V is the Cayley transform of a selfadjoint operator A, then $I - V$ is injective, and*

$$A = i(I + V)(I - V)^{-1}$$

where the domain of the right hand side, which is defined to be $\operatorname{Im}(I - V)$, and the domain $D(A)$ of A, are equal.

Proof. Since $i \notin \sigma(A)$, for every $h \in H$ there is some $g \in D(A)$ such that

$$h = (A + i)g = Ag + ig, \tag{13.7}$$

and then

$$Vh = (A - i)(A + i)^{-1}(A + i)g = Ag - ig. \tag{13.8}$$

Subtracting (13.8) from (13.7) gives $(I - V)h = 2ig$, which shows that $I - V = 2i(A + i)^{-1}$. This means that $I - V$ is injective and its inverse $(I - V)^{-1} = \frac{1}{2i}(A + i)$ has domain equal to $D(A) = \operatorname{Im}(I - V)$.

Next, by adding (13.7) and (13.8) we obtain $(I + V)h = 2Ag$. Putting things together we find that

$$i(I + V)(I - V)^{-1}g = (I + V)\frac{1}{2}h = Ag$$

for all $g \in D(A) = \operatorname{Im}(I - V)$, as required. $\square$

13.3.2   The multiplication operator formulation of the spectral theorem

Theorem 13.3.4 (The spectral theorem for selfadjoint operators). *Let A be a selfadjoint operator in a Hilbert space H. Then there exists a locally compact Hausdorff space X, a semi-finite regular Borel measure μ on X, a measurable function $f\colon X \to \mathbb{R}$ and a unitary operator $U\colon L^2(X, \mu) \to H$, such that $D(A) = U(D(M_f))$ and*

$$U^*AU = M_f$$

where $M_f\colon h \mapsto fh$ is the multiplication operator on the domain

$$D(M_f) = \left\{ h \in L^2(X, \mu) : fh \in L^2(X, \mu) \right\}.$$

Moreover

$$\sigma(A) = \text{ess-range}(f).$$

Proof. Let V be the Cayley transform of A. By Proposition 13.3.2, V is a unitary. By the spectral theorem for bounded normal operators (Theorem 11.1.10), there exists a unitary $U\colon L^2(X, \mu) \to H$ where X and μ are as in the statement of the theorem, and a function $v \in L^\infty(X, \mu)$ such that

$$U^*VU = M_v.$$

Since V is unitary, $M_v M_v^* = M_{|v|^2} = I$, so $|v| = 1$ almost everywhere. By Proposition 13.3.3, $I - V$ is injective, whence 1 is not an eigenvalue of V and therefore it is not an eigenvalue of M_v, and it follows that the set where $v = 1$ is of measure zero. We can therefore define (at almost every point of X) a function $f = \psi \circ v\colon X \to \mathbb{R}$, where ψ is given by (13.6). Then we get back

$v = \phi \circ f$, where ϕ is given by (13.4). Let M_f be the multiplication operator with the domain

$$D(M_f) = \{h \in L^2(X, \mu) : fh \in L^2(X, \mu)\}$$

as in Example 13.1.26. The elementary Equation (13.5) shows that M_v is the Cayley transform of M_f. By Proposition 13.3.3, $D(M_f) = \mathrm{Im}(I - M_v)$ and $D(A) = \mathrm{Im}(I - V)$, so $D(A) = U(D(M_f))$. Proposition 13.3.3 also gives that

$$M_f = i(I + M_v)(I - M_v)^{-1}.$$

Plugging $U^*VU = M_v$ into the above equation, we find that for $h \in D(M_f)$

$$
\begin{aligned}
M_f h &= i(I + U^*VU)(I - U^*VU)^{-1}h \\
&= iU^*(I + V)UU^*(I - V)^{-1}Uh \\
&= U^*AUh,
\end{aligned}
$$

where the final equality follows from Proposition 13.3.3. Finally, since μ is semi-finite, the spectrum of M_f is equal to the essential range of f (see Example 13.2.5). □

Remark 13.3.5. There is also a spectral theorem for *normal* operators (see Exercise 13.5.9 for a definition; see [42] for statement and proof).

Let us briefly examine how the spectral theorem manifests itself in a couple of examples.

Example 13.3.6. Consider $A_2 = i\frac{d}{dx}$ from Example 13.2.6 (with domain corresponding to periodic boundary conditions) in the Hilbert space $H = L^2[0, 1]$. We saw that $\sigma(A_2) = \{2\pi n : n \in \mathbb{Z}\}$. In this case letting $X = \mathbb{Z}$, μ the counting measure and $U: \ell^2(\mathbb{Z}) \to H$ given on the standard basis vectors by $Ue_n = e^{-2\pi i n x}$, we see that U implements a unitary equivalence between A_2 and the multiplication operator M_f with $f: \mathbb{Z} \to \mathbb{R}$ given by $f(n) = 2\pi n$. We see that in this case our selfadjoint operator is unitarily equivalent to an unbounded diagonal operator. Note that the measure space (X, μ) is determined uniquely only up to some sort of weak *measure equivalence* that we shall not go into here[2].

Example 13.3.7. Let $A = i\frac{d}{dx}$ be the differentiation operator in $L^2(\mathbb{R})$ with domain

$$D(A) = \{f \in L^2(\mathbb{R}) : f \in AC \text{ and } f' \in L^2(\mathbb{R})\},$$

where by $f \in AC$ we mean that the restriction of f to any finite interval is absolutely continuous. One can show that the Fourier transform $\mathcal{F}: L^2(\mathbb{R}) \to L^2(\mathbb{R})$, given by

$$\mathcal{F}(f)(x) = \hat{f}(x) = \frac{1}{\sqrt{2\pi}} \int_{\mathbb{R}} f(y)e^{-iyx}\,dy,$$

---

[2]This lack of uniqueness occurred already in Theorem 11.1.10 — recall that in the separable case, we showed how the measure could be changed into a probability measure.

implements a unitary equivalence of A with the multiplication operator M_ξ for $\xi(x) = x$, which is simply denoted M_x. It follows that $\sigma(A) = \mathbb{R}$, which is not evident at first sight since A does not have eigenvectors. The domain of A can be described as

$$\mathcal{F}^{-1}(D(M_x)) = \left\{ f \in L^2(\mathbb{R}) : x\hat{f}(x) \in L^2(\mathbb{R}) \right\}.$$

Similarly, the *Laplacian operator*

$$L = -\Delta = -\sum_{j=1}^{n} \frac{\partial^2 u}{\partial x_j^2}$$

in $L^2(\mathbb{R}^n)$ is unitarily equivalent to the multiplication with the function $|x|^2 = \sum x_j^2$.

The above two examples are meant to illustrate what the conclusion of the spectral theorem looks like, but in neither case did the spectral theorem reveal something we did not know. In the first example, we used the existence of an orthonormal basis of eigenvectors (diagonalizability) to *prove* selfadjointness. In the second example, the unitary equivalence to a multiplication operator was the consequence of Fourier analysis, and it is from this that we obtained both the spectrum of A and that A is selfadjoint. The power of the spectral theorem is in giving a description of what a general selfadjoint operator does, in providing guidance on how to approach complicated selfadjoint operators, and in showing what to look for.

13.3.3   The Borel functional calculus and the spectral measure formulation of the spectral theorem

Once we have the spectral theorem in multiplication operator form in our hands, we can follow the ideas presented in Sections 11.2 and 11.3 to obtain the bounded Borel functional calculus and the spectral measure formulation of the spectral theorem. Since no significant new ideas are needed, we shall state the results and leave the proofs to the reader.

Let $\mathcal{B}(\mathbb{R})$ denote the $*$-algebra of bounded Borel functions on $\mathbb{R}$ equipped with the supremum norm. Let us write ξ_n for the function

$$\xi_n(x) = \begin{cases} x & x \in [-n, n] \\ 0 & \text{else} \end{cases}.$$

Theorem 13.3.8 (Bounded Borel functional calculus). *Let A be a selfadjoint operator in a separable Hilbert space H. There exists a contractive $*$-homomorphism from $\mathcal{B}(\mathbb{R})$ into $B(H)$, denoted $\phi \mapsto \phi(A)$, which is natural in the sense that $1(A) = I$ and*

$$\lim_{n \to \infty} \xi_n(A)h = Ah \qquad (13.9)$$

for all $h \in D(A)$. *If* ϕ_n *is a bounded sequence in* $\mathcal{B}(\mathbb{R})$ *that converges pointwise to* ϕ, *then* $\phi_n(A) \to \phi(A)$ *in the strong operator topology.*

Exercise 13.3.9. Prove Theorem 13.3.8. One can follow the proof of Theorem 11.2.8, noting that the function f in Theorem 13.3.4 is continuous.

From the bounded Borel functional calculus, we obtain the existence of a spectral measure and recover A as a spectral integral, as described in Section 11.3. Indeed, given a Borel set $S \subseteq \sigma(A)$, we can apply the bounded Borel functional calculus to define

$$E(S) = \chi_S(A).$$

Then $E(S)$ is an orthogonal projection, the **spectral projection** corresponding to S. The projection valued function $S \mapsto E(S)$ is a spectral measure as defined in Definition 11.3.2. Following Section 11.3 one can make sense of integration against a spectral measure and obtain that for every bounded Borel function on X, the functional calculus is given by

$$\phi(A) = \int \phi dE = \int \phi(\lambda) dE_\lambda. \tag{13.10}$$

Remark 13.3.10. The reader should be aware that in some texts the term **spectral resolution** is used instead of spectral measure. In other treatments, the emphasis is put on the projections $E_\lambda := E((-\infty, \lambda])$ for $\lambda \in \mathbb{R}$. The family $\{E_\lambda\}_{\lambda \in \mathbb{R}}$ is an increasing family of projections, referred to as a **resolution of identity**. The integral in (13.10) can then be interpreted as a *Stieltjes type integral*, e.g., for a continuous function ϕ of compact support $\int \phi(\lambda) dE_\lambda$ is the limit of "Riemann sums" $\sum_{j=1}^{N} \phi(\lambda_j)(E_{\lambda_j} - E_{\lambda_{j-1}})$, justifying the notation.

The spectral integrals (13.10) can also be defined in a weak sense as in Section 11.3. Given any spectral measure E, one defines for every $g, h \in H$ and every Borel set S

$$\mu_{g,h}(S) = \langle E(S)g, h \rangle.$$

The properties of the spectral measure E imply that $\mu_{g,h}$ is a regular Borel measure. With some work, $\int \phi(\lambda) dE_\lambda$ can be defined to be the unique operator in $B(H)$ such that

$$\int \phi d\mu_{g,h} = \left\langle \left(\int \phi(\lambda) dE_\lambda \right) g, h \right\rangle$$

for all $g, h \in H$. When E is the spectral measure corresponding to a selfadjoint operator A, the existence of such a unique operator is easy to justify, and it is readily verified that then one recovers the functional calculus as $\phi(A) = \int \phi(\lambda) dE_\lambda$.

From (13.9) we see that A can also be represented as

$$Ah = \lim_{n \to \infty} \xi_n(A)h = \lim_{n \to \infty} \int_{-n}^{n} \lambda dE_\lambda h,$$

for all h in the domain, which is given by

$$D(A) = \left\{ h \in H : \int \lambda^2 d\mu_h < \infty \right\},$$

where we put $\mu_h := \mu_{h,h}$. The above integral representation of A is abbreviated as

$$A = \int \lambda dE_\lambda$$

which is referred to as the **spectral decomposition** of A.

Exercise 13.3.11. Justify that the spectral integrals $\int \phi(\lambda) dE_\lambda$ and $\int \lambda dE_\lambda$ converge weakly to $\phi(A)$ and to A in the sense that was explained above, and that the domain of A has the spectral integral description as claimed.

Example 13.3.12. Recalling the operator A_2 from Example 13.3.6 (let's just call it A), the spectral decomposition is given by

$$A = \int \lambda dE_\lambda = \sum_{n \in \mathbb{Z}} 2\pi n E_n$$

where $E_n = E(\{n\})$ is the projection of $L^2[0,1]$ onto the space spanned by the basis vector $e^{-2\pi i n x}$. The sum above converges only in the sense that for $h \in D(A)$, the following limit exists

$$Ah = \lim_{N \to \infty} \sum_{n=-N}^{N} 2\pi n E_n h.$$

Bounded functions of A are then given by

$$\phi(A) = \int \phi(\lambda) dE_\lambda = \sum_{n \in \mathbb{Z}} \phi(2\pi n) E_n$$

where now this sum converges in the strong operator topology.

Example 13.3.13. The spectral decomposition of the operator A from Example 13.3.7 can be understood as follows. For every $S \subseteq \mathbb{R}$, the spectral projection $E(S)$ is the projection onto the subspace of $L^2(\mathbb{R})$ that Fourier transforms into $M_{\chi_S} L^2(\mathbb{R}) = L^2(S)$. Thus the spectral projection acts on $h \in L^2(\mathbb{R})$ as $E(S)h = \mathcal{F}^{-1}(\chi_S \hat{h})$, or, in all its glory,

$$(E(S)h)(x) = \frac{1}{\sqrt{2\pi}} \int_S \hat{h}(\lambda) e^{i\lambda x} d\lambda.$$

The spectral decomposition of A is given by

$$Ah = \lim_{R \to \infty} \frac{1}{\sqrt{2\pi}} \int_{-R}^{R} \lambda \hat{h}(\lambda) e^{i\lambda x} d\lambda.$$

For every $h \in H$, the measure μ_h (sometimes also referred to as "spectral measure") is a probability measure, and gives every set S the following probability:

$$\mu_h(S) = \langle E(S)h, h \rangle = \int_S |\hat{h}(\lambda)|^2 d\lambda.$$

An element $h \in H$ is in the domain of A if and only if $\int_{\mathbb{R}} \lambda^2 |\hat{h}(\lambda)|^2 d\lambda < \infty$, and then $\langle Ah, h \rangle$ is given by $\int_{\mathbb{R}} \lambda |\hat{h}(\lambda)|^2 d\lambda$, which may be interpreted as the *expected value of the random variable* λ under the measure $d\mu_h = |\hat{h}(\lambda)|^2 d\lambda$.

13.4   Application: Stone's theorem and quantum mechanics

At the beginning of the chapter, we discussed how the study of unbounded operators is motivated by the need to find a functional analytic framework for investigating differential operators. There are other branches of science and mathematics where unbounded operators arise, one of the most important being quantum mechanics. The mathematical framework of quantum mechanics requires unbounded selfadjoint operators for its very formulation; we shall briefly describe this framework below. Another part of mathematics where unbounded operators naturally arise is in the analysis of dynamical systems in continuous time, or, more specifically, the study of *strongly continuous one-parameter semigroups of operators*. This section showcases the use of unbounded operators in both topics. We shall focus on one-parameter unitary groups, and prove the foundational theorem of Stone that describes the relationship of such a group with a selfadjoint operator that generates it. The reader can find further discussion of one-parameter semigroups of general operators in [17] or [42]. See also Exercises 13.5.13–13.5.15.

One-parameter groups of unitaries and Stone's theorem are fundamental to the mathematical formalism of quantum mechanics. We now briefly sketch a small part of this formalism without any pretense of explaining the underlying physics, and the reader will have to take all this on faith (see [24] for a solid mathematical introduction to quantum theory).

Briefly, the state of a quantum mechanical system can be described by a unit vector ψ in a Hilbert space H. The nature of the space H and the interpretation of ψ depends on the system under study. For example, the position of a particle in space can be described by a state $\psi \in H = L^2(\mathbb{R}^3)$. Since ψ is a unit vector in L^2, it gives rise to a probability measure $|\psi(x)|^2 dx$ on $\mathbb{R}^3$. In quantum mechanics, the position does not have a definite value, rather, for every region $\Omega \subset \mathbb{R}^3$ the probability that the particle is in Ω is given by $\int_\Omega |\psi(x)|^2 dx$.

An *observable quantity* in that system is described by selfadjoint operator A in H, which is typically unbounded. For example, the operator corresponding to the energy of a particle in space under the influence of a (bounded and continuous, say), potential function $V \colon \mathbb{R}^3 \to \mathbb{R}$ might be given by the following operator, called the **Hamiltonian** of the system:

$$\mathsf{H} = -\Delta + M_V \tag{13.11}$$

where $-\Delta$ is the Laplacian and M_V is the operator of multiplication by V (by Example 13.3.7 and Proposition 13.1.27, H is selfadjoint). The possible measurable quantities, such as energy levels of a system or position of a particle, are points in the spectrum $\sigma(A)$ of the selfadjoint operator A corresponding to the observable. Statistical information can be obtained by applying a vector state to a function of the operator, for example, $\langle A\psi, \psi \rangle$ is the expected value of the observable corresponding to A when the system is in the state ψ.

Supposing that the state at time $t = 0$ is given by $\psi_0 \in H$, we let the state of the system at time $t > 0$ be denoted by $\psi(t) = U_t \psi_0$, where U_t is the operator of letting the system evolve t units of time. General considerations lead to the postulate that for all t, the operator U_t should be linear and invertible, and it should send unit vectors to unit vectors. Further considerations lead (under additional restrictive but realistic physical assumptions) to postulate that an application of the evolution operator U_t followed by an application of the evolution operator U_s is equivalent to the application of the evolution operator U_{s+t}. Finally, it is reasonable to postulate also that the evolution of the state of the system is continuous in time, therefore $t \mapsto U_t \psi_0$ should be a continuous path in H. The requirements of the time evolution in a quantum mechanical system are formalized in the following definition.

Definition 13.4.1. A *one-parameter unitary group* (or *unitary group*, for short) is a family $U = \{U_t\}_{t \in \mathbb{R}}$ of unitaries on a Hilbert space H such that:

1. $U_0 = I_H$ and $U_s U_t = U_{s+t}$ for all $s, t \in \mathbb{R}$,

2. $\lim_{s \to t} \|U_s h - U_t h\| = 0$ for all $t \in \mathbb{R}$ and $h \in H$.

Thus, one-parameter unitary groups are unitary representations of the group $\mathbb{R}$ that are continuous with respect to the strong operator topology. It turns out that requiring norm continuity (i.e. $\lim_{s \to t} \|U_s - U_t\| = 0$) would be too stringent a condition – many interesting evolutions arising in practice do not satisfy this condition. On the other hand, unitary groups that are not SOT continuous are so pathological, that there are necessarily $g, h \in H$ such that the path $t \mapsto \langle U_t h, g \rangle$ is not even measurable[3].

Stone's theorem gives a full description of the way in which unitary groups arise, establishing a one-to-one correspondence between selfadjoint operators and unitary groups. One half of this correspondence is a direct consequence of the functional calculus, as follows.

---

[3]See [15, Theorem 6.8.3].

Proposition 13.4.2. *Let A be a selfadjoint operator in a Hilbert space H. Define*

$$U_t = \exp(itA) \tag{13.12}$$

for every $t \in \mathbb{R}$, by applying the bounded Borel functional calculus. Then $U = \{U_t\}_{t \in \mathbb{R}}$ is a one-parameter unitary group. Moreover, for all $h \in D(A)$,

$$Ah = \lim_{t \to 0} \frac{1}{i} \frac{U_t h - h}{t}. \tag{13.13}$$

Proof. By the algebraic properties of the exponential functions $\phi_s(x) = e^{isx}$, and since the functional calculus is a $*$-homomorphism, we obtain that $U_0 = I_H$, U_t is a unitary and $U_s U_t = U_{s+t}$ for all $s, t \in \mathbb{R}$. The functions ϕ_s are a bounded family that converges pointwise to ϕ_t as $s \to t$, so by the continuity of the Borel functional calculus we have that $U_s \to U_t$ in the strong operator topology. This shows that $\{U_t\}_{t \in \mathbb{R}}$ is a unitary group.

We want to show that A can be recovered from U via (13.13). By the spectral theorem, we may assume that $A = M_f$ is a multiplication operator on $L^2(X, \mu)$ and

$$D(M_f) = \left\{ h \in L^2(X, \mu) : \int |fh|^2 < \infty \right\},$$

thus $U_t = M_{\exp(itf)}$. To prove the convergence in (13.13), note that the difference quotients

$$\frac{1}{i} \frac{U_t h - h}{t} = \frac{e^{itf(x)} h(x) - h(x)}{it}$$

converge pointwise to $f(x)h(x)$ as $t \to 0$ for every $x \in X$. To show norm convergence for $h \in D(A)$, we use the elementary inequality $|e^{ia} - 1| \le |a|$ for all $a \in \mathbb{R}$, to find that

$$\left| \frac{e^{itf(x)} h(x) - h(x)}{it} \right| \le |f(x)h(x)|.$$

But if $h \in D(A)$ then $fh \in L^2$, thus

$$\left| f(x)h(x) - \frac{e^{itf(x)} h(x) - h(x)}{it} \right|^2 \le 4|f(x)h(x)|^2 \in L^1$$

so by the dominated convergence theorem we conclude that (13.13) holds. $\square$

Definition 13.4.3. Given a unitary group U, we define its ***infinitesimal generator*** (or ***generator***, for short) to be the operator (L, D) with domain D consisting of all $h \in H$ such that the limit

$$\lim_{t \to 0} \frac{1}{i} \frac{U_t h - h}{t}$$

exists, and we define Lh to be this limit for all $h \in D$.

Proposition 13.4.2 shows that every selfadjoint A generates a unitary group U by (13.12), but we do not yet know that A is recovered by (13.13) as the generator L of U, only that $A \subseteq L$. This subtlety concerning domains will be resolved after the following lemma.

Lemma 13.4.4. *Let U be a unitary group, and let (L, D) be its generator. Then*

1. *L is densely defined, closed and Hermitian.*

2. *If $h \in D$ then $U_s h \in D$ and*

$$\frac{d}{dt} U_t h\Big|_{t=s} := \lim_{t \to s} \frac{U_t h - U_s h}{t - s} = iL U_s h = iU_s L h \qquad (13.14)$$

   for all $s \in \mathbb{R}$.

Proof. Let (L, D) be the generator as defined in Definition 13.4.3. First, (13.14) follows from writing

$$U_s \frac{U_{t-s} - I}{t - s} h = \frac{U_t - U_s}{t - s} h = \frac{U_{t-s} - I}{t - s} U_s h.$$

If $h \in D$, then the left hand side converges to $iU_s L h$, and therefore the right hand side converges to $iL U_s h$ and $U_s h \in D$.

Our next goal is to prove that D is dense. To this end, for any $h \in H$ and $T > 0$, we define

$$h_T = \int_0^T U_s h \, ds$$

where the integral is defined as in Section 8.6.2. Since $s \mapsto U_s h$ is continuous, we have that

$$\frac{1}{T} h_T \xrightarrow{T \to 0} h.$$

To see that D is dense, it remains to show that $h_T \in D$ for all $T > 0$. But

$$\frac{1}{i} \frac{U_t h_T - h_T}{t} = \frac{1}{i} \left(\frac{1}{t} \int_T^{T+t} U_s h \, ds - \frac{1}{t} \int_0^t U_s h \, ds \right)$$

and by continuity the right hand side converges to $-i(U_T h - h)$ as $t \to 0$. This shows that D is dense and that $L h_T = -i(U_T h - h)$. A slightly more involved computation shows that if $h \in D$, then

$$\int_0^T U_s L h \, ds = -i(U_T h - h).$$

We'll use this to show that L is closed. Suppose that $\{h_n\}$ is a sequence in D such that $h_n \to h$ and $Lh_n \to k$. Then $\int_0^T U_s Lh_n ds = -i(U_T h_n - h_n)$, and letting $n \to \infty$ we find that

$$\frac{1}{i} \frac{U_T h - h}{T} = \frac{1}{T} \int_0^T U_s k ds.$$

Letting $T \to 0$, we obtain that $h \in D(L)$ and that $Lh = k$. This shows that A is closed.

Next, for every $h, g \in D$, we use $U_t^* = U_{-t}$ to see that

$$\langle Lh, g \rangle = \lim_{t \to 0} \left\langle \frac{U_t - I}{it} h, g \right\rangle = \lim_{t \to 0} \left\langle h, \frac{U_{-t} - I}{-it} g \right\rangle = \langle h, Lg \rangle$$

so L is Hermitian. $\qquad\square$

Corollary 13.4.5. *Let A be a selfadjoint operator and define a unitary group $U_t = \exp(itA)$ as in Proposition 13.4.2. Then A is the infinitesimal generator of U.*

Proof. Let L denote the infinitesimal generator of U. By Proposition 13.4.2, $A \subseteq L$. This implies, using just the definition of the adjoint, that $L^* \subseteq A^*$. But since L is Hermitian, $L \subseteq L^*$. Putting everything together,

$$A^* = A \subseteq L \subseteq L^* \subseteq A^*,$$

and we conclude that $L = A$. $\qquad\square$

Lemma 13.4.6. *Let V and U be two unitary groups that have the same infinitesimal generator L. Then $V_t = U_t$ for all $t \in \mathbb{R}$.*

Proof. Let $h \in D(L)$ be a unit vector and consider the function $\phi(t) = \langle V_t h, U_t h \rangle$. By Lemma 13.4.4 and an easy to establish "Leibnitz product rule" for differentiation of inner products,

$$\frac{d\phi(t)}{dt} = \left\langle \frac{d}{dt} V_t h, U_t h \right\rangle + \left\langle V_t h, \frac{d}{dt} U_t h \right\rangle$$
$$= \langle iLV_t h, U_t h \rangle + \langle V_t h, iLU_t h \rangle$$
$$= i\langle LV_t h, U_t h \rangle - i\langle LV_t h, U_t h \rangle = 0.$$

Thus $\phi(t) = \phi(0) = 1$ for all t, from which it follows that the unit vectors $U_t h$ and $V_t h$ are equal for all t. Since $D(L)$ is dense, $U_t = V_t$ for all t. $\qquad\square$

Theorem 13.4.7 (Stone's theorem). *Let A be a selfadjoint operator in a Hilbert space H. For $t \in \mathbb{R}$, let*

$$U_t = \exp(itA).$$

Then $U = \{U_t\}_{t \in \mathbb{R}}$ is a one-parameter unitary group with infinitesimal generator A.

Conversely, every one-parameter unitary group arises this way: if U is a unitary group and A is its generator, then $U_t = \exp(itA)$ for all $t \in \mathbb{R}$.

Proof. The direct implication is Proposition 13.4.2 together with Corollary 13.4.5. Now we prove the converse. Let U be a unitary group, and let A be its generator. By Lemma 13.4.4, A is a closed, densely defined Hermitian operator. To prove A is selfadjoint, we invoke Theorem 13.2.10 to reduce this to showing that $\mathrm{Im}(A\pm i) = H$. Let us show, for example, that $\mathrm{Im}(A+i) = H$. The identity $\|Ah + ih\|^2 = \|Ah\|^2 + \|h\|^2 \geq \|h\|^2$ shows that $A + i$ is bounded below on its domain. Since A is closed, it follows that $\mathrm{Im}(A + i)$ is closed, so for $\mathrm{Im}(A + i) = H$ it suffices to show that $\mathrm{Im}(A + i)$ is dense. By Lemma 13.2.8 it suffices to show that $\ker(A^* - i) = \{0\}$. Let $g \in D(A^*)$ be such that $A^*g = ig$. We need to show that $g = 0$. To this end, let $h \in D(A)$, and define $\phi \colon \mathbb{R} \to \mathbb{C}$ by

$$\phi(t) = \langle g, U_t h \rangle.$$

By Lemma 13.4.4, ϕ is continuously differentiable and

$$\phi'(t) = \langle g, iAU_t h \rangle = -i \langle A^* g, U_t h \rangle = \phi(t).$$

It follows that $\phi(t) = \langle g, h \rangle e^t$ for all t. But $|\phi(t)| \leq \|g\| \|h\|$ for all t, so to avoid a contradiction we must have $\langle g, h \rangle = 0$. Since $D(A)$ is dense, we conclude that $g = 0$.

Now that we know that A is selfadjoint, we can use the first part of the theorem to construct a unitary group $V_t = \exp(itA)$ with generator A. Since U and V have the same generator, Lemma 13.4.6 implies that $U = V$. ☐

Example 13.4.8 (Schrödinger equation). Let $\mathsf{H} = -\Delta + M_V$ be the selfadjoint operator discussed around Equation (13.11), acting in the Hilbert space $L^2(\mathbb{R}^3)$. We know that $-\mathsf{H}$ generates a unitary group $U_t = \exp(-it\mathsf{H})$ which corresponds to the evolution[4] of a certain system. If the system starts from a state $\psi_0 \in L^2(\mathbb{R}^3)$, then its state at time t is given by $\psi_t = U_t\psi_0$. Assuming further that $\psi_0 \in D(\mathsf{H})$, we can differentiate $t \mapsto \psi_t = U_t\psi_0$, and thus we obtain

$$\frac{d\psi_t}{dt} = \frac{dU_t\psi_0}{dt} = -i\mathsf{H}U_t\psi_0 = -i\mathsf{H}\psi_t.$$

Considering that ψ_t is an element in $L^2(\mathbb{R}^3)$ that depends on time, we write $\psi(t, x) = \psi_t(x)$. We conclude that (with a liberal interpretation of derivative) the dynamics of the system are governed by the celebrated *Schrödinger equation*

$$i\frac{\partial}{\partial t}\psi(t, x) = -\Delta\psi(t, x) + V(x)\psi(t, x).$$

Thus, we see that the assumption that the time evolution of a system is described by the unitary group generated by the Hamiltonian forces on us the conclusion that the state of the system evolves according to the Schrödinger equation. *Why* should the time evolution of the system be described like that? That is a question whose answer lies beyond the scope of functional analysis.

---

[4] We chose $-\mathsf{H}$ instead of H so as to get equations that comply with standard conventions.

13.5   Additional exercises

Exercise 13.5.1. Let A be a closed operator in a Banach space X. Recall that for every $\lambda \in \rho(A)$, the resolvent R_λ is defined to be $R_\lambda = (A - \lambda)^{-1}$.

1. Prove that the resolvent set $\rho(A)$ is an open subset of $\mathbb{C}$.

2. Prove the **resolvent identity**: for every $\lambda, \mu \in \rho(A)$

$$R_\lambda - R_\mu = (\lambda - \mu)R_\mu R_\lambda.$$

3. Prove that $z \mapsto R_z$ is a holomorphic function from $\rho(A)$ into $B(X)$.

Exercise 13.5.2. Let A be a densely defined closed operator in a Hilbert space H. Prove that A^* is closed and densely defined and that $A^{**} = A$.

Exercise 13.5.3. Let $Q = M_x$ and $P = \frac{d}{dx}$ be the unbounded operators in $L^2(\mathbb{R})$ defined on the dense domain $C_c^\infty(\mathbb{R})$.

1. The *commutator* of P and Q is "defined" to be the operator $[P, Q] := PQ - QP$. Make the definition precise.

2. Show that $[P, Q] = I$, and explain carefully what the equality means.

3. Prove that there exist no two bounded operators $A, B \in B(H)$ such that $[A, B] = I$.

4. Is it possible to find a pair of operators A, B in a Hilbert space such that one of them is bounded and such that $[A, B] = I$?

Exercise 13.5.4. In this exercise, we will see that a closed, densely defined Hermitian operator may have infinitely many different selfadjoint extensions.

1. Let $H = L^2[0, 1]$ and consider the operators B defined by $Bf = if'$ with domain

$$D(B) = \{f \in L^2[0, 1] : f \in AC \text{ and } f' \in L^2 \text{ and } f(0) = f(1) = 0\}.$$

Prove that for every $\theta \in [0, 2\pi)$ the operator $B_\theta f = if'$ on the domain

$$D(B_\theta) = \{f \in L^2[0, 1] : f \in AC \text{ and } f' \in L^2 \text{ and } f(0) = e^{i\theta} f(1)\},$$

is selfadjoint. Is B a core for B_θ?

2. Find the spectral decomposition of B_θ, i.e., express it as a multiplication operator or use any other version of the spectral theorem to describe it.

3. Prove that there is no selfadjoint operator A that is a proper extension of B_θ.

Exercise 13.5.5. Let A be a selfadjoint operator in a Hilbert space H.

1. Prove that $\langle Ah, h \rangle \geq 0$ for all $h \in D(A)$ if and only if $\sigma(A) \subseteq [0, \infty)$. In this case, we say that A is **positive**, and we write $A \geq 0$.

2. Prove that if A is a positive operator in H then there exists a unique selfadjoint operator $B \geq 0$ such that $A = B^2$ (take care to define the domain of B correctly).

Exercise 13.5.6 (Diagonal operators). In this exercise, we explore a simple criterion for establishing that an operator is essentially selfadjoint (we used this in Example 13.2.6).

1. Let A be a Hermitian operator in H. Suppose that there exists an orthonormal basis $\{e_n\}_{n=1}^{\infty}$ for H consisting of eigenvectors of A, in the sense that for all n the basis vector e_n is in $D(A)$ and there exists λ_n such that $Ae_n = \lambda_n e_n$. Assume moreover that $\lambda_n \in \mathbb{R}$ for all n. Prove that A is essentially selfadjoint and describe its closure and its spectrum.

2. Use the above to prove that the operator $Af = -f''$ in $L^2[0, 2\pi]$ with domain
$$D(A) = \{f \in C^2([0, 1]) : f(0) = f(1) = 0\},$$
is essentially selfadjoint. What is its closure?

3. Compare with Examples 13.1.20 and 13.1.24.

Exercise 13.5.7 (Criterion for essential selfadjointness). Let A be a Hermitian operator in a Hilbert space H. Then A is essentially selfadjoint if and only if
$$\ker(A^* - i) = \ker(A^* + i) = \{0\}.$$

Exercise 13.5.8. Show that a selfadjoint operator in a Hilbert space is unitarily equivalent to a direct sum of multiplication operators $M_x : h \to xh$ on a family of L^2 spaces on $\mathbb{R}$.

Exercise 13.5.9. A densely defined operator N in a Hilbert space H is said to be **normal** if $NN^* = N^*N$.

1. Prove that the multiplication operator M_f determined by a measurable function $f : X \to \mathbb{C}$ on a measure space (X, μ) is a normal operator in $L^2(X, \mu)$.

2. State and prove (or read) the spectral theorem for normal operators in a Hilbert space.

Exercise 13.5.10. Prove that a one-parameter unitary group U on H is **uniformly continuous** in the sense that $\lim_{t \to s} \|U_t - U_s\| = 0$ for all $s \in \mathbb{R}$, if and only if there exists a *bounded* selfadjoint operator $A \in B(H)$ such that $U_t = \exp(itA)$.

Exercise 13.5.11. For each of the following, prove that it is a strongly continuous, but not uniformly continuous, one-parameter unitary group, and find its generator.

1. The shift $U = \{U_t\}_{t \in \mathbb{R}}$ on $L^2(\mathbb{R})$ given by $[U_t f](x) = f(x - t)$.

2. The rotation $V = \{V_t\}_{t \in \mathbb{R}}$ on $L^2(\mathbb{T})$ given by $[V_t f](z) = f(e^{it}z)$.

Exercise 13.5.12. Let U and V be unitary groups on H with generators A and B, respectively. Assume that there exists a constant K such that

$$\|U_t - V_t\| \leq K|t|$$

for all t in a neighborhood of the origin. Prove[5] that $D(A) = D(B)$ and that there exists a bounded operator $C \in B(H)$ such that $A = B + C$.

Exercise 13.5.13. A one-parameter semigroup is a family $T = \{T_t\}_{t \geq 0}$ of bounded operators on a Banach space X such that

1. $T_0 = I$, and

2. $T_s T_t = T_{s+t}$ for all $s, t \geq 0$.

A C_0-**semigroup** (also called a **strongly continuous semigroup**) is a one-parameter semigroup such that

$$\lim_{t \to 0^+} \|T_t x - x\| = 0$$

for all $x \in X$. Prove that a C_0-semigroup satisfies the following.

1. There exist constants a, b such that $\|T_t\| \leq a e^{bt}$ for all $t \geq 0$.

2. For all $x \in X$ the map $t \mapsto T_t x$ is continuous.

3. Define the generator of T to be $Ax = \lim_{t \to 0^+} t^{-1}(T_t x - x)$ for all x such that this limit exists. Prove that A is a densely defined closed operator.

4. Prove that if two C_0-semigroups have the same generator then they are equal.

Exercise 13.5.14. A one-parameter semigroup $T = \{T_t\}_{t \geq 0}$ is said to be *strongly continuous* if for all $x \in X$ the path $t \mapsto T_t x$ is continuous in the norm topology, and it is said to be *weakly continuous* if for all $x \in X$ the path $t \mapsto T_t x$ is continuous in the weak topology. Prove that a semigroup of operators is strongly continuous if and only if it is weakly continuous.

---

[5]This simple exercise has interesting consequences; see [ ].

Exercise 13.5.15. Let X be a Banach space and let $F \subseteq X^*$ be a linear subspace that separates the points of X. Let $\mathcal{S}$ be a dense subsemigroup of $\mathbb{R}_+$ such that $0 \in \mathcal{S}$ (for example, $\mathcal{S} = \mathbb{Q}_+ := \mathbb{Q} \cap [0, \infty)$). Suppose that $\{T_s\}_{s \in \mathcal{S}}$ is a semigroup of bounded operators on X, in that $T_s \circ T_t = T_{s+t}$ for all $s, t \in \mathcal{S}$ and that $T_0 = I_X$. We assume in addition that

$$\langle T_s(x), y \rangle \xrightarrow{\mathcal{S} \ni s \to 0} \langle x, y \rangle \quad \text{for all } x \in X, y \in F. \tag{13.15}$$

In this exercise, we explore the question[6] of whether $\{T_s\}_{s \in \mathcal{S}}$ can be extended to a strongly continuous semigroup $\{T_t\}_{t \in \mathbb{R}_+}$ parameterized by $\mathbb{R}_+$.

1. Prove that if $\|T_s(x) - x\| \xrightarrow{\mathcal{S} \ni s \to 0} 0$ for all $x \in X$ then $\{T_s\}_{s \in \mathcal{S}}$ can be extended to a strongly continuous semigroup $\{T_t\}_{t \in \mathbb{R}_+}$ parameterized by $\mathbb{R}_+$.

2. Prove for X separable and reflexive and $F = X^*$, (13.15) implies that $\{T_s\}_{s \in \mathcal{S}}$ can be extended to a strongly continuous semigroup $\{T_t\}_{t \in \mathbb{R}_+}$ parameterized by $\mathbb{R}_+$.

3. Show that there exist X and F such that (13.15) holds but $\{T_s\}_{s \in \mathcal{S}}$ cannot be extended to a strongly continuous semigroup $\{T_t\}_{t \in \mathbb{R}_+}$ parameterized by $\mathbb{R}_+$.

---

[6] See [33] if you can't solve it.

Bibliography

[1] C. Anantharaman and S. Popa. An Introduction to II_1 Factors. *preprint*, 8, 2017.

[2] W.B. Arveson. *A Short Course on Spectral Theory*. Springer, 2002.

[3] W.B. Arveson. *An Invitation to C*-Algebras*, volume 39. Springer Science & Business Media, 2012.

[4] S. Bochner and R.S. Phillips. Absolutely convergent Fourier expansions for non-commutative normed rings. *Ann. of Math.*, 43(3):409–418, 1942.

[5] J. Bourgain and H. Brezis. On the equation $divY = f$ and application to control of phases. *Journal of the American Mathematical Society*, 16(2):393–426, 2003.

[6] A. Bowers and N.J. Kalton. *An Introductory Course in Functional Analysis*. Springer, 2014.

[7] H. Brezis. *Functional Analysis, Sobolev Spaces and Partial Differential Equations*. Springer, 2011.

[8] N.P. Brown and N. Ozawa. *C*-Algebras and Finite-Dimensional Approximations*. American Mathematical Society, 2008.

[9] L. Carleson. Interpolations by bounded analytic functions and the corona problem. *Ann. of Math.*, 76(3):547–559, 1962.

[10] L. Carleson. On convergence and growth of partial sums of Fourier series. *Acta Math.*, 116(4):135–157, 1966.

[11] E.W. Cheney. *Introduction to Approximation Theory*. McGraw-Hill, 1966.

[12] N.E. Cotter. The Stone–Weierstrass theorem and its application to neural networks. *IEEE transactions on neural networks*, 1(4):290–295, 1990.

[13] G. Cybenko. Approximation by superpositions of a sigmoidal function. *Mathematics of control, signals and systems*, 2(4):303–314, 1989.

[14] K.R. Davidson. *C*-algebras by Example*. American Mathematical Society, 1996.

[15] E.B. Davies. *Spectral Theory and Differential Operators*, volume 42. Cambridge University Press, 1995.

[16] R.G. Douglas. *Banach Algebra Techniques in Operator Theory*. Springer, 2013.

[17] Klaus-Jochen Engel and Rainer Nagel. *A short course on operator semigroups*. Springer Science & Business Media, 2006.

[18] L.C. Evans. *Partial Differential Equations*. American Mathematical Society, 2010.

[19] G.B. Folland. *Real Analysis: Modern Techniques and their Applications*. Wiley, 1999.

[20] I.M. Gelfand. *Commutative Normed Rings*. Chelsea Publishing Company, 1964.

[21] M. Gerhold and O.M. Shalit. Bounded perturbations of the Heisenberg commutation relation via dilation theory. *Proceedings of the American Mathematical Society*, 151(09):3949–3957, 2023.

[22] I. Gohberg and S. Goldberg. *Basic Operator Theory*. Birkhäuser, 2001.

[23] I. Gohberg, S. Goldberg, and M.A. Kaashoek. *Classes of Linear Operators. Volume 1*. Operator Theory, Advances and Applications. Springer Basel AG, Basel, 1990.

[24] B.C. Hall. *Quantum Theory for Mathematicians*, volume 267. Springer Science & Business Media, 2013.

[25] P.R. Halmos. What does the spectral theorem say? *The American Mathematical Monthly*, 70(3):241–247, 1963.

[26] K. Hornik, M. Stinchcombe, and H. White. Multilayer feedforward networks are universal approximators. *Neural networks*, 2(5):359–366, 1989.

[27] F. Jones. *Lebesgue Integration on Euclidean Space*. Jones & Bartlett Learning, 2001.

[28] R.V. Kadison and J.R. Ringrose. *Fundamentals of the Theory of Operator Algebras, Vol. I*. American Mathematical Society, 1997.

[29] R.V. Kadison and J.R. Ringrose. *Fundamentals of the Theory of Operator Algebras, Vol. II*. American Mathematical Society, 1997.

[30] Y. Katznelson. *An Introduction to Harmonic Analysis*. Cambridge University Press, 2004.

[31] P. Lax. *Functional Analysis*. Wiley Interscience, 2002.

[32] M. Leshno, V.Y. Lin, A. Pinkus, and S. Schocken. Multilayer feedforward networks with a nonpolynomial activation function can approximate any function. *Neural Networks*, 6(6):861–867, 1993.

[33] E. Levy and O.M. Shalit. Continuous extension of a densely parameterized semigroup. *Semigroup Forum*, 78(2):276–284, 2009.

[34] J. Lindenstrauss, A. Pazy, and B. Weiss. *Functional Analysis: Hilbert and Banach Spaces (in Hebrew)*. Academon Press, 1981.

[35] J. Munkres. *Topology*. Pearson, 2000.

[36] F.J. Murray and J. von Neumann. On rings of operators. *Annals of Mathematics*, 37(1):116–229, 1936.

[37] M.A. Nielsen. *Neural Networks and Deep Learning*, volume 25. Determination press San Francisco, CA, USA, 2015. The original online book is available at http://neuralnetworksanddeeplearning.com.

[38] G.K. Pedersen. *Analysis Now*, volume 118. Springer Science & Business Media, 2012.

[39] A. Pinkus. Approximation theory of the MLP model in neural networks. *Acta Numerica*, 8:143–195, 1999.

[40] M. Reed and B. Simon. *Functional Analysis (Methods of Modern Mathematical Physics, Vol. I)*. Academic Press, 1980.

[41] F. Riesz and B. Sz.-Nagy. *Functional Analysis*. Frederick Ungar Publishing Co., 1956.

[42] W. Rudin. *Functional Analysis*. McGraw-Hill, 1973.

[43] W. Rudin. *Real and Complex Analysis*. McGraw-Hill, 1986.

[44] O.M. Shalit. *A First Course in Functional Analysis*. CRC Press, 2017.

[45] B. Simon. *A Comprehensive Course in Analysis, Part 4: Operator Theory*. American Mathematical Soc., 2015.

[46] A.D. Sokal. A really simple elementary proof of the uniform boundedness theorem. *The American Mathematical Monthly*, 118(5):450–452, 2011.

[47] R.S. Strichartz. *A Guide to Distribution Theory and Fourier Transforms*. World Scientific Publishing Company, 2003.

[48] F.G. Tricomi. *Integral Equations*. Dover Publications, 1985.

[49] J. von Neumann. Zur algebra der funktionaloperationen und theorie der normalen operatoren. *Mathematische Annalen*, 102(1):370–427, 1930.

[50] N. Wiener. Tauberian theorems. *Ann. of Math.*, 33(1):1–100, 1932.

Index

For Product Safety Concerns and Information please contact our EU
representative GPSR@taylorandfrancis.com
Taylor & Francis Verlag GmbH, Kaufingerstraße 24, 80331 München, Germany